Novel Electrode Materials for Rechargeable Batteries

Novel Electrode Materials for Rechargeable Batteries

Guest Editors

Jian Peng
Zhangxiang Hao
Zhe Hu

MDPI

Basel • Beijing • Wuhan • Barcelona • Belgrade • Novi Sad • Cluj • Manchester

Guest Editors
Jian Peng
Eastern Institute for Advanced Study
Eastern Institute of Technology
Ningbo
China

Zhangxiang Hao
School of Science
Hubei University of Technology
Wuhan
China

Zhe Hu
College of Materials Science and Engineering
Shenzhen University
Shenzhen
China

Editorial Office
MDPI AG
Grosspeteranlage 5
4052 Basel, Switzerland

This is a reprint of the Special Issue, published open access by the journal *Molecules* (ISSN 1420-3049), freely accessible at: https://www.mdpi.com/journal/molecules/special_issues/7T8U60FOY2.

For citation purposes, cite each article independently as indicated on the article page online and as indicated below:

Lastname, A.A.; Lastname, B.B. Article Title. *Journal Name* **Year**, *Volume Number*, Page Range.

ISBN 978-3-7258-5269-7 (Hbk)
ISBN 978-3-7258-5270-3 (PDF)
https://doi.org/10.3390/books978-3-7258-5270-3

Contents

molecules

MDPI

Article

Study on Extraction Valuable Metal Elements by Co-Roasting Coal Gangue with Coal Gasification Coarse Slag

Jincheng Zhao [1,2,3], Tao Yu [1,2,3,*], Huan Zhang [1], Yu Zhang [1], Lanting Ma [1,3], Jinling Li [2,3], Chengtun Qu [1,2] and Te Wang [2,3]

1 Yan'an Key Laboratory of Low Carbon Synergistic Control Technology and Reservoir Protection for Oil and Gas Field Environmental Pollution, Shaanxi Fuquan Environmental Protection Technology Co., Ltd., Yan'an 727500, China; 21211070886@stumail.xsyu.edu.cn (J.Z.); zhuan@wnu.edu.cn (H.Z.); zhangyu@xynu.edu.cn (Y.Z.); malanting2020@xsyu.edu.cn (L.M.); xianquct@xsyu.edu.cn (C.Q.)

2 State Key Laboratory of Petroleum Pollution Control, College of Chemistry and Chemical Engineering, Xi'an Shiyou University, Xi'an 710065, China; lijinling@xsyu.edu.cn (J.L.); 21212070982@stumail.xsyu.edu.cn (T.W.)

3 Shaanxi Oil and Gas Pollution Control and Reservoir Protection Key Laboratory, College of Chemistry and Chemical Engineering, Xi'an Shiyou University, Xi'an 710065, China

* Correspondence: 180708@xsyu.edu.cn

Abstract: Coal gangue (CG) and coal gasification coarse slag (CGCS) possess both hazardous and resourceful attributes. The present study employed co-roasting followed by H_2SO_4 leaching to extract Al and Fe from CG and CGCS. The activation behavior and phase transformation mechanism during the co-roasting process were investigated through TG, XRD, FTIR, and XPS characterization analysis as well as Gibbs free energy calculation. The results demonstrate that the leaching rate of total iron (TFe) reached 79.93%, and Al^{3+} achieved 43.78% under the optimized experimental conditions (co-roasting process: CG/CGCS mass ratio of 8/2, 600 °C, 1 h; H_2SO_4 leaching process: 30 wt% H_2SO_4, 90 °C, 5 h, liquid to solid ratio of 5:1 mL/g). Co-roasting induced the conversion of inert kaolinite to active metakaolinite, subsequently leading to the formation of sillimanite (Al_2SiO_5) and hercynite ($FeAl_2O_4$). The iron phases underwent a selective transformation in the following sequence: hematite (Fe_2O_3) → magnetite (Fe_3O_4) → wustite (FeO) → ferrosilite ($FeSiO_3$), hercynite ($FeAl_2O_4$), and fayalite (Fe_2SiO_4). Furthermore, we found that acid solution and leached residue both have broad application prospects. This study highlights the significant potential of co-roasting CG and CGCS for high-value utilization.

Keywords: co-roasting; H_2SO_4 leaching; phase transformation; Al; Fe extraction

1. Introduction

Coal gangue (CG) is a solid waste generated in the process of coal mining and washing, constituting approximately 10–15% of the total coal production [1,2]. Currently, China's coal gangue reserves amount to around 7 billion tons, with its production continuously increasing each year, making it one of the largest industrial wastes [3,4]. Coal gasification slag (CGS) is a solid waste generated by the coal gasification process. It primarily consists of residual materials derived from raw coal, which undergo a series of processes including crushing, melting, sintering, gasification, deposition, and accumulation. Ultimately, CGS comprises residues that have not undergone redox and gasification reactions. Depending on the particle size, it can be further classified into two categories: coal gasification coarse slag (CGCS) and coal gasification fine slag (CGFS) [5]. The coal gasification process consumes approximately 100 million tons of coal annually, with the national production of CGS exceeding 60 million tons per year and showing exponential growth [6]. The deposition of CG and CGS typically occurs in open piles on vacant land. Despite the distance from urban areas, the accumulation of substantial quantities of coal-based waste residues still engrosses significant land resources and poses a grave ecological threat. The environmental concerns encompass air pollution caused by dust, water pollution, and soil contamination from heavy metals (Cu, Cd, Cr, Pb, Zn, Hg, and Sn), as well as various geological hazards [7,8]. Additionally, the oxidative spontaneous combustion during stacking of CG and CGS results in the emission of detrimental gases such as CO_2, CO, H_2S, SO_2, and NO_x [9]. In addition, both CG and CGS exhibit high concentrations of silicon, aluminum, iron, and titanium. The primary constituents of CG include quartz, clay minerals, and carbonaceous materials [10]. On the other hand, the mineral composition of CGS is predominantly composed of amorphous alumino-silicates, crystalline phase minerals, and residual carbon. Amorphous alumino-silicates account for over 60% of the total composition while quartz and calcite dominate the crystalline phase minerals [11].

At present, the comprehensive utilization of CG and CGS is primarily concentrated in areas such as combustion and power generation, construction materials, fertilizers, chemical product extraction and preparation, boiler blending, and ecological restoration [12–14]. However, the comprehensive utilization rate of both CG and CGS in China significantly lags behind that of developed countries. Therefore, it is strategically imperative to undertake extensive utilization of CG and CGS. Leaching is a widely used method for metal extraction and can be performed using various types of solvents and reagents. It involves dissolving the metal from the raw material into a liquid solution, which can then be further processed to isolate the metal. Leaching represents an energy-efficient process that can be optimized to minimize waste generation and mitigate environmental impact. Nevertheless, it is crucial to consider the potential environmental consequences associated with any extraction method and opt for the most environmentally friendly alternative whenever feasible.

Currently, the main methods regarding the extraction of aluminum from coal gangue (CG) and coal gasification slag (CGS) are alkali and acid methods [15,16]. Alkali methods, including sintering and leaching [16], have been extensively studied for the extraction of alumina from CG and CGS. However, the sintering method requires high temperatures (around 1100 °C to 1200 °C), resulting in increased energy consumption and significant emission of dicalcium silicate residue [17]. In addition, the alkali leaching method enables the alumina in CG and CGS to transform into a sodium aluminate solution, while iron and titanium become insoluble residue, resulting in a waste of resources. Additionally, alkaline solutions dissolve silica in CG and CGS, necessitating an additional desilication process [18]. Consequently, the application of alkali methods for aluminum extraction is not promising.

Compared with alkali methods, acid extraction offers competitiveness and suitability for treating CG and CGS due to its advantages such as a shorter process duration, lower energy consumption, easy separation of silica and aluminum components, as well as thorough separation. Although there are fewer studies on iron extraction from CG and CGS at present time, common methods include acid leaching extraction or reductive roasting-magnetic separation. Reduction roasting-magnetic separation involves roasting under a reducing atmosphere (e.g., H_2 [19,20], methane, or natural gas [21]), which facilitates rapid conversion of hematite to magnetite in CG and CGS followed by magnetic separation to obtain magnetite. Similar to acid extraction of aluminum, acid extraction of iron also possesses these advantages. However, it should be noted that acid leaching extraction usually requires pre-activation when extracting iron and aluminum from CG and CGS.

Various methods, including mechanical activation and thermal activation (roasting activation, and microwave activation) [22–24], have been reported as effective pre-activation approaches for acid leaching to extract the available components in CG and CGS. However, a single activation method often fails to achieve efficient extraction, leading to the adoption of composite activation methods in current research for extracting its active ingredients [25,26]. Qin et al. [23] employed a combination of mechanical grinding, low-temperature roasting, and deionized water leaching to extract aluminum and other trace elements from CG. The fully ground CG was sieved through a 100-mesh sieve and mixed according to the mass ratio of CG:NH_4Cl:$(NH_4)_2SO_4$ = 1:2:1. Subsequently, the mixture was subjected to roasting at 400 °C for 1 h in a muffle furnace. The roasted sample was placed in a conical flask of 20 mL of 0.4 m deionized water and shaken at 60 °C for 1 h. The final leaching yields of Al, Li and Fe were 56.35%, 80.83% and 32.77%, respectively. Han et al. [27] used an acid-base combination method, wherein a 2.0 mol/L KOH solution was utilized as an activator to immerse the activated CG in a hydrochloric acid solution (2.0 mol/L) for 2 h. The resulting acid leach solution was subsequently evaporated at 110 °C for 2 h, yielding extraction rates of Al^{3+} and Si^{4+} after drying that reached as high as 78.9% and 69.2%, respectively. Shao et al. [2] proposed a recovery technology using a combination of thermal activation + HNO_3 leaching to treat CG. Following milling, the CG was subjected to thermal activation at 550 °C for 0.5 h in a muffle furnace. Under optimized leaching conditions (150 °C, theoretical HNO_3 dosage, liquid-solid ratio of 5:1 mL/g, and 2 h), extraction rates of aluminum, gallium, lithium, and iron were achieved at 95.2%, 56.4%, 80.5%, and 2.1%, respectively. Additionally, flotation is also a valuable method to upgrade solid particles before leaching [28,29], which can not only improve the efficiency and quality of aluminum and iron extraction by acid leaching, but also minimizes the process cost and environmental pollution. Moreover, it finds extensive application in the efficient recovery of minerals, solid waste, as well as heavy metal ions present in water.

In these methods, the leaching rate of valuable metal elements is influenced not only by process conditions but also by the composition of CG and CGS. These existing techniques are difficult to directly replicate and apply. Furthermore, most studies focus on single coal-based waste residues such as CG, CGS, and fly ash, with limited research on extracting valuable metal elements from two or more coal-based waste residues. Therefore, it is necessary to explore a leaching process condition that can simultaneously extract valuable metal elements from both CG and CGS. It should be noted that both CG and CGCS contain fixed carbon which can act as a reducing agent during the reduction roasting process. Based on current research progress, an innovative approach involving co-roasting and H_2SO_4 leaching was proposed for extracting valuable metal elements (Al, Fe) from CG and CGCS. This study addresses the gap in synergistic extraction of valuable metal elements from both CG and CGCS; compared to separate extraction methods, it improves the leaching rate of valuable metal elements while reducing energy consumption and H_2SO_4 dosage. The aim of this method is to provide technical support for achieving synergistic extraction of valuable metal elements from two or more coal-based waste residues.

2. Results and Discussion

2.1. Optimization of Co-Roasting Parameter

The influence of CGCS content on the extraction efficiency of valuable metal elements was investigated by thoroughly mixing CG with CGCS in a specific ratio, followed by flat placement in a corundum crucible and co-roasting in a tubular pyrolysis furnace at 700 °C for 1 h. Simultaneously, the leaching conditions were set as follows: leaching temperature of 50 °C, H_2SO_4 concentration of 30 wt%, liquid to solid ratio of 5:1 mL/g, and leaching time of 5 h.

As depicted in Figure 1a, in the absence of CGCS addition, the leaching rates of Al^{3+} and TFe were merely 23.74% and 57.22%, respectively. With an increase in CGCS content, the leaching of Al^{3+} exhibited a non-linear trend with an initial decrease followed by an increase before declining again, reaching its peak at 32.69% for a content of 20 wt%. Conversely, the leaching rate of TFe gradually increased and reached its maximum value of 69.52% when the CGCS content was at 25 wt%. This behavior can be attributed to the lower percentage of Al^{3+} content compared to that of TFe within CGCS, resulting in an enhanced release of TFe as more CGCS is added. Additionally, the decrease in the extraction rate of Al^{3+} can be attributed to the relatively low initial Al content in the CGCS sample compared to CG. As the amount of CGCS added increases, there is a gradual reduction in the overall Al content in the mixed raw material, ultimately leading to a decline in the extraction rate of Al^{3+}. Furthermore, there was a notable rise in carbon content which facilitated sufficient CO generation for hematite (Fe_2O_3) reduction. Considering both the high proportion of

CG present in the co-roasting mixture and elevated aluminum concentration within CG itself, it was deemed appropriate to select a CGCS addition level at 20 wt%.

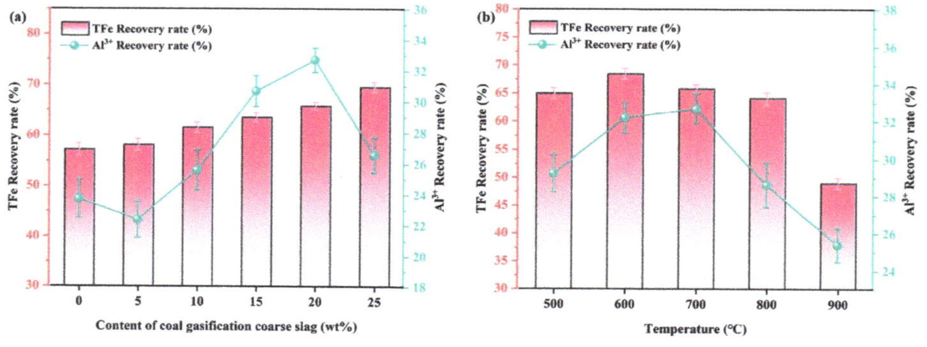

Figure 1. (**a**) The effect of different CGCS content on the leaching rates of Al^{3+} and TFe, (**b**) the effect of different co-roasting temperature on the leaching rates of Al^{3+} and TFe (co-roasting process: 1 h; H_2SO_4 leaching process: 30 wt% H_2SO_4, 50 °C, 5 h, liquid to solid ratio of 5:1 mL/g).

As shown in Figure 1b, the effect of co-roasting temperature on the extraction of valuable metal elements was investigated by incorporating 20 wt% CGCS, while keeping other conditions such as liquid to solid ratio constant. With the increase in co-roasting temperature, the leaching rates of Al^{3+} and TFe initially exhibited an upward trend followed by a decline. At 600 °C, the leaching rate of TFe reached a peak of 68.47%, indicating that the reduction effect of co-roasting was optimized at this time. This can be attributed to the temperature-induced promotion of hematite (Fe_2O_3) reduction to magnetite (Fe_3O_4) and wustite (FeO). However, as the temperature continued to rise, the leaching rate of TFe gradually decreased due to potential formation of other substances that hindered iron phase reduction in the co-roasting process. The leaching rates of Al^{3+} were measured at 32.21% and 32.69% at temperatures of 600 °C and 700 °C, respectively, which could be attributed to metakaolinite becoming inert material beyond 700 °C resulting in minimal or even negative impact on leaching rates' enhancement. Therefore, considering both leaching rates' performance and economic factors, a co-roasting temperature of 600 °C was selected.

2.2. Raw Material Characterization

The XRD pattern of CG raw material and mixed raw material are presented in Figure 2a, revealing predominant mineral constituents including kaolinite, quartz, hematite, muscovite, and calcite. Figure 2b displays the FTIR spectra of CG raw material and mixed raw material. The bands observed at 914, 3620 cm^{-1} and 3653, 3678, 3693 cm^{-1} can be attributed to the internal and external OH stretching vibrations in the layered structure of kaolinite octahedron [30,31]. Additionally, characteristic bands corresponding to Al-O-Si and Al-O vibrations in kaolinite were identified at 534 and 692 cm^{-1}, respectively [32,33]. Furthermore, absorption peaks at 428 and 1114 cm^{-1} along with double peaks at 1029 and 1006 cm^{-1} correspond to the Si-O vibration in kaolinite. Lastly, absorption peaks observed at 796 and 752 cm^{-1} are indicative of the Si-O-Si stretching vibration in kaolinite [34,35].

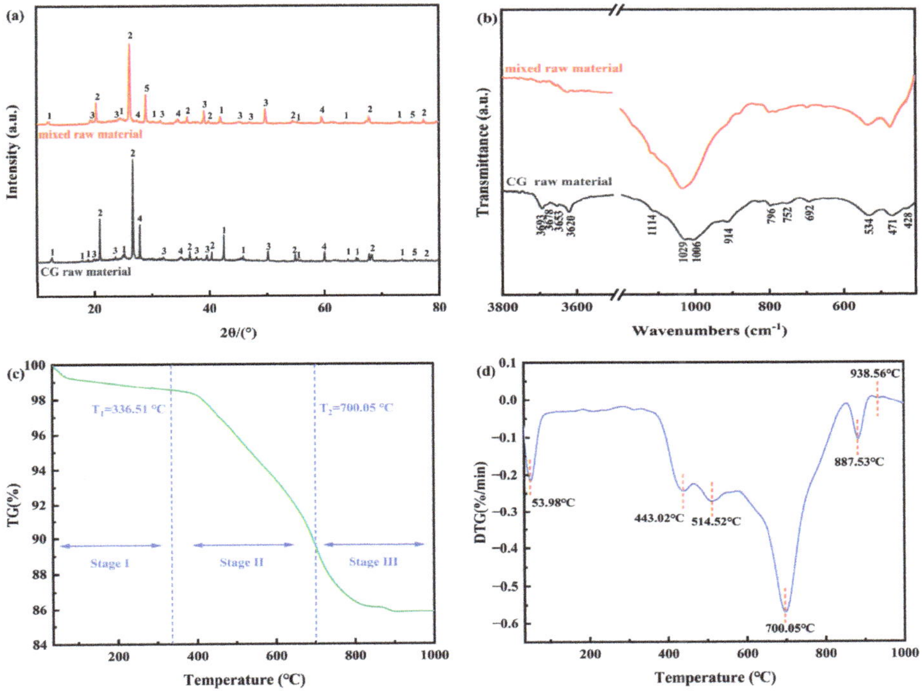

Figure 2. (**a**) XRD patterns of CG raw material, mixed raw material; 1—kaolinite; 2—quartz; 3—hematite; 4—muscovite; 5—calcite; (**b**) FTIR spectra of CG raw material and mixed raw material, (**c**,**d**) TG and DTG curves of co-roasting process.

The co-roasting of coal gangue and coal gasification coarse slag was divided into three stages based on the variations in TG and DTG curves, as shown in Figure 2c,d. Stage I occurred at an onset temperature of 35 °C–336.51 °C, accompanied by moisture evaporation. Stage II, ranging from 336.51 °C to 700.05 °C, involved the release of volatiles, carbon burning, and observed structural hydroxyl desorption of kaolinite [31,36]. This stage exhibited a maximum DTG peak with a value of 0.569%/min at a corresponding temperature of 700.05 °C. Stage III existed within the range of 700.05 °C–1000 °C and displayed two DTG peaks at temperatures of 887.53 °C and 938.56 °C, respectively; this finding may be attributed to macromolecular organic matter degradation, fixed carbon decomposition, and mullite formation [37].

2.3. *Phase Transformation during Co-Roasting Process*

The XRD patterns of the co-roasting products obtained at different co-roasting temperatures are presented in Figure 3a. At 500 °C, the diffraction peaks of kaolinite still persist but with reduced intensity compared to the mixed raw material, indicating partial decomposition of kaolinite [38]. Upon increasing the temperature from 500 °C to 600 °C, the diffraction peaks of kaolinite disappear completely, suggesting successful transformation from inert kaolinite to active metakaolinite. Notably, active metakaolinite appears amorphous and does not exhibit any discernible features in XRD analysis [39]. Additionally, some hematite diffraction peaks vanish at 500 °C while a new magnetite phase is detected.

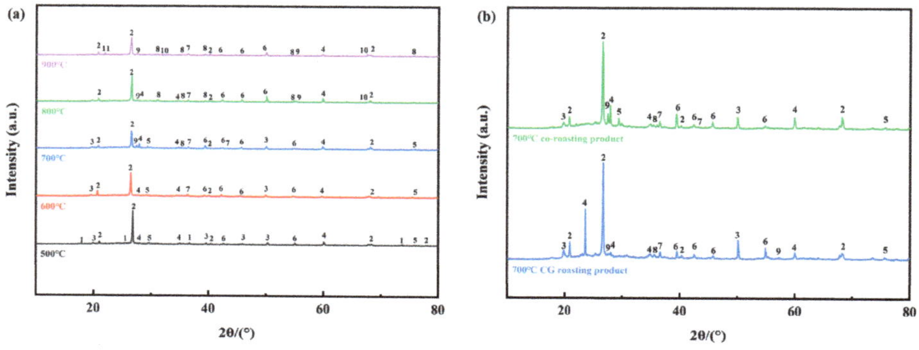

Figure 3. (a) XRD patterns of 500 °C–900 °C co-roasting products; (b) XRD patterns of 700 °C CG roasting product, 700 °C co-roasting products; 1—kaolinite; 2—quartz; 3—hematite; 4—muscovite; 5—calcite; 6—magnetite; 7—wustite; 8—fayalite; 9—sillimanite; 10—hercynite; 11—ferrosilite.

Further increasing temperature to 600 °C, a significant decomposition of hematite occurred, leading to an increase in the diffraction peaks of magnetite. Meanwhile, the presence of wustite was observed following co-roasting at 600 °C. Subsequently, at temperatures ranging from 700 °C–900 °C, distinct diffraction peaks corresponding to fayalite, sillimanite, hercynite, and ferrosilite were identified [40–43]. This observation elucidates why the reduction of iron phases during co-roasting was impeded and resulted in a minimal or even decreased leaching rate of Al^{3+}. Notably, at a co-roasting temperature of 800 °C, calcite underwent decomposition while muscovite remained remarkably stable even when subjected to elevated temperatures up to 900 °C.

Figure 3b displays the XRD patterns of the CG roasting product and co-roasting product at 700 °C. As depicted in the figure, compared to the solely CG roasting product at 700 °C, the diffraction peaks of hematite are weaker while those of magnetite are stronger in the co-roasting product at 700 °C, and there is an increase in wustite's number of diffraction peaks. This can be attributed to a higher carbon content in mixed raw material that provides a better reducing atmosphere and promotes further reduction of iron phases [44,45]. Additionally, there are more diffraction peaks for sillimanite in CG roasting products than co-roasting products which result in a lower leaching rate for Al^{3+}.

The aforementioned results demonstrate that the temperature increase and carbon presence contribute to the reduction of iron phase in the co-roasting process. However, excessively high temperatures lead to sintering of wustite with Si and Al in the mixed raw material, resulting in the formation of other iron phases which impede iron phase reduction. Simultaneously, as temperature increases, metakaolinite reacts with Si and Fe in the mixed raw material to form sillimanite and hercynite, consequently reducing the leaching rate of Al^{3+}.

The co-roasting products were characterized using FTIR analysis. As depicted in Figure 4, at a co-roasting temperature of 500 °C, the gradual removal of the hydroxyl group from kaolinite resulted in the absence of absorption peaks associated with OH stretching vibrations (3693, 3678, 3653, 3620, and 914 cm^{-1}), indicating the initiation of kaolinite transformation into metakaolinite [18,31]. Simultaneously, a decrease in the intensity of Al-O-Si stretching vibration at 534 cm^{-1} suggested that the dehydroxylation process was accompanied by the disruption of Al-O-Si bonds and exacerbation of crystal defects.

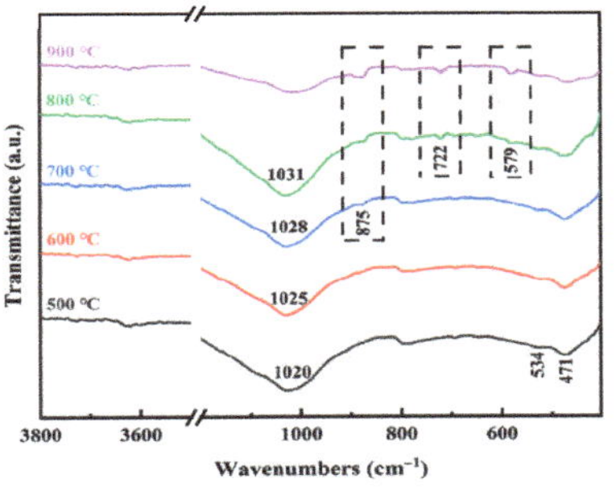

Figure 4. FTIR patterns of 500 °C–900 °C co-roasting product.

With the further increase in co-roasting temperature, the Si-O vibration peaks at 1006 and 1029 cm^{-1} merge into a broadband peak at 1020 cm^{-1}, which subsequently shifts to shorter wavelengths between 500 °C–800 °C. This shift is attributed to the enhanced orderliness of the silicate structure [45]. During the transformation of kaolinite into metakaolinite, as metakaolinite gradually loses its ordering and average bond angle increases within the Si-O-Si network, the stretching vibration peak at 471 cm^{-1} assigned to Si-O-Si weakens and broadens with increasing temperatures from 500 °C–800 °C [46].

Significantly, the adsorption bond located at 875 cm^{-1} associated with inert sillimanite (Al_2SiO_5) was detected after co-roasting at temperatures ranging from 700 °C to 900 °C [45]. Furthermore, the characteristic peaks of hercynite ($FeAl_2O_4$) related to Fe-O at 579 cm^{-1} and Al-O bond at 722 cm^{-1} were observed after co-roasting within the temperature range of 800 °C–900 °C [47]. These findings are consistent with the XRD results, providing further evidence for the formation of new inert components contributing to the decrease in leaching rates of Al^{3+} and TFe.

In addition to XRD and FTIR, XPS analysis was conducted to characterize the mixed raw material and co-roasting products. The effect of co-roasting on the atomic valence characteristics of mineral surfaces was investigated through XPS analysis in Figure 5. Figure 5a illustrates the comprehensive survey scans of the six samples, revealing that Si, Al, Fe, O, C, K, Ca, Mg, and Na were the predominant elements at the surface. These findings align with the chemical composition and proximate analysis of raw materials.

In Figure 5b, the high-resolution XPS survey of Fe2p for these six samples reveals that the Fe2p3/2 peak exhibits a narrower width, higher intensity, and larger area compared to the Fe2p1/2 peak due to spin-orbit (j-j) coupling [48–50]. The Fe2p spectrum of the mixed raw material indicates that the two photoelectron peaks at binding energies of 726.9 eV (Fe2p1/2) and 712.79 eV (Fe2p3/2) can be attributed to Fe^{3+}, which appears to correspond with the presence of Fe^{3+} in the hematite phase [51,52]. Additionally, the observed photoelectron peaks related to Fe(II) minerals in trace amounts seem to represent contributions from Fe^{2+} species.

According to Figure 5b, the Fe^{2+} content of the co-roasting product at 500 °C exhibited an increase, with the ratio of Fe^{2+}/Fe^{3+} rising from 1.06 to 1.12. Additionally, lower binding energy revealed a low intensity peak of Fe^{2+}, indicating the transformation of hematite after co-roasting and contributing to the presence of magnetite in the product at 500 °C [51,53]. This suggests that reduction from the Fe^{3+} state to Fe^{2+} has occurred [54]. Moreover, these findings align with those obtained from XRD analysis as evidenced by satellite peaks (714.98 eV and 729.18 eV) observed in the co-roasting product at 600 °C for Fe^{2+}. The ratio of Fe^{2+}/Fe^{3+} further increased to 1.18 during this stage. Since magnetite does not exhibit satellite peaks [49,55], it can be inferred that these satellite peaks are attributed to wustite based on XRD analysis results, signifying that the transition from magnetite to wustite has already taken place. Specifically, with the increase in co-roasting temperature, the binding energy positions of the maxima of the Fe2p3/2 and Fe2p1/2 peaks were shifted from 710.4 eV and 724.2 eV to 709.72 eV and 723.32 eV, with negative shifts of 0.68 eV and 0.88 eV, respectively, compared to the mixed raw material. Moreover, there was an increase in the Fe^{2+}/Fe^{3+} ratio to 2.43, indicating a change in the chemical environment where more divalent iron is present (i.e., alterations in types and quantities of elements bound to it [56]). Combining these findings with XRD analyses, it can be attributed that within the co-roasting product at temperatures ranging from 700 °C–900 °C, Fe^{2+} originates from two sources: a minor contribution from magnetite formation and transformation of wustite into fayalite, hercynite, and ferrosilite phases.

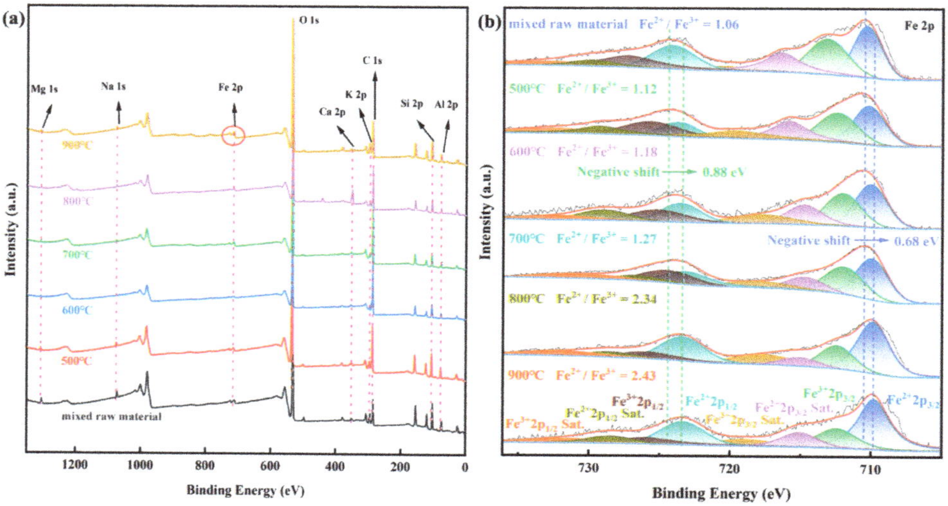

Figure 5. XPS analysis (**a**) survey scan XPS spectra for mixed raw material, and 500 °C–900 °C co-roasting products; XPS survey of Fe2p for (**b**) mixed raw material, and 500 °C–900 °C co-roasting products.

The XPS results mentioned above provide robust evidence supporting the transformation of hematite phase into magnetite, wustite, fayalite, hercynite, and ferrosilite phases during the co-roasting process. Additionally, this phase transformation process resulted in a reduction in both TFe and Al^{3+} leaching rates.

2.4. Co-Roasting Reaction Mechanism

The reactions that may occur during the co-roasting process of mixed raw material are enumerated as Equations (1)–(11) [57,58]. Thermodynamic calculations for some of the reactions were performed using HSC 9.0 software, and the temperature dependency of $\Delta_r G^\theta$ is illustrated in Figure 6.

$$3Fe_2O_3\ (s) + C\ (s) \rightarrow 2Fe_3O_4\ (s) + CO\ (g) \quad (1)$$

$$3Fe_2O_3\ (s) + CO\ (g) \rightarrow 2Fe_3O_4\ (s) + CO_2\ (g) \quad (2)$$

$$6Fe_2O_3\ (s) + C\ (s) \rightarrow 4Fe_3O_4\ (s) + CO_2\ (g) \quad (3)$$

$$Fe_3O_4\ (s) + C\ (s) \rightarrow 3FeO\ (s) + CO\ (g) \quad (4)$$

$$Fe_3O_4\ (s) + CO\ (g) \rightarrow 3FeO\ (s) + CO_2\ (g) \quad (5)$$

$$FeO\ (s) + Al_2O_3\ (s) \rightarrow FeAl_2O_4(s) \tag{6}$$

$$FeO\ (s) + SiO_2\ (s) \rightarrow FeSiO_3(s) \tag{7}$$

$$2FeO\ (s) + SiO_2\ (s) \rightarrow Fe_2SiO_4(s) \tag{8}$$

$$Al_2O_3{\cdot}2SiO_2{\cdot}2H_2O\ (s) \rightarrow Al_2O_3{\cdot}2SiO_2\ (s) + 2H_2O\ (g) \tag{9}$$

$$Al_2O_3{\cdot}2SiO_2(s) + 0.5Fe_2O_3(s) + 0.5C\ (s) \rightarrow FeAl_2O_4(s) + 2SiO_2(s) + 0.5CO\ (g) \tag{10}$$

$$2[Al_2O_3{\cdot}2SiO_2]\ (s) \rightarrow 0.5Al_2SiO_5(s) + 1.5SiO_2\ (s) + [Al_2O_3{\cdot}2SiO_2]\ (s) + 0.5[\gamma - Al_2O_3]\ (s) \tag{11}$$

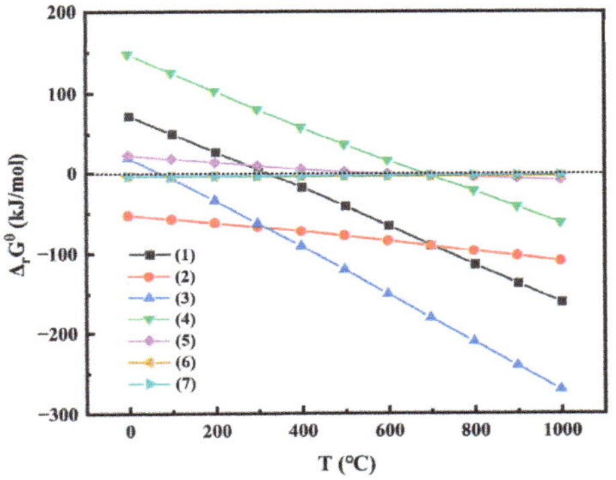

Figure 6. Diagram of Gibbs free energy (Δ_rG^{θ}) versus temperature for reactions occurring during the co-roasting process of mixed raw material.

Due to the unavailability of data on metakaolinite, the phase transformation temperature of kaolinite during thermal activation is inferred from experimental findings reported by literature [59,60]. The transition temperature from kaolinite to metakaolinite ranges between 500 °C and 700 °C [Equation (9)].

The diagram in Figure 7 illustrates the mechanism of co-roasting CG and CGCS. During the co-roasting process, hematite (Fe_2O_3) is reduced to magnetite (Fe_3O_4), which further undergoes reduction to wustite (FeO) due to the presence of fixed carbon in CG and CGCS. Subsequently, wustite (FeO) reacts with Si and Al elements in the mixed raw material, resulting in the formation of iron phases such as ferrosilite ($FeSiO_3$), hercynite ($FeAl_2O_4$), and fayalite (Fe_2SiO_4). Additionally, kaolinite present in the mixed raw material undergoes dehydration into metakaolinite, which then transforms into sillimanite (Al_2SiO_5) and hercynite ($FeAl_2O_4$). In summary, CGCS enables efficient reduction of iron minerals in CG while also enhancing leaching rates of Al^{3+} and TFe.

2.5. Optimization of H_2SO_4 Leaching Parameter

After conducting the co-roasting experiment, a significant enhancement in the activity of aluminum and iron was observed. Consequently, batch experiments were conducted to investigate the optimal leaching conditions, encompassing H_2SO_4 concentration and leaching temperature.

As can be seen from Figure 8a, the effect of H_2SO_4 concentration (10–50 wt%) on the extraction of valuable metal elements were investigated under this conditions, including a CGCS content of 20 wt%, co-roasting in a tubular pyrolysis furnace at 600 °C for 1 h, leaching temperature of 50 °C, liquid to solid ratio of 5:1 mL/g, and leaching time of 5 h.

With the increase in H_2SO_4 concentration, the leaching rates of Al^{3+} and TFe exhibited an initial upward trend followed by a plateau. Specifically, at an H_2SO_4 concentration of 30 wt%, the leaching rate of TFe was measured to be 68.47%, while that of Al^{3+} was found to be 32.21%. Subsequently, further increases in H_2SO_4 concentration only resulted in marginal changes in the leaching rates of Al^{3+} and TFe, indicating near-complete reaction between Al^{3+}, TFe, and H_2SO_4 at a concentration of 30 wt%. Therefore, it can be inferred that co-roasting pretreatment with CG and CGCS achieves optimal leaching efficiency when using an H_2SO_4 concentration of 30 wt%.

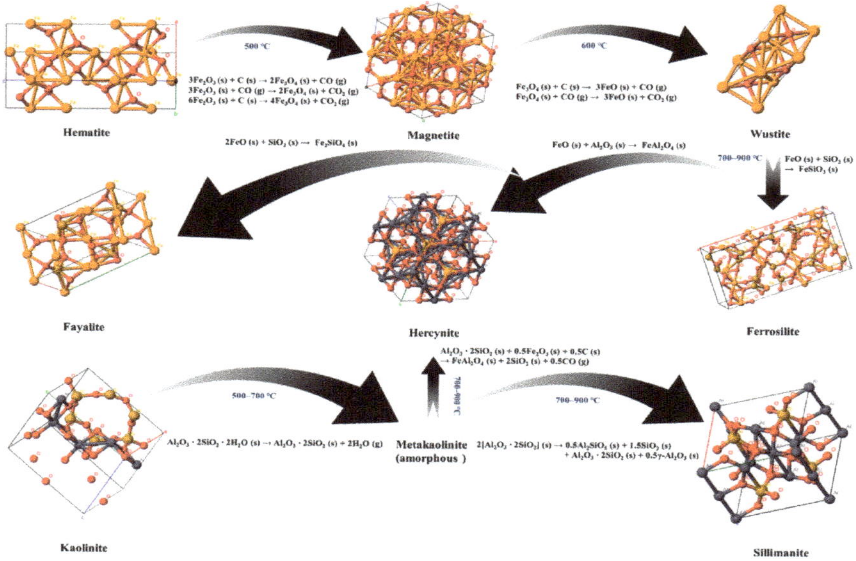

Figure 7. Mechanism of co-roasting of CG and CGCS.

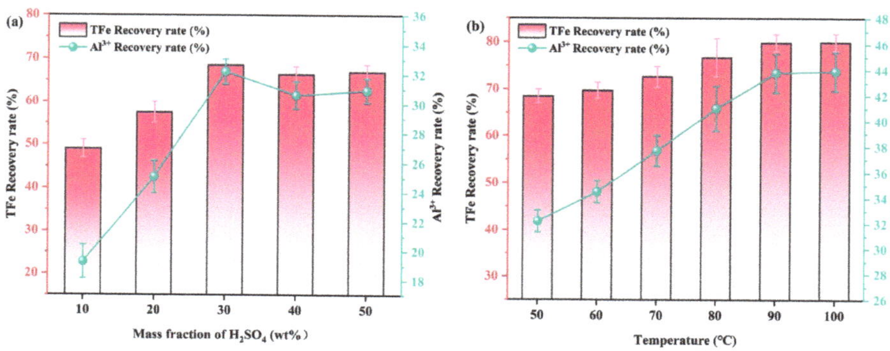

Figure 8. (**a**) The effect of H_2SO_4 concentration on the leaching rates of Al^{3+} and TFe, (**b**) the effect of leaching temperature on the leaching rates of Al^{3+} and TFe (co-roasting process: CG/CGCS mass ratio of 8/2, 600 °C, 1 h; H_2SO_4 leaching process: 5 h, liquid to solid ratio of 5:1 mL/g).

The effect of leaching temperature (50 °C–100 °C) on the extraction of valuable metal elements were investigated, as depicted in Figure 8b. In this study, the H_2SO_4 concentration was fixed at 30 wt%, while other conditions such as the liquid to solid ratio were maintained constant. As shown in Figure 8b, the leaching rates of Al^{3+} and TFe at 50 °C were only 32.21% and 68.47%, respectively. As the temperature increased, both the leaching rates of

Al^{3+} and TFe gradually improved; when the temperature reached 90 °C, the leaching rate of TFe reached 79.93%, while that of Al^{3+} was recorded as 43.78%. This can be attributed to enhanced diffusion processes and chemical reaction rates with increasing temperature, leading to more complete reactions. Moreover, beyond a temperature of 90 °C, further increases had minimal impact on enhancing the leaching rates of Al^{3+} and TFe significantly. Therefore, a leaching temperature of 90 °C was selected.

2.6. Analysis of Mixed Raw Material, Co-Roasting Product and Leached Residue

The mixed raw material, as well as the co-roasting product and leached residue obtained under optimal conditions, were characterized to analyze their differences. The SEM-EDS images of the three samples are recorded and presented in Figure 9. As depicted in Figure 9a, the particle size distribution of the mixed raw material exhibits non-uniformity, while displaying a typical lamellar structure of kaolinite. In Figure 9b, it can be observed that despite dehydroxylation of kaolinite into metakaolinite during co-roasting, the lamellar structure still exists, which may be attributed to the fact that although the $AlO_2(OH)_4$ octahedral layer is damaged, the SiO_4 tetrahedral layer can still maintain the lamellar structure. Furthermore, Figure 9c presents a scanning electron microscope image of the leached residue revealing evident signs of erosion. This indicates that after leaching aluminum and iron with sulfuric acid, the microstructure of aluminosilicates is disrupted. Additionally, both co-roasting product and leached residue exhibit some pore structures.

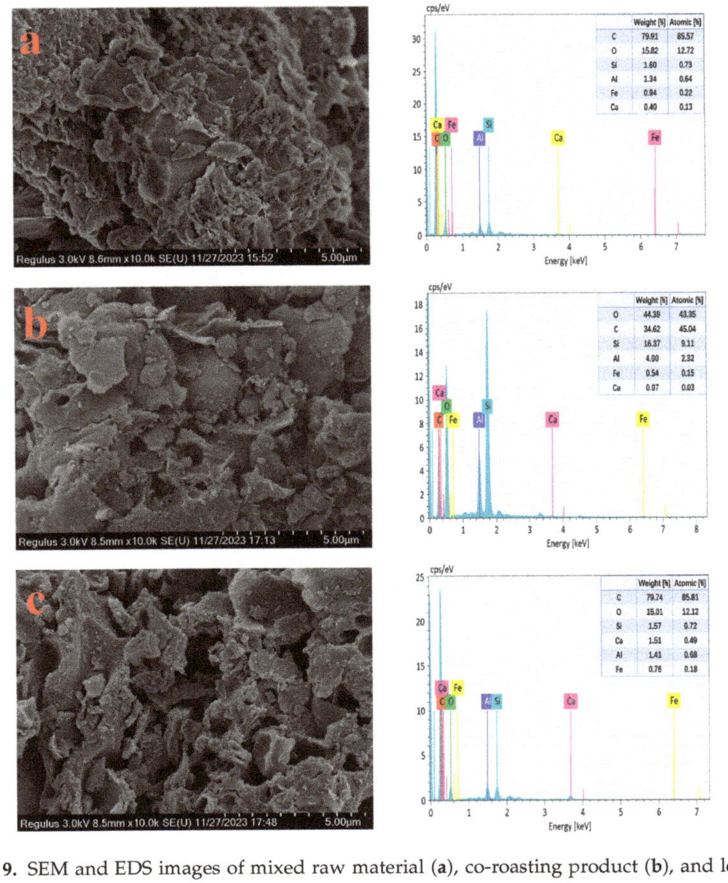

Figure 9. SEM and EDS images of mixed raw material (**a**), co-roasting product (**b**), and leached residue (**c**).

The EDS diagram in Figure 9 reveals a significantly higher mass percentage of elemental Ca in the leached residue compared to the other two samples, which may be due to the reaction of sulphuric acid with CaO in the samples to generate calcium sulphate, resulting in the enrichment of calcium ions in the leached residue.

The particle size distribution of the mixed raw material, co-roasting product, and leached residue is illustrated in Figure 10. The average particle sizes of the mixed raw material, co-roasting product, and leached residue were determined to be 151.10 μm, 103.87 μm, and 42.86 μm, respectively. Notably, the leached residue exhibited the smallest average particle size among all samples investigated, consistent with SEM observations and further confirming its abundant pore structure.

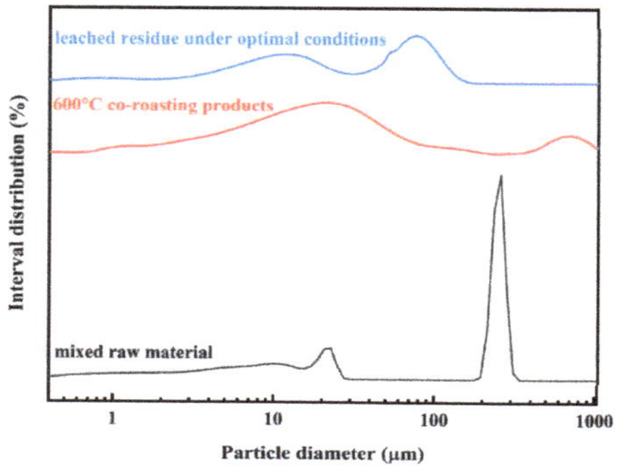

Figure 10. The particle size distribution of the mixed raw material, co-roasting product, and leached residue.

2.7. Prospects and Challenges in the Application of Acid Solution and Leached Residue

Table 1 illustrates the Chemical composition (wt%) of leached residue and chemical content (g/L) of an acid solution. As given in Table 1, alumina, potassium oxide, iron oxide and sodium oxide from the co-roasting product were partially dissolved in the acid solution, while calcium oxide and magnesium oxide reacted with sulphuric acid and were deposited in the leached residue, and the leached residue had the largest percentage of SiO_2; this is in agreement with the results of XRD analyses (Figure S3) of the leached residue.

Table 1. Chemical composition (wt%) of leached residue and chemical content (g/L) of an acid solution.

	SiO_2	Al_2O_3	K_2O	Fe (Fe_2O_3, Fe_3O_4, FeO, etc.)	S ($CaSO_4$, $MgSO_4$, etc.)	Na_2O	Others
The chemical composition (wt%) of leached residue	74.29	13.44	3.31	1.34	0.524	0.492	6.604
	$Al_2(SO_4)_3$		$FeSO_4$	$Fe_2(SO_4)_3$	K_2SO_4	Na_2SO_4	Others
The chemical content (g/L) of an acid solution	0.82		0.92	0.09	0.02	0.01	<0.01

Most of the existing coal-based waste treatment technologies focus solely on extracting valuable metal elements from a single type of coal-based waste residue, neglecting the synergistic extraction potential of multiple types of coal-based waste residues. In this study, a combination of CG and CGCS co-roasting + H_2SO_4 leaching was employed to extract valuable components (Al, Fe) from the acid solution, resulting in a leached residue with high silicon content and abundant pore structures. This indicates that both the acid solution and leached residue obtained in this study possess rich usable resources; however, effectively harnessing their value while minimizing environmental pollution remains a significant challenge. The technologies for disposal of coal-based waste residues have been systematically summarized and compared, with results presented in Table 2.

Table 2. Summary and comparison of coal-based waste residues disposal technologies.

Substance	Methods	Reagents	Extraction	Application	Ref.
Acid solution (high-iron CG)	Calcination + acid leaching	HCl	Al: 90% Fe: 91%	Preparation of PAFC flocculants	[61]
Acid solution (high-alumina fly ash)	Acid leaching + direct-electricity conversion technology + roasting	HCl	No reported	Preparation of Al_2O_3	[62]
Aacid solution (coal-bearing kaolinite)	Mechanical grinding + acid leaching	H_2SO_4	Al: 100%	Preparation of γ-Al_2O_3 powder	[63]
Leached residue (CG)	Calcination + acid leaching	HCl	No reported	Preparation of NaA zeolite	[64]
Leached residue (CGFS)	Calcination + acid leaching	HCl	The total leaching rate of all metal oxides: 80%	Preparation of mesoporous silica	[65]
Leached residue (CGS)	Non-hydrothermal sol–gel method	No reported	No reported	Preparation of MCM-41	[66]
Acid solution and leached residue (CG)	High temperature acid leaching	HCl	Al: 92.54% SiO_2: 96.01%	Preparation of Al_2O_3 and SiC	[15]
Acid solution and leached residue (CG and CGCS)	Co-roasting + acid leaching	H_2SO_4	Al: 43.78% Fe: 79.93%	Preparation of aluminium-iron flocculants and mesoporous silica	This work

We believe that the findings of these studies highlight the potential applications and benefits of utilizing H_2SO_4 acid solution and coal-based waste leached residue, providing technical support for the proper handling of acid solution and leached residue. It is crucial to note that adherence to regulations and guidelines is essential in handling these leaching wastes to minimize their environmental impact. Furthermore, our next step involves utilizing aluminum and iron ions present in the acid solution for fabricating coal-based waste residues-based aluminum-iron flocculants, which can be applied in treating industrial wastewater such as coal slurry water and dye wastewater, with a view to achieving the purpose of treating wastes with wastes. Simultaneously, we also focus on preparing mesoporous materials from acid-leached residue, with subsequent systematic investigations planned.

3. Materials and Methods

3.1. Materials

The CG and CGCS used in this study were obtained from Yan'an City, Shaanxi Province, China. The CG and CGCS were crushed and then ground with a rod mill for 2–3 min, passing through a 100-mesh sieve. The used H_2SO_4 was of analytical reagent and was purchased from Xi 'an Jiuheng Huaye Equipment Co., Ltd. The experiments were conducted using deionized water.

The chemical compositions of CG and CGCS, as well as their proximate analysis, are presented in Table 3.

Table 3. Chemical composition and proximate analysis of CG and CGCS (mass, %).

	Substance	SiO_2	Al_2O_3	Fe_2O_3	K_2O	MgO	CaO	Others	LOI
Chemical composition	Content in CG	58.02	20.87	10.67	4.41	2.17	1.34	2.52	20.10
	Content in CGCS	28.80	10.51	20.23	1.37	2.36	31.93	4.80	36.75
	Substance	Moisture		Ash		Volatiles		Fixed carbon	
Proximate analysis	Content in CG	1.7		85.81		9.93		2.56	
	Content in CGCS	2.15		74.88		5.78		17.19	

The Sulphur content of CG and CGCS both are <0.1%.

3.2. Experimental Procedures

The schematic in Figure 11 illustrates the co-roasting experiment of CG and CGCS, consisting of two main steps: co-roasting and H_2SO_4 leaching. In a typical co-roasting experiment, the as-received CG was thoroughly mixed with CGCS in a specific ratio, placed flat in a corundum crucible, and subjected to co-roasting at predetermined temperatures (500 °C–900 °C) and duration (1 h) using a tubular pyrolysis furnace (model GSL-1500X). The tubular pyrolysis furnace is filled with N_2 gas to provide a reduced atmosphere. To investigate the effect of co-roasting+H_2SO_4 leaching conditions on the extraction efficiency of valuable metal elements, the leaching process was conducted under the following conditions: leaching temperatures ranging from 50 °C to 100 °C, H_2SO_4 concentration varying from 10 wt% to 50 wt%, liquid to solid ratio set at 5:1 mL/g, and a leaching time of 5 h. The optimization process of the above experimental conditions is shown in Figures S1 and S2. The co-roasting product was subjected to H_2SO_4 leaching in a flat-bottomed flask equipped with a condensing device.

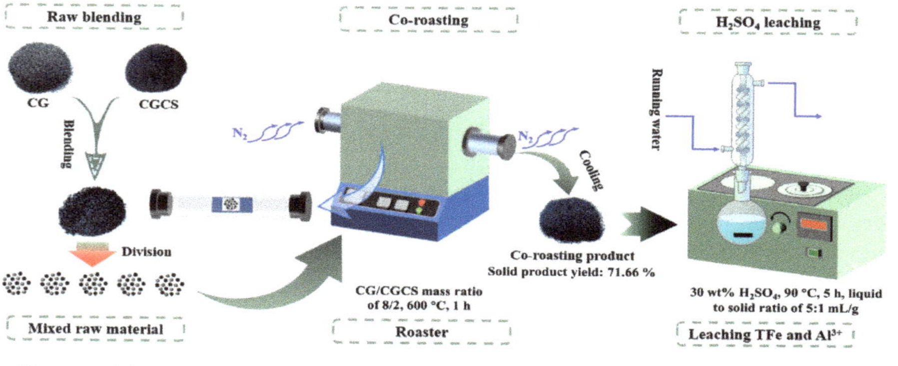

Figure 11. Schematic of CG and CGCS co-roasting experiment.

Firstly, the co-roasting product obtained under optimal co-roasting conditions was mixed with predetermined amounts of H_2SO_4 at a specified liquid to solid ratio. Subsequently, the resulting slurry was transferred to a flat-bottomed flask equipped with a condensing device and subjected to reaction in a thermostatic magnetic water bath pot, while maintaining a stirring speed of 150 r/min. Once the temperature of the mixed slurry reaches the predetermined temperature, the timing for the reaction begins. After the reaction, the slurry is cooled to room temperature and filtered. Evaluate the concentration of Al^{3+} and total iron (TFe) in the leaching solution, determine the ratio of Al^{3+} and TFe in the leaching solution to that in the mixed raw material, and subsequently calculate the leaching efficiency. These indicators of aluminum and iron leaching rates in the leaching solution serve as performance assessment metrics for evaluating the effectiveness of the co-roasting process.

The equation for determining the leaching rate is as follows:

$$\eta_{Al^{3+}/TFe} = \frac{m}{M} \times 100\% \quad (12)$$

where η represents the recovery rates of Al^{3+} and TFe, while m and M denote the Al^{3+} and TFe contents in the leaching solution and mixed raw material, respectively.

3.3. Characterization Methods

The chemical compositions of both the CG, CGCS, and leached residue were analyzed by X-ray fluorescence (XRF, Tokyo Rigaku Co., Simultix 12, Tokyo, Japan). XRF analysis was conducted with an Rh target, Si (Li) crystal detector, and calibrated based on fundamental parameters. Proximate analysis of the CG and CGCS was performed according to GB/T 212-2008 standards [67]. The pyrolysis characteristics of CG and mixed raw material were recorded using a thermogravimetric analyzer (TGA 8000 Perkin Elmer, Waltham, MA, USA) from 35 °C to 1000 °C in a nitrogen atmosphere at a flow rate of 20 mL/min. The heating rates of 10 °C/min were set to research the pyrolysis characteristics. The mineral phase-evolution features of the samples were conducted by X-ray diffraction technique (XRD; Bruker D8 ADVANCE, Ettlingen, Germany) with Cu Kα radiation from 10° to 80° at a scan speed of 10 °/min. The chemical bonds and functional groups in the samples were analyzed through a Fourier transform infrared spectrometer (FTIR; Bruker VERTEX 70, Ettlingen, Germany) within the wavenumber range of 400–3800 cm^{-1}. XPS analysis was performed on a ULVAC Inc. spectrometer equipped (XPS, ULVAC PHI 5000 VersaProbe 4, Chigasaki, Japan) with monochromatic Al Kα radiation as an excitation source, and the analyzed area of 400 μm in diameter. Raw data obtained were processed by the XPS Peak Fits software package (version number: XPSpeak 41) to fit all spectra. A scanning electron microscope (JSM-6390A, Jeol, Tokiyo, Japan) was used to analyze the microscopic morphology and the type and percentage of elements of samples. The particle size distribution of the samples was determined using a laser particle size analyzer (EyeTech, S/N 60296, Carlton, VIC, Australia). Analysis of the composition of the leaching solution was performed by using spectrophotometry.

4. Conclusions

The present study proposes an innovative approach of co-roasting and H_2SO_4 leaching to extract valuable metal elements (Al, Fe) from CG and CGCS, aiming at their treatment. Through thermodynamic analyses and systematic experimental investigations, the following outcomes were obtained:

(1) The effects of various factors on the extraction efficiency of valuable metal elements (Al, Fe) were investigated through single factor experiments. Under optimal conditions including 20 wt% CGCS content, 600 °C co-roasting temperature for 1 h followed by leaching at 90 °C with a liquid to solid ratio of 5:1 mL/g using a H_2SO_4 concentration of 30 wt%, a TFe leaching rate of 79.93% was achieved along with an Al^{3+} leaching rate of 43.78% after a leaching time of 5 h. Furthermore, we found that acid solution and leached residue both have broad application prospects.

(2) The activation behavior and phase transformation mechanism during the co-roasting process were investigated through Gibbs free energy calculation, as well as XRD, FTIR, and XPS characterization analysis. Inert kaolinite in CG and CGCS was converted to active metakaolinite after co-roasting. The reaction of metakaolinite with Si and Al elements from the mixed raw material produced sillimanite (Al_2SiO_5) and hercynite ($FeAl_2O_4$). Hematite (Fe_2O_3) was reduced to magnetite (Fe_3O_4) and wustite (FeO) by fixed carbon in CG and CGCS. Subsequently, wustite reacted with Si and Al elements from the mixed raw material to form iron phases such as ferrosilite ($FeSiO_3$), hercynite ($FeAl_2O_4$), and fayalite (Fe_2SiO_4).

(3) Co-roasting+H_2SO_4 leaching provides a novel method for extensive utilization of CG and CGCS while also offering a new approach for treating two or more types of coal-based solid waste that can alleviate industry development's pressure on the environment.

Supplementary Materials: The following supporting information can be downloaded at: https://www.mdpi.com/article/10.3390/molecules29010130/s1, Figure S1: Coal gangue H_2SO_4 leaching experiments; Figure S2: Coal gangue calcination+H_2SO_4 leaching experiments; Figure S3: XRD patterns of leached residue; 1-kaolinite; 2-quartz; 3-hematite; 4-muscovite; 5-calcite; 6-magnetite; 7-wustite; 8-fayalite; 9-sillimanite; 10-hercynite; 11-ferrosilite; 12-magnesium sulfate; 13-calcium sulfate; 14-potassium oxide; 15-alumina oxide; 16-sodium oxide.

Author Contributions: J.Z.: data curation, investigation, and writing—original draft; H.Z. and J.L.: validation, conceptualization, and supervision; Y.Z. and T.W.: data curation, investigation, and formal analysis; L.M.: investigation, methodology, and data curation; C.Q.: project administration and funding acquisition; T.Y.: writing—review and editing, validation, and supervision. All authors have read and agreed to the published version of the manuscript.

Funding: The work was supported financially by the Shaanxi Provincial Department and Municipal Linkage Key Project: Key Technology Development and Engineering Demonstration of Coal-Based Waste Co-utilization in Extremely Thin Coal Seam (No. 2022GD-TSLD-69), the Special Program for Serving Localities of Shaanxi Provincial Department of Education: Research on Reuse Technology of Return Fluid from Oil and Gas Field Measures in North Shaanxi Region(No. 22JC054), the open subject of State Key Laboratory of Petroleum Pollution Control (PPC2019001), and the Xi'an Shiyou University Postgraduate Innovation and Practical Ability Training Project (No. YCS23113051).

Institutional Review Board Statement: Not applicable.

Informed Consent Statement: Not applicable.

Data Availability Statement: Data are contained within the article and Supplementary Materials.

Acknowledgments: The authors also thank the support of The Youth Innovation Team of Shaanxi University and the work of the Modern Analysis and Testing Center of Xi'an Shiyou University.

Conflicts of Interest: Authors J.Z., T.Y., H.Z., Y.Z., L.M., C.Q. were employed by the company Shaanxi Fuquan Environmental Protection Technology Co., Ltd. The remaining authors declare that the research was conducted in the absence of any commercial or financial relationships that could be construed as a potential conflict of interest.

References

1. Li, M.; Zhang, J.; Li, A.; Zhou, N. Reutilisation of coal gangue and fly ash as underground backfill materials for surface subsidence control. *J. Clean. Prod.* **2020**, *254*, 120113. [CrossRef]
2. Shao, S.; Ma, B.; Wang, C.; Chen, Y. Extraction of valuable components from coal gangue through thermal activation and HNO_3 leaching. *J. Ind. Eng. Chem.* **2022**, *113*, 564–574. [CrossRef]
3. Li, J.; Wang, J. Comprehensive utilization and environmental risks of coal gangue: A review. *J. Clean. Prod.* **2019**, *239*, 117946. [CrossRef]
4. Wang, H.; Fang, X.; Du, F.; Tan, B.; Zhang, L.; Li, Y.; Xu, C. Three-dimensional distribution and oxidation degree analysis of coal gangue dump fire area: A case study. *Sci. Total Environ.* **2021**, *772*, 145606. [CrossRef] [PubMed]
5. Shu, Q.; Sun, Z.; Zhu, G.; Wang, C.; Li, H.; Qi, F.; Zhang, Q.; Li, S. Highly efficient synthesis of ZSM-5 zeolite by one-step microwave using desilication solution of coal gasification coarse slag and its application to VOCs adsorption. *Process Saf. Environ. Prot.* **2022**, *167*, 173–183. [CrossRef]
6. Zhu, D.; Zuo, J.; Jiang, Y.; Zhang, J.; Zhang, J.; Wei, C. Carbon-silica mesoporous composite in situ prepared from coal gasification fine slag by acid leaching method and its application in nitrate removing. *Sci. Total Environ.* **2020**, *707*, 136102. [CrossRef] [PubMed]
7. Li, M.; Li, A.; Zhang, J.; Huang, Y.; Li, J. Effects of particle sizes on compressive deformation and particle breakage of gangue used for coal mine goaf backfill. *Powder Technol.* **2020**, *360*, 493–502. [CrossRef]
8. Mjonono, D.; Harrison, S.; Kotsiopoulos, A. Supplementing structural integrity of waste rock piles through improved packing protocols to aid acid rock drainage prevention strategies. *Miner. Eng.* **2019**, *135*, 13–20. [CrossRef]

9. Wu, Y.; Yu, X.; Hu, S.; Shao, H.; Liao, Q.; Fan, Y. Experimental study of the effects of stacking modes on the spontaneous combustion of coal gangue. *Process Saf. Environ. Prot.* **2019**, *123*, 39–47. [CrossRef]
10. Shi, S.; Ma, B.; Zhao, D.; Li, X.; Shao, S.; Wang, C.; Chen, Y. Thermal Behavior of $Al(NO_3)_3{\cdot}9H_2O$ and its Application in preparing Al_2O_3 and Regenerating HNO_3. *Process Saf. Environ. Prot.* **2023**, *178*, 728–738. [CrossRef]
11. Chen, Y.; Zhou, X.; Wan, S.; Zheng, R.; Tong, J.; Hou, H.; Wang, T. Synthesis and characterization of geopolymer composites based on gasification coal fly ash and steel slag. *Constr. Build. Mater.* **2019**, *211*, 646–658. [CrossRef]
12. Li, Z.; Zhang, Y.; Zhao, H.; Chen, H.; He, R. Structure characteristics and composition of hydration products of coal gasification slag mixed cement and lime. *Constr. Build. Mater.* **2019**, *213*, 265–274. [CrossRef]
13. Liu, T.; Awasthi, M.K.; Awasthi, S.K.; Ren, X.; Liu, X.; Zhang, Z. Influence of fine coal gasification slag on greenhouse gases emission and volatile fatty acids during pig manure composting. *Bioresour. Technol.* **2020**, *316*, 123915. [CrossRef] [PubMed]
14. Yuan, N.; Zhao, A.; Hu, Z.; Tan, K.; Zhang, J. Preparation and application of porous materials from coal gasification slag for wastewater treatment: A review. *Chemosphere* **2021**, *287*, 132227. [CrossRef] [PubMed]
15. Xiao, J.; Li, F.; Zhong, Q.; Bao, H.; Wang, B.; Huang, J.; Zhang, Y. Separation of aluminum and silica from coal gangue by elevated temperature acid leaching for the preparation of alumina and SiC. *Hydrometallurgy* **2015**, *155*, 118–124. [CrossRef]
16. Yang, Q.; Zhang, F.; Deng, X.; Guo, H.; Zhang, C.; Shi, C.; Zeng, M. Extraction of alumina from alumina rich coal gangue by a hydro-chemical process. *R. Soc. Open Sci.* **2020**, *7*, 192132. [CrossRef] [PubMed]
17. Dong, L.; Li, Y.; Yan, J.; Shu, X.Q. Efficient Extraction of SiO_2 and Al_2O_3 from Coal Gangue by Means of Acidic Leaching. *Adv. Mater. Res.* **2014**, *878*, 149–156. [CrossRef]
18. Cheng, F.; Cui, L.; Miller, J.D.; Wang, X. Aluminum Leaching from Calcined Coal Waste Using Hydrochloric Acid Solution. *Miner. Process. Extr. Metall. Rev.* **2012**, *33*, 391–403. [CrossRef]
19. Gaviría, J.; Bohé, A.; Pasquevich, A.; Pasquevich, D. Hematite to magnetite reduction monitored by Mössbauer spectroscopy and X-ray diffraction. *Phys. B Condens. Matter* **2007**, *389*, 198–201. [CrossRef]
20. Pineau, A.; Kanari, N.; Gaballah, I. Kinetics of reduction of iron oxides by H_2: Part I: Low temperature reduction of hematite. *Thermochim. Acta* **2006**, *447*, 89–100. [CrossRef]
21. Strangway, P.; Ross, H. The Kinetics of Hematite Reduction by Partially Reformed Natural Gas. *Can. Metall. Q.* **1966**, *5*, 221–235. [CrossRef]
22. Han, R.; Guo, X.; Guan, J.; Yao, X.; Hao, Y. Activation Mechanism of Coal Gangue and Its Impact on the Properties of Geopolymers: A Review. *Polymers* **2022**, *14*, 3861. [CrossRef] [PubMed]
23. Qin, Q.; Deng, J.; Geng, H.; Bai, Z.; Gui, X.; Ma, Z.; Miao, Z. An exploratory study on strategic metal recovery of coal gangue and sustainable utilization potential of recovery residue. *J. Clean. Prod.* **2022**, *340*, 130765. [CrossRef]
24. Guan, X.; Chen, J.; Zhu, M.; Gao, J. Performance of microwave-activated coal gangue powder as auxiliary cementitious material. *J. Mater. Res. Technol.* **2021**, *14*, 2799–2811. [CrossRef]
25. Guo, Y.; Yan, K.; Cui, L.; Cheng, F. Improved extraction of alumina from coal gangue by surface mechanically grinding modification. *Powder Technol.* **2016**, *302*, 33–41. [CrossRef]
26. Li, C.; Wan, J.; Sun, H.; Li, L. Investigation on the activation of coal gangue by a new compound method. *J. Hazard. Mater.* **2010**, *179*, 515–520. [CrossRef]
27. Han, L.; Ren, W.; Wang, B.; He, X.; Ma, L.; Huo, Q.; Wang, J.; Bao, W.; Chang, L. Extraction of SiO_2 and Al_2O_3 from coal gangue activated by supercritical water. *Fuel* **2019**, *253*, 1184–1192. [CrossRef]
28. Zhao, W.; Yang, B.; Yi, Y.; Feng, Q.; Liu, D. Synergistic activation of smithsonite with copper-ammonium species for enhancing surface reactivity and xanthate adsorption. *Int. J. Min. Sci. Technol.* **2023**, *33*, 519–527. [CrossRef]
29. Shen, Z.; Wen, S.; Wang, H.; Miao, Y.; Wang, X.; Meng, S.; Feng, Q. Effect of dissolved components of malachite and calcite on surface properties and flotation behavior. *Int. J. Miner. Metall. Mater.* **2023**, *30*, 1297–1309. [CrossRef]
30. Wang, Y.; Zhang, Y.; Zhou, Q.; Zhang, Y.; Sun, J. Thermal kinetics analysis of coal-gangue selected from Inner Mongolia in China. *J. Therm. Anal. Calorim.* **2018**, *131*, 1835–1843. [CrossRef]
31. Xu, B.; Liu, Q.; Ai, B.; Ding, S.; Frost, R.L. Thermal decomposition of selected coal gangue. *J. Therm. Anal. Calorim.* **2018**, *131*, 1413–1422. [CrossRef]
32. Madejová, J. FTIR techniques in clay mineral studies. *Vib. Spectrosc.* **2003**, *31*, 1–10. [CrossRef]
33. Hu, H.Y.P. Controlled coating of antimony-doped tin oxide nanoparticles on kaolinite particles. *Appl. Clay Sci.* **2010**, *48*, 368–374. [CrossRef]
34. Ion, R.M.; Radovici, C.; Fierascu, R.C.; Fierascu, I. Thermal and mineralogical investigations of iron archaeological materials. *J. Therm. Anal. Calorim.* **2015**, *121*, 1247–1253. [CrossRef]
35. Cheng, H.; Li, K.; Liu, Q.; Zhang, S.; Li, X.; Frost, R.L. Insight into the thermal decomposition of kaolinite intercalated with potassium acetate: An evolved gas analysis. *J. Therm. Anal. Calorim.* **2014**, *117*, 1231–1239. [CrossRef]
36. Shao, S.; Ma, B.; Wang, C.; Chen, Y. Thermal behavior and chemical reactivity of coal gangue during pyrolysis and combustion. *Fuel* **2023**, 331. [CrossRef]
37. Bi, H.; Wang, C.; Lin, Q.; Jiang, X.; Jiang, C.; Bao, L. Pyrolysis characteristics, artificial neural network modeling and environmental impact of coal gangue and biomass by TG-FTIR. *Sci. Total Environ.* **2021**, *751*, 142293. [CrossRef] [PubMed]
38. Guo, Y.; Zhao, Q.; Yan, K.; Cheng, F.; Lou, H.H. Novel Process for Alumina Extraction via the Coupling Treatment of Coal Gangue and Bauxite Red Mud. *Ind. Eng. Chem. Res.* **2014**, *53*, 4518–4521. [CrossRef]

39. Magdy, A.; Fouad, Y.; Abdel-Aziz, M.; Konsowa, A. Synthesis and characterization of Fe_3O_4/kaolin magnetic nanocomposite and its application in wastewater treatment. *J. Ind. Eng. Chem.* **2017**, *56*, 299–311. [CrossRef]
40. Yuan, S.; Zhou, W.; Han, Y.; Li, Y. Selective enrichment of iron particles from complex refractory hematite-goethite ore by coal-based reduction and magnetic separation. *Powder Technol.* **2020**, *367*, 305–316. [CrossRef]
41. Zhu, D.; Cui, Y.; Vining, K.; Hapugoda, S.; Douglas, J.; Pan, J.; Zheng, G. Upgrading low nickel content laterite ores using selective reduction followed by magnetic separation. *Int. J. Miner. Process.* **2012**, *106–109*, 1–7. [CrossRef]
42. Yuan, S.; Zhou, W.; Han, Y.; Li, Y. Efficient enrichment of low-grade refractory rhodochrosite by preconcentration-neutral suspension roasting-magnetic separation process. *Powder Technol.* **2020**, *361*, 529–539. [CrossRef]
43. Yuan, S.; Zhou, W.; Han, Y.; Li, Y. Selective enrichment of iron from fine-grained complex limonite using suspension magnetization roasting followed by magnetic separation. *Sep. Sci. Technol.* **2019**, *55*, 3427–3437. [CrossRef]
44. Li, X.-B.; Xiao, W.; Liu, W.; Liu, G.-H.; Peng, Z.-H.; Zhou, Q.-S.; Qi, T.-G. Recovery of alumina and ferric oxide from Bayer red mud rich in iron by reduction sintering. *Trans. Nonferrous Met. Soc. China* **2009**, *19*, 1342–1347. [CrossRef]
45. Qiao, X.C.; Si, P.; Yu, J.G. A Systematic Investigation into the Extraction of Aluminum from Coal Spoil through Kaolinite. *Environ. Sci. Technol.* **2008**, *42*, 8541–8546. [CrossRef] [PubMed]
46. Liu, X.; Liu, X.; Hu, Y. Investigation of the thermal behaviour and decomposition kinetics of kaolinite. *Clay Miner.* **2015**, *50*, 199–209. [CrossRef]
47. Omsinsombon, J.; Chaiyasat, A.; Busabok, C.; Chaiyasat, P. A novel iron aluminate composite polymer particle for high-efficiency self-coating solar heat reflection. *Sol. Energy Mater. Sol. Cells* **2021**, *230*, 111248. [CrossRef]
48. Jin, J.; Liu, X.; Yuan, S.; Gao, P.; Li, Y.; Zhang, H.; Meng, X. Innovative utilization of red mud through co-roasting with coal gangue for separation of iron and aluminum minerals. *J. Ind. Eng. Chem.* **2021**, *98*, 298–307. [CrossRef]
49. Yamashita, T.; Hayes, P. Analysis of XPS spectra of Fe^{2+} and Fe^{3+} ions in oxide materials. *Appl. Surf. Sci.* **2008**, *254*, 2441–2449. [CrossRef]
50. Fang, S.; Xu, L.; Wu, H.; Xu, Y.; Wang, Z.; Shu, K.; Hu, Y. Influence of surface dissolution on sodium oleate adsorption on ilmenite and its gangue minerals by ultrasonic treatment. *Appl. Surf. Sci.* **2020**, *500*, 144038. [CrossRef]
51. Omran, M.; Fabritius, T.; Elmahdy, A.M.; Abdel-Khalek, N.A.; El-Aref, M.; Elmanawi, A.E.-H. XPS and FTIR spectroscopic study on microwave treated high phosphorus iron ore. *Appl. Surf. Sci.* **2015**, *345*, 127–140. [CrossRef]
52. Salama, W.; El Aref, M.; Gaupp, R. Spectroscopic characterization of iron ores formed in different geological environments using FTIR, XPS, Mössbauer spectroscopy and thermoanalyses. *Spectrochim. Acta Part A Mol. Biomol. Spectrosc.* **2015**, *136*, 1816–1826. [CrossRef]
53. Dong, H.; Du, W.; Dong, J.; Che, R.; Kong, F.; Cheng, W.; Ma, M.; Gu, N.; Zhang, Y. Depletable peroxidase-like activity of Fe_3O_4 nanozymes accompanied with separate migration of electrons and iron ions. *Nat. Commun.* **2022**, *13*, 5365. [CrossRef] [PubMed]
54. Grosvenor, A.; Kobe, B.; McIntyre, N. Examination of the oxidation of iron by oxygen using X-ray photoelectron spectroscopy and QUASESTM. *Surf. Sci.* **2004**, *565*, 151–162. [CrossRef]
55. Liu, H.; Zhang, S.; Yang, W.; Yu, N.; Liu, C.; Chai, Y.; Dong, B. Directional Reconstruction of Iron Oxides to Active Sites for Superior Water Oxidation. *Adv. Funct. Mater.* **2023**, *33*, 2303776. [CrossRef]
56. Gu, A.; Chen, K.; Zhou, X.; Gong, C.; Wang, P.; Jiao, Y.; Mao, P.; Chen, K.; Lu, J.; Yang, Y. Trimetallic MOFs-derived Fe-Co-Cu oxycarbide toward peroxymonosulfate activation for efficient trichlorophenol degradation via high-valent metal-oxo species. *Chem. Eng. J.* **2023**, *468*, 143444. [CrossRef]
57. Li, X.; Wang, H.; Zhou, Q.; Qi, T.; Liu, G.; Peng, Z. Efficient separation of silica and alumina in simulated CFB slag by reduction roasting-alkaline leaching process. *Waste Manag.* **2019**, *87*, 798–804. [CrossRef]
58. Zhou, Q.-S.; Li, C.; Li, X.-B.; Peng, Z.-H.; Liu, G.-H.; Qi, T.-G. Reaction behavior of ferric oxide in system Fe_2O_3–SiO_2–Al_2O_3 during reductive sintering process. *Trans. Nonferrous Met. Soc. China* **2016**, *26*, 842–848. [CrossRef]
59. Zhang, Y.; Xu, L.; Seetharaman, S.; Liu, L.; Wang, X.; Zhang, Z. Effects of chemistry and mineral on structural evolution and chemical reactivity of coal gangue during calcination: Towards efficient utilization. *Mater. Struct.* **2014**, *48*, 2779–2793. [CrossRef]
60. Zhang, C.; Yang, X.Y.; Li, Y.F. Mechanism and Structural Analysis of the Thermal Activation of Coal-Gangue. *Adv. Mater. Res.* **2011**, *356–360*, 1807–1812. [CrossRef]
61. Kong, D.; Zhou, Z.; Song, S.; Feng, S.; Lian, M.; Jiang, R. Preparation of Poly Aluminum-Ferric Chloride (PAFC) Coagulant by Extracting Aluminum and Iron Ions from High Iron Content Coal Gangue. *Materials* **2022**, *15*, 2253. [CrossRef] [PubMed]
62. Zhang, T.A.; Han, X.; Lv, G.; Pan, X.; Agajan, S.; Fu, D.; Liu, J.; Zhang, J. Research on Alumina Preparation from Aluminium Chloride Solution by Electrolysis Process. In *Light Metals 2018, Proceedings of the TMS 2018, Phoenix, AZ, USA, 11–15 March 2018*; Martin, O., Ed.; Springer International Publishing: Cham, Switzerland, 2018; pp. 97–103. [CrossRef]
63. Asuha, S.; Talintuya, T.; Han, Y.; Zhao, S. Selective extraction of aluminum from coal-bearing kaolinite by room-temperature mechanochemical method for the preparation of γ-Al_2O_3 powder. *Powder Technol.* **2018**, *325*, 121–125. [CrossRef]
64. Kong, D.; Jiang, R. Preparation of NaA Zeolite from High Iron and Quartz Contents Coal Gangue by Acid Leaching—Alkali Melting Activation and Hydrothermal Synthesis. *Crystals* **2021**, *11*, 1198. [CrossRef]
65. Liu, S.; Chen, X.; Ai, W.; Wei, C. A new method to prepare mesoporous silica from coal gasification fine slag and its application in methylene blue adsorption. *J. Clean. Prod.* **2019**, *212*, 1062–1071. [CrossRef]

66. Li, C.-C.; Qiao, X.-C.; Yu, J.-G. Large surface area MCM-41 prepared from acid leaching residue of coal gasification slag. *Mater. Lett.* **2016**, *167*, 246–249. [CrossRef]
67. *GB/T 212-2008*; Proximate Analysis of Coal. China Coal Industry Association: Beijing, China, 2008.

molecules

MDPI

Article

Boosting Photocatalytic Performance of ZnO Nanowires via Building Heterojunction with g-C_3N_4

Yayang Wang [1,2], Ziyi Liu [1], Yuesheng Li [1,*], Xiaojie Yang [1,*], Lingfei Zhao [3] and Jian Peng [3,*]

1 Hubei Key Laboratory of Radiation Chemistry and Functional Materials, School of Nuclear Technology and Chemistry & Biology, Hubei University of Science and Technology, Xianning 437100, China; wyy934313@163.com (Y.W.); yiyi3092208242@163.com (Z.L.)

2 School of Chemistry and Chemical Engineering, Wuhan University of Science and Technology, Wuhan 430074, China

3 Institute for Superconducting and Electronic Materials, Australian Institute for Innovative Materials, Innovation Campus, University of Wollongong, Squires Way, North Wollongong, NSW 2522, Australia; lingfeiz@uow.edu.au

* Correspondence: frank78929@163.com (Y.L.); mailyangxiaojie@126.com (X.Y.); jianp@uow.edu.au (J.P.)

Abstract: The development of a stable and highly active photocatalyst has garnered significant attention in the field of wastewater treatment. In this study, a novel technique involving a facile stirring method was devised to fabricate an array of g-C_3N_4/ZnO nanowire (ZnO NW) composites. Through the introduction of g-C_3N_4 to augment the generation of electron-hole pairs upon exposure to light, the catalytic efficacy of these composites was found to surpass that of the pristine ZnO NWs when subjected to simulated sunlight. The photocatalytic performance of a 20 mg·L^{-1} methylene blue solution was found to be highest when the doping rate was 25 wt%, resulting in a degradation rate of 99.1% after 60 min. The remarkable enhancement in catalytic efficiency can be ascribed to the emergence of a captivating hetero-junction at the interface of g-C_3N_4 and ZnO NWs, characterized by a harmoniously aligned band structure. This alluring arrangement effectively curtailed charge carrier recombination, amplified light absorption, and augmented the distinct surface area, culminating in a notable boost to the photocatalytic prowess. These findings suggest that the strategic engineering of g-C_3N_4/ZnO NW heterostructures holds tremendous promise as a pioneering avenue for enhancing the efficacy of wastewater treatment methodologies.

Keywords: g-C_3N_4; ZnO NW; heterostructure; photodegradation

1. Introduction

The swift growth of the socio-economy and chemical industry as resulted in the release of large volumes of wastewater containing antibiotics and dyes into water bodies, leading to water pollution. This poses a threat to ecosystems, human health, and the environment, making it a pressing issue that requires attention [1–5]. Over time, various methods for the removal of toxic components in wastewater have continued to be devised, including membrane separation, adsorption, chemical methods, and photocatalytic technology [6–11]. Among these techniques, semiconductor photocatalysis had emerged as one of the most crucial and promising method due to its ease of handling, good reproducibility, simplicity, outstanding cost-effective performance, and environmental friendliness [12,13]. Therefore, designing and developing a novel semiconductor material with broad spectral response, low cost, outstanding photoactivity, and excellent stability is a key research area in the quest to improve the efficient utilization of solar energy [14].

Previous studies had investigated numerous semiconductors as efficient photocatalysts, such as TiO_2 [15], ZnO, $SrTiO_3$ [16], Fe_2O_3 [17], CdS [18], WO_3 [19], and ZnS [20]. Among numerous multifunctional semiconductor metal oxide materials, ZnO had been widely applied in fields such as luminescent materials, gas sensing, optical devices, solar cells, and photocatalysis due to its characteristics including non-toxicity, ease of substitution, low cost, and stability. As an emerging semiconductor material, ZnO had received significant attention worldwide for its potential in degrading organic pollutants and is generally considered one of the most attractive semiconductor materials [21]. While ZnO offers numerous benefits, its usefulness in practical applications is restricted by its wide band gap of 3.37 eV and high recombination efficiency of electron-hole pairs. To overcome these limitations, researchers had employed various strategies, such as doping, surface modification, and forming nanostructures to improve the optical and electronic characteristics of the materials [22]. So far, fabricating ZnO-based semiconductor composites has been proved to be an effective route to improving the photocatalytic activity of ZnO. By utilizing the difference between the conduction bands and valence of various photocatalytic materials, this approach successfully segregates photogenerated electrons and holes, thereby increasing the photocatalytic capability of ZnO. Moreover, semiconductors with narrow band gaps can absorb visible light, thereby expanding the light-absorption range of ZnO [23–26]. Li et al. developed the chemical vapor deposition method to synthesis ZnO nanowires (ZnO NWs) decorated with Au NPs on silicon substrates [27]. The photocatalytic activity of Au/ZnO NWs was found to be significantly higher than that of pristine ZnO NWs in the degradation of methylene blue (MB) solution under simulated sunlight irradiation. In order to improve the photocatalytic efficiency of ZnO NWs, Yamina et al. [28] successfully synthesized iron-doped ZnO NWs (Fe/ZnO NWs) by a two-step hydrothermal method. The Fe/ZnO NWs were tested for their photocatalytic ability to break down organic pollutants in water using simulated pollutants like MB, methyl orange (MO), and acid red 14 (AR14). The results showed that doping 1 wt% Fe/ZnO NWs led to a 9%, 20%, and 5% increase in photodegradation rates of MB, MO, and AR14, respectively. Z. Braiek et al. [29] synthesized ZnO/In_2S_3 core/shell NWs on indium tin oxide substrates through a simple and economical two-step electrochemical method, with successful results. Experimental results showed that the ZnO/In_2S_3 core/shell NWs could completely degrade pollutants within 120 min, with photodegradation rates that were five times faster than those of pure ZnO NWs. Nevertheless, the existing precursor selections in these state-of-the-art methods remain inadequate for the practical implementation of ZnO-based catalysis in wastewater treatment. Major hurdles include their exorbitant costs, intricate synthesis pathways, and inherent toxicity. Thus, the development of innovative synthesis techniques becomes imperative in order to fabricate ZnO-based catalytic materials that can foster the advancement of low-cost, highly efficient, and safe methods for wastewater purification.

Although the heterojunction between ZnO and g-C_3N_4 is not new, it is still of paramount importance in developing a simple, efficient, and low-cost method to synthesize a high-stability and high-performance catalyst. In this work, a remarkable set of g-C_3N_4/ZnO NW composites, showcasing exceptional catalytic performance, remarkable stability, and effortless recyclability, were successfully synthesized through the meticulous selection of precursors and the implementation of a novel and straightforward stirring technique. In the preparation process of ZnO NWs, the samples demonstrated a remarkable level of reproducibility. This could be attributed to several factors, including the simplicity of the operation, meticulous control over specific experimental conditions and relevant parameters, and the excellent reproducibility of the synthesis method employed. Additionally, the stability and uniformity maintained throughout the synthesis process further contributed to the exceptional reproducibility observed in the prepared ZnO NW samples. Due to the slender band gap (E_g = 2.7 eV), ample specific surface area, robust photostability, and non-toxic, non-polluting nature of the elaborately selected g-C_3N_4, the as-prepared g-C_3N_4/ZnO NW composites showed enhanced photocatalytic performance. The crystal phase, morphology, chemical composition, valence, optical properties, and specific surface area of the samples were analyzed by multiple techniques. In addition, the photocatalytic performance of the prepared catalyst was evaluated through light-induced degradation experiments of MB. The stability of the catalyst was explored by cycling experiments and possible degradation mechanisms were proposed and discussed by analyzing active species trapping experiments. This study could provide a valuable roadmap for harnessing the potential of g-C_3N_4/ZnO semiconductor photocatalysts in the eradication of organic pollutants from wastewater, paving the way for groundbreaking advancements in the field of wastewater treatment.

2. Results and Discussion

2.1. XRD and FT-IR Analysis

Figure 1a presented the Fourier transform infrared (FT-IR) spectrum of the g-C_3N_4, ZnO NWs, and 25 wt% g-C_3N_4/ZnO NWs. The peaks observed at 1245.19 cm^{-1} and 1639.51 cm^{-1} in the g-C_3N_4 spectrum were attributed to the C-N and C=N stretching vibrations in the aromatic carbon–nitrogen heterocycles. Furthermore, a vibration peak originating from three-s-triazine rings appeared near 812.14 cm^{-1}, which corresponded to the XRD results and provided further evidence of the graphene-like structure of g-C_3N_4. The relatively weaker absorption peak at 546.42 cm^{-1} could be assigned to the symmetric stretching vibration of Zn-O, indicating the formation of ZnO crystals. The FT-IR spectrum of 25 wt% g-C_3N_4/ZnO NWs was similar to the characteristic spectra of g-C_3N_4 and ZnO NWs, providing evidence that g-C_3N_4 was effectively composited with ZnO NWs while retaining its typical graphite structure.

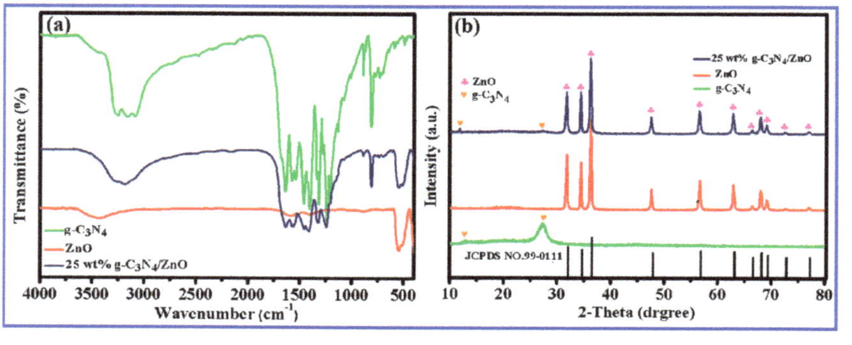

Figure 1. Characterization diagrams of g-C_3N_4, ZnO NWs, and 25 wt% g-C_3N_4/ZnO NWs: (**a**) IR pattern and (**b**) XRD pattern.

Figure 1b shows the X-ray powder diffraction (XRD) patterns of the g-C_3N_4, ZnO NWs, and 25 wt% g-C_3N_4/ZnO NW composites. The XRD pattern of the g-C_3N_4 showed two obvious diffraction peaks at 13.0° and 27.3°. The diffraction peak near 13.0° belonged to the (100) plane and represented the distance of the repeating unit in the plane of the conjugated layer, corresponding to the macrocyclic structure between the three-s-triazine units. The characteristic peak near 27.3° could be attributed to the (002) plane of typical graphitic layered structure, which was attributed to the interlayer stacking structure of aromatic hydrocarbons [30]. The characteristic diffraction peaks of the ZnO NWs at 2θ of 31.77°, 34.42°, 36.25°, 47.54°, 56.59°, 62.85°, 66.37°, 67.94°, 69.08°, 72.56°, and 76.95° were assigned to (100), (002), (101), (102), (110), (103), (200), (112), (201), (004), and (202) crystal faces (JCPDS NO.99-0111), indicating that ZnO with wurtzite structure was successfully prepared. The XRD patterns of 25 wt% g-C_3N_4/ZnO NW composites exhibited the characteristic peaks of g-C_3N_4 and ZnO at the same time, and each diffraction peak was sharp without impurity peaks. The successful preparation of g-C_3N_4/ZnO NWs was confirmed, and it was also shown that the resulting samples exhibited excellent crystallinity.

2.2. XPS Analysis

XPS measurements were conducted to analyze the superficial composition and chemical state of the samples (Figure 2). The full survey spectrum in Figure 2a confirmed the presence of Zn, O, C, and N elements in the 25 wt% g-C_3N_4/ZnO NW composite, providing evidence for the co-existence of 25 wt% g-C_3N_4 and ZnO NWs in the composite. As reported previously [31], the two sharp peaks at 1045.2 eV (Zn $2p_{1/2}$) and 1022.0 eV (Zn $2p_{3/2}$) in Figure 2b matched well with that of Zn^{2+}. In Figure 2c, the characteristic peak of O 1s at 531.1 eV corresponded to O^{2-} in ZnO. The observed binding energies at 284.7 eV and 288.5 eV in the C 1s sub-peak spectrum of Figure 2d corresponded to the C-C and C=N bonds, respectively. In Figure 2e, the strong characteristic peaks of N 1s appeared at 399.1 eV and 400.1 eV, corresponding to C-N=C and N-$(C)_3$, respectively, which further verified the successful composite of ZnO NWs and g-C_3N_4. The XPS results provided additional evidence to support the successful formation of the heterojunction, which was consistent with the other characterization findings.

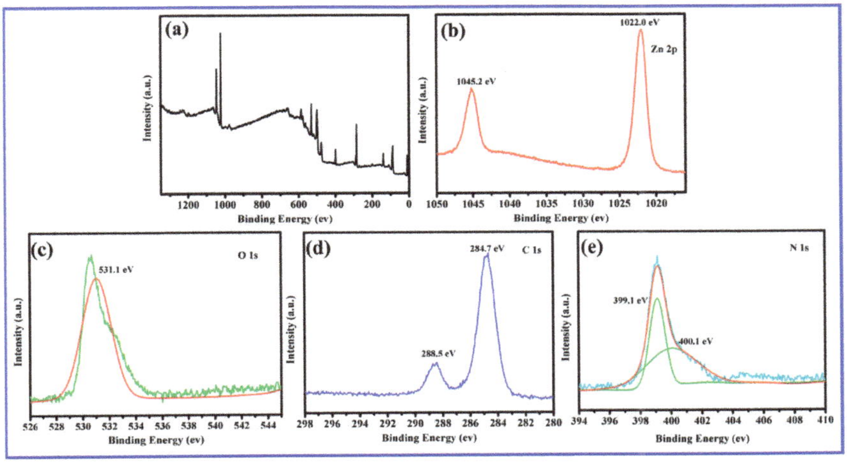

Figure 2. XPS spectra of 25 wt% g-C_3N_4/ZnO NWs: (**a**) survey spectrum, (**b**) Zn 2p, (**c**) O 1s, (**d**) C 1s, and (**e**) N 1s.

2.3. Morphology and Microstructure Analysis

The morphology of g-C_3N_4, ZnO NWs, and 25 wt% g-C_3N_4/ZnO NW samples was analyzed by scanning electron microscope (SEM). Figure 3a,b show the multilayer stacked structure of g-C_3N_4. As shown in Figure 3c,d, the prepared ZnO NWs exhibited a linear structure with a diameter of approximately 100 nm and a length of approximately 2.433×10^3 nm. The surface of these nanowires presented fine and dense features, without any apparent orientation or pointed structure. Figure 3e,f clearly indicate that, in addition to the stacking of g-C_3N_4 sheets in 25 wt% g-C_3N_4/ZnO NWs, the linear structure of ZnO was uniformly distributed on the surface of g-C_3N_4. This structure could be beneficial to the formation of heterojunctions to enhance the photocatalytic activity. The lamellar structure of g-C_3N_4, the disordered linear structure of ZnO NWs, and the microstructure of g-C_3N_4/ZnO NW composites were further analyzed by transmission electron microscopy. The results of both Figures 3 and 4 illustrate that ZnO was dispersed on the surface of g-C_3N_4, which was in line with the findings of XRD and SEM analyses.

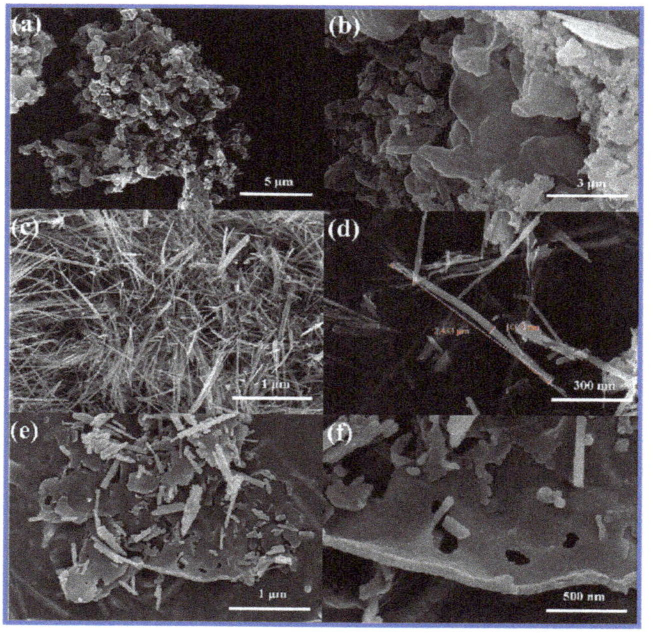

Figure 3. (**a**,**b**) SEM images of g-C_3N_4, (**c**,**d**) SEM images of ZnO NWs, and (**e**,**f**) SEM images of 25 wt% g-C_3N_4/ZnO NWs.

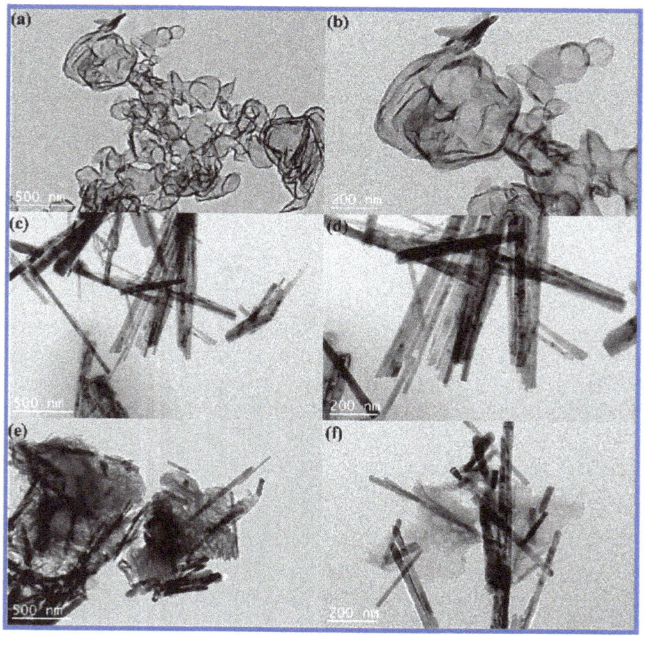

Figure 4. (**a**,**b**) TEM images of g-C_3N_4, (**c**,**d**) TEM images of ZnO NWs, and (**e**,**f**) TEM images of 25 wt% g-C_3N_4/ZnO NWs.

2.4. Analysis of Optical Properties

In order to assess the rate of recombination of photoinduced charge carriers, photoluminescence (PL) emission spectra were measured. The optical properties of g-C_3N_4, ZnO NWs, and 25 wt% g-C_3N_4/ZnO NWs were investigated with an excitation wavelength of 365 nm. As shown in Figure 5a, the PL intensity of 25 wt% g-C_3N_4/ZnO NWs was much weaker than that of single g-C_3N_4. This result indicated that the recombination rate of photoinduced carriers in the 25 wt% g-C_3N_4/ZnO NW composite was greatly reduced, thereby enhancing the photocatalytic activity.

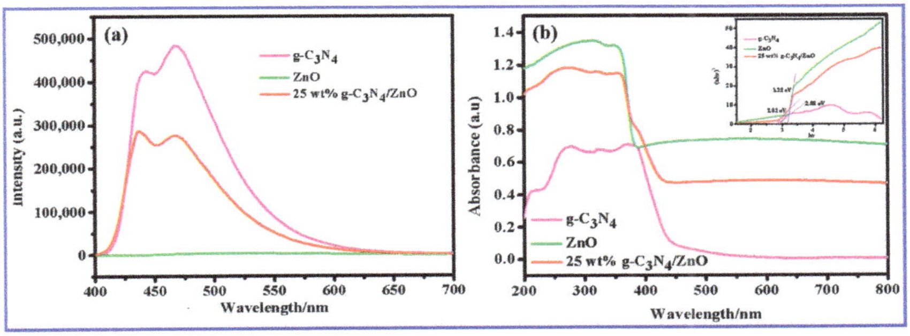

Figure 5. (**a**) Photoluminescence (PL) emission spectra of g-C_3N_4, ZnO NWs, and 25 wt% g-C_3N_4/ZnO NWs and (**b**) UV-Vis spectra of g-C_3N_4, ZnO NWs, and 25 wt% g-C_3N_4/ZnO NWs.

The optical properties of the photocatalyst were analyzed by UV-Vis DRS. Figure 5b shows the UV-Vis spectra and bandgap diagrams (upper right inset) of g-C_3N_4, ZnO NWs, and 25 wt% g-C_3N_4/ZnO NWs, with E_g values of 2.88 eV, 3.22 eV, and 2.81 eV, respectively. The successful recombination of g-C_3N_4 with ZnO NWs resulted in a narrow band gap and better photo response.

2.5. N_2 Adsorption–Desorption Isotherm Analysis

The surface area of the catalysts was measured by testing their specific surface area. The nitrogen adsorption–desorption isotherms for three different samples are presented in Figure 6, including g-C_3N_4, ZnO NWs, and a combination of 25 wt% g-C_3N_4/ZnO NWs. According to the data, the measured specific surface areas of g-C_3N_4, ZnO NWs, and 25 wt% g-C_3N_4/ZnO NWs were 22.5233 $m^2 \cdot g^{-1}$, 10.2677 $m^2 \cdot g^{-1}$, and 15.4558 $m^2 \cdot g^{-1}$, respectively. The analysis of the data showed that the specific surface area of ZnO NWs was significantly increased after g-C_3N_4 was combined with ZnO NWs. As the specific surface area increases in a photocatalytic reaction, contact with the reactants can be achieved more fully. This is highly beneficial for the later stages of the photocatalytic process.

2.6. Evaluation of Photocatalytic Degradation Performance

MB was used as a simulated pollutant to explore the photocatalytic performance of the catalyst under visible light irradiation ($\lambda \geq 400$ nm). To ensure that MB did not undergo self-degradation, a blank control experiment was incorporated in the study. Specifically, 20 $mg \cdot L^{-1}$ MB was subjected to photodegradation without the presence of any catalyst. Based on the subsequent experimental findings, it can be concluded that MB does not possess the ability to self-degrade and is suitable for use in subsequent photocatalytic degradation experiments. It could be seen from Figure 7a that the photocatalytic performance of g-C_3N_4/ZnO series composites was enhanced when compared with single ZnO. The activity of the photocatalysts followed the order g-C_3N_4/ZnO NWs, pure ZnO NWs, and g-C_3N_4. When the doping ratio of g-C_3N_4 was 25 wt%, the formed composite material showed the best performance in degrading MB, and the degradation rate of MB solution was as high as 99.07% within 60 min. The g-C_3N_4/ZnO NW composite exhibited excellent photocatalytic performance, which can be attributed to two factors. Firstly, the combination of g-C_3N_4 and ZnO NWs increased light absorption and specific surface area. Secondly, the formation of the heterostructure facilitated the transfer and separation of charge carriers. The results further indicated that the as-prepared composite was of high photocatalytic activity and potential ability to efficiently degrade organic pollutants.

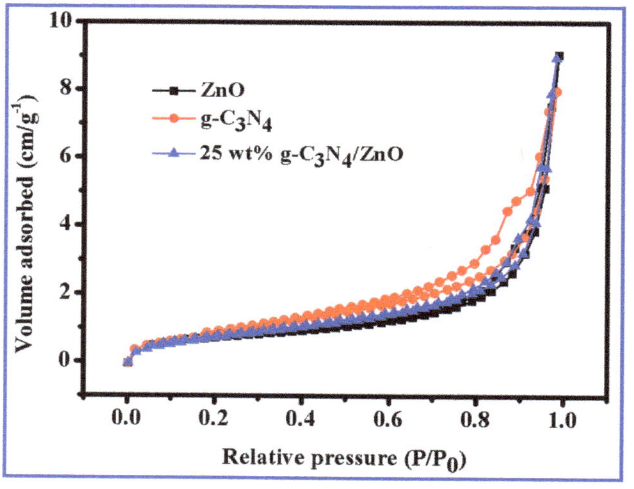

Figure 6. Nitrogen multilayer adsorption–desorption curves of g-C_3N_4, ZnO NWs, and 25 wt% g-C_3N_4/ZnO NWs.

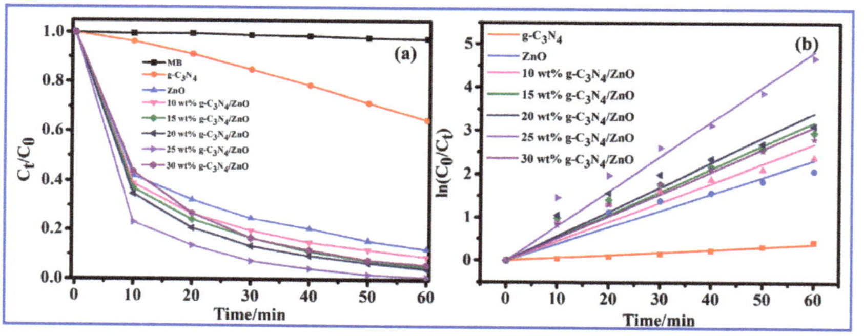

Figure 7. (**a**) Photocatalytic degradation curves of MB on different prepared samples and (**b**) degradation kinetics curves of MB with different photocatalysts.

In order to gain a deeper understanding of the mechanism in the photocatalytic degradation activity of various catalysts for MB, the experimental data in Figure 7a were fitted with the first-order kinetics according to the formula $-\ln (C_t/C_0) = kt$, as shown in Figure 7b. In this formula, "C_0" refers to the initial concentration of MB, "C_t" represents the concentration of MB after a given illumination time "t", and "k" denotes the rate constant of pseudo first order. By plotting the graph of $-\ln (C_t/C_0)$ versus time "t", the rate constants for all photocatalysts were determined from the slope of the straight line. Table 1 lists the corresponding kinetic rate constants and the corresponding MB degradation rates when using different catalysts. From the analysis of the data, it is worth noting that all samples exhibited an almost linear curve ($R^2 > 0.96$), which confirmed that the degradation of MB dye followed pseudo-first-order reaction kinetics. It is well-known that a higher photocatalytic activity is achieved with a higher first order constant [32]. The catalytic reaction rate of the composite catalysts was higher than that of pure ZnO. Moreover, the k value of 25 wt% g-C_3N_4/ZnO was the largest (0.08045), and 25 wt% g-C_3N_4/ZnO also produced the best degradation effect, which is consistent with the experimental results.

Table 1. Kinetic rate constants of MB degradation rate with different catalysts.

	Y = ln(C_0/C_t)	**R^2**	**Degradation Rate (%)**
g-C_3N_4	Y = 0.00653X	0.98	35.21
ZnO	Y = 0.03886X	0.96	87.53
10 wt% g-C_3N_4/ZnO	Y = 0.04498X	0.96	90.82
15 wt% g-C_3N_4/ZnO	Y = 0.05332X	0.98	94.80
20 wt% g-C_3N_4/ZnO	Y = 0.05691X	0.98	95.55
25 wt% g-C_3N_4/ZnO	Y = 0.08045X	0.99	99.07
30 wt% g-C_3N_4/ZnO	Y = 0.05156X	0.98	93.99

2.7. Reusability and Stability of the Composites

Apart from exhibiting excellent photocatalytic performance, assessing the reusability and cycle stability of photocatalysts is a necessary and crucial step in their practical application. Therefore, 25 wt% g-C_3N_4/ZnO NWs were subjected to continuous degradation of MB to evaluate their recovery availability under identical conditions. After each cycle test, the catalyst was first subjected to high-speed centrifugation, followed by several washes with ethanol and water. Finally, the catalyst was dried in an oven at 60 °C, ground, and reused in the next cycle. As shown in Figure 8a, the catalytic degradation efficiency for MB of 25 wt% g-C_3N_4/ZnO NWs decreased slightly when the number of degradation cycles increased. The photocatalyst was tested for reusability and cycle stability by conducting three cycle experiments, where 25 wt% g-C_3N_4/ZnO NWs were continuously used to degrade MB under the same conditions. After each cycle, the used catalyst was centrifuged, washed with ethanol and water, dried at 60 °C, and ground before reuse. The results showed that the degradation rates of MB by the photocatalyst were 97.77%, 94.44%, and 87.67% in the first, second, and third cycles, respectively. Despite this gradual decline in performance, the removal efficiency of MB remained high at 80.2% after three degradation–regeneration runs, indicating the good reusability and cycle stability of the composite photocatalyst.

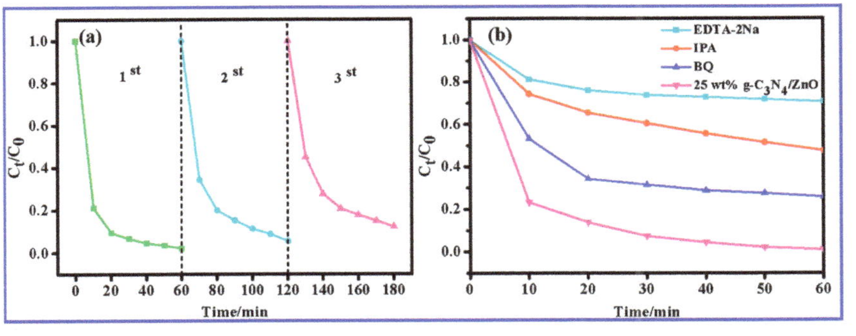

Figure 8. MB degradation by 25 wt% g-C_3N_4/ZnO NWs: (**a**) cycle experiment diagram and (**b**) capture experiment diagram of related active substances.

2.8. Possible Photodegradation Mechanism

To identify the active species involved in the photocatalytic process, radical species trapping experiments were carried out by adding quenchers, including EDTA-2Na (h^+ scavenger), IPA (·OH scavenger), and BQ (·O^{2-} scavenger). As shown in Figure 8b, the photocatalytic performance of 25 wt% g-C_3N_4/ZnO NW was greatly affected by the addition of EDTA-2Na. Its degradation rate decreased from 99.07% to 29.34%, indicating that h^+

was the main contributor to MB degradation. Adding IPA inhibited the degradation of MB to some extent, and the degradation rate was reduced from 99.07% to 52.53%, indicating that ·OH also participated in the photocatalytic reaction process. The introduction of BQ as a capture agent for $\cdot O^{2-}$ had little effect on MB degradation, and the degradation rate decreased from 99.07% to 74.32%. Therefore, ·OH and h^+ were the main active species of the 25 wt% g-C_3N_4/ZnO NWs in the catalytic degradation of MB under visible light.

Based on the relevant characterization data and experimental results, Figure 9 reveals the possible mechanism of complex photocatalytic degradation of MB by 25 wt% g-C_3N_4/ZnO NWs. Since the band gaps of g-C_3N_4 and ZnO NWs were 2.88 eV and 3.22 eV, respectively, the valence and conduction band positions of g-C_3N_4 and ZnO NWs were roughly calculated from the forbidden band width E_g. The CB potential values of g-C_3N_4 and ZnO NWs were −1.21 eV and −0.32 eV, respectively, and the VB potential values of g-C_3N_4 and ZnO NWs were 1.44 eV and 2.9 eV, respectively. Both the CB potential and VB potential values of g-C_3N_4 were more negative than those of ZnO NWs, therefore, this potential difference enabled the formation of a unique heterostructure inside the composite. Under simulated sunlight illumination, both g-C_3N_4 and ZnO NWs were activated and generated photogenerated electron-hole pairs. The e^- on the CB of g-C_3N_4 migrated rapidly to the CB of ZnO, while the h^+ of VB in ZnO migrated to the VB potential of g-C_3N_4. This well-matched energy-band structure was conducive to the effective separation of photogenerated electron-hole pairs between g-C_3N_4 and ZnO NWs, which improved the photocatalytic activity of g-C_3N_4/ZnO NW composites [33].

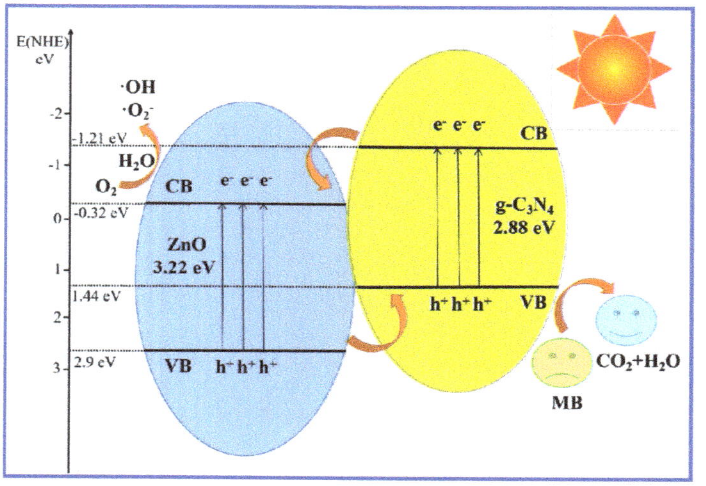

Figure 9. Schematic diagram for the proposed photocatalytic reaction mechanism of 25 wt% g-C_3N_4/ZnO NWs.

Compared to $O_2/\cdot O^{2-}$ with a reduction potential of −0.33 eV/NHE, the CB position of g-C_3N_4 was more negative than $O_2/\cdot O^{2-}$ and the potential of CB of ZnO NWs was more positive. Therefore, the e^- at the CB position of g-C_3N_4 could easily react with dissolved oxygen to form $\cdot O^{2-}$ while the e^- at the CB position of ZnO NWs could not reduce O_2 to $\cdot O^{2-}$. Compared to O_2/H_2O_2 with a reduction potential of 0.695 eV/NHE, the e^- at the CB position of ZnO NWs could react with O_2 and h^+ in water to form ·OH. The VB of g-C_3N_4 was lower than that of OH/H_2O (+2.72 eV/NHE), so the h^+ on the VB of g-C_3N_4 could not oxidize H_2O to ·OH and directly react with MB. However, the potential of ZnO NWs was higher than that of the OH/H_2O, so the h^+ on its CB could oxidize H_2O to ·OH, which could also be used in the following oxidation reaction. ·OH, h^+, and $\cdot O^{2-}$ free-radical active substances will undergo redox reaction with MB, and then be degraded into small molecular organic compounds, and finally be degraded into non-toxic and harmless H_2O and CO_2 [34].

3. Experimental

3.1. Materials

Zinc acetate dihydrate ($(CH_3COO)_2Zn{\cdot}2H_2O$), urea ($CN_2H_4O$), isopropanol ($C_3H_8O$, IPA), disodium edetate ($C_{10}H_{14}Na_2O_8$, EDTA-2Na), potassium bromide (KBr), and absolute ethanol (C_2H_6O) were purchased from Shanghai Sinopharm Chemical Reagent Co., Ltd. Shanghai, China. Methylene blue ($C_{16}H_{18}N_3ClS$, MB) and p-benzoquinone ($C_6H_4O_2$, BQ) were purchased from Shanghai Aladdin Biochemical Technology Co., Ltd. Shanghai, China. The chemical reagents used in this experiment were of analytical grade and were not further purified before use.

3.2. Preparation of g-C_3N_4

We weighed 70 g of urea (CH_4N_2O) into an alumina crucible, and transferred the covered crucible to a muffle furnace. Calcination was performed at a constant temperature of 550 °C for 180 min, and the heating rate was set at 5 $°C{\cdot}min^{-1}$. The sample was calcined and grounded to obtain light-yellow g-C_3N_4 powder.

3.3. Preparation of ZnO NWs

ZnO NWs were prepared by direct solid-state thermal decomposition of $(CH_3COO)_2Zn$ $2H_2O$. We took a 14 g sample of $(CH_3COO)_2Zn$ $2H_2O$ and ground it in an agate mortar for 10 min, then transferred it to a 30 mL alumina crucible. The crucible was placed in a muffle furnace and heated under an atmosphere of air to a constant temperature of 300 °C for 120 min. The heating rate was set to 2 $°C{\cdot}min^{-1}$. After the muffle furnace was cooled to room temperature, the gray ZnO NW powder was taken out, and the ZnO NWs were fully ground for later use.

3.4. Preparation of g-C_3N_4/ZnO NWs

We weighed 50 mg of g-C_3N_4 and added it to a reagent bottle containing the desired mass of ZnO NWs (the doping amounts of g-C_3N_4 were 10 wt%, 15 wt%, 20 wt%, 25 wt%, and 30 wt%), and added 40 mL of absolute ethanol and magnetons to the reagent bottle. After stirring for 8 h, the mixture was transferred to an oven at 60 °C for 48 h to remove absolute ethanol. The resulting solid was ground for later use.

3.5. Characterization

Fourier transform infrared (FT-IR) spectra were acquired using a spectrometer (Nicolet 5700, Thermo Fisher Scientific, MA, USA) to verify the presence and identify the specific vibration modes associated with functional groups. The sample's crystal structure was determined using X-ray diffraction (XRD, Lab X XRD-6100, Shimazdu, Kyoto, Japan) with Cu-Kα radiation source. The measurements were taken with a working voltage of 40 kV and working current of 20 mA, and the scanning range spanned 2θ = 10~80°. The shapes and structures of the materials were examined using advanced imaging techniques, including field-emission scanning electron microscopy (FE-SEM, SU8220, Hitachi, Tokyo, Japan) and field emission transmission electron microscopy (FE-TEM, FEI Tecnai G2 F30, Hillsboro, OR, USA). X-ray photoelectron spectrometer (XPS) technology (Thermo ESCALAB 250XI, Shanghai Yuzhong Industrial Co., Ltd., Shanghai, China) was used to obtain the relevant information of the constituent elements, valence states, and chemical bonds of the samples. The binding energies of the elements in the samples were all based on the C 1s binding energy (284.6 eV) as the corrected binding energy. The Brunauer–Emmett–Teller method was used to measure the specific surface areas (BET, NOVA TOUCH LX1, Quantachrome, FL, USA). Ultraviolet-visible diffuse reflectance spectra were carried out by the spectrophotometer (UV-Vis DRS, UV-3600i Plus, Shimadzu Corporation, Kyoto, Japan) in the range of 200–800 nm. The Edinburgh FLS1000 fluorescence spectrophotometer was used to obtain the photoluminescence spectra of the samples at room temperature with the excitation laser wavelength of 365 nm. Additionally, the ultraviolet-visible (UV-Vis) spectrophotometric method (TU-1950, Persee, Nanjing, China) was employed to study the dye degradation.

3.6. Photocatalytic Experiment

The ability of a photocatalyst to decompose MB under visible light was used as a measure of its photocatalytic activity. All relevant experiments in this work were carried out in an open system at room temperature. In order to confirm that MB did not have the possibility of self-degradation, a blank control experiment was added. In the absence of a catalyst, the control experiment was carried out on the initial concentration of 20 mg·L^{-1} of MB. The detailed steps were as follows: 50 mg of photocatalyst was added to 50 mL MB solution with an initial concentration of 20 mg·L^{-1}, and gently stirred for 30 min in the dark to obtain adsorption–desorption equilibrium. Next, using a xenon lamp as the light source, the photocatalytic degradation experiment was conducted under visible light irradiation, and 2 mL of the reaction solution was taken at regular intervals and placed it a centrifuge tube until the reaction was over. Finally, the centrifuge tube containing the reaction solution was placed in a centrifuge at 10,000 r·min^{-1}, and the supernatant was taken after high-speed centrifugation. The absorbance values of MB (λ_{max} = 554 nm) supernatant under different reaction times were determined by UV-visible spectrophotometer. The degradation rate of MB (%) was calculated as follows:

$$\text{Degradation Rate}(\%) = \frac{C_0 - C_t}{C_t} \times 100\% = \frac{A_0 - A_t}{A_0} \times 100\%$$

where C_t and C_0 represented the concentration of MB corresponding to time t and t_0, respectively, and A_0 and A_t represented the absorbance of the MB solution at time t and t_0, respectively.

4. Conclusions

We have successfully designed a low-cost, intrinsically safe and facile "one-pot stirring" method to prepare a series of composite photocatalysts containing g-C_3N_4-doped ZnO NWs with enhanced photocatalytic performance by the meticulous selection of precursors. The photocatalytic degradation of MB was used to assess the catalytic performance of the photocatalyst under simulated sunlight irradiation. The 25 wt% g-C_3N_4/ZnO NWs showed the best photocatalyst performance, achieving an outstanding MB removal rate of 99.1% within 60 min. The improved performance of the composite could be attributed to several factors, including the improved separation of photogenerated electron-hole pairs, the large surface area providing more reaction sites, and the enhanced light-absorption capability. Furthermore, the composites exhibited excellent stability, maintaining their morphology, crystal phase, and catalytic activity after three cycles of testing. This work not only provides a more economical solution for the efficient purification of industrial wastewater, but also provides a theoretical basis for the study of ZnO-based composite materials, which is of great significance for opening up broad prospects for environmental protection applications.

Author Contributions: Formal analysis and writing—original draft, Y.W.; resources and investigation, Z.L.; funding acquisition, Y.L.; supervision and writing—review and editing, X.Y.; data analysis and interpretation, L.Z.; guidance support and final review of proposed thesis, J.P. All authors have read and agreed to the published version of the manuscript.

Funding: This work was funded by Professor Yuesheng Li and the projects are as follows: the Xianning City Program of Science & Technology (No. 2022ZRKX051), the Hubei University of Science and Technology Doctoral Research Initiation Project (No. BK202217), the Science Development Foundation of Hubei University of Science & Technology (Nos. 2021F005, 2021ZX14, 2020TD01, 2021ZX01, and 2022FH09), the Hubei Provincial Colleges and Universities Outstanding Young and

Middle-Aged Technological Innovation Team Project (No. T2020022), and the Xianning City Key Program of Science & Technology (No. 2021GXYF021).

Conflicts of Interest: The authors declare no conflict of interest.

Sample Availability: Not applicable.

References

1. Feng, L.; Cheng, Y.; Zhang, Y.; Li, Z.; Yu, Y.; Feng, L.; Zhang, S.; Xu, L. Distribution and human health risk assessment of antibiotic residues in large-scale drinking water sources in Chongqing area of the Yangtze River. *Environ. Res.* **2020**, *185*, 109386. [CrossRef]
2. Jiang, X.; Qu, Y.; Zhong, M.; Li, W.; Huang, J.; Yang, H.; Yu, G. Seasonal and spatial variations of pharmaceuticals and personal care products occurrence and human health risk in drinking water—A case study of China. *Sci. Total Environ.* **2019**, *694*, 133711. [CrossRef]
3. Guo, X.; Duan, J.; Li, C.; Zhang, Z.; Wang, W. Highly efficient Z-scheme g-C_3N_4/ZnO photocatalysts constructed by co-melting-recrystallizing mixed precursors for wastewater treatment. *J. Mater. Sci.* **2020**, *55*, 2018–2031. [CrossRef]
4. Peng, J.; Zhang, B.; Hua, W.; Liang, Y.; Zhang, W.; Du, Y.; Peleckis, G.; Indris, S.; Gu, Q.; Cheng, Z.; et al. A Disordered Rubik's Cube-Inspired Framework for Sodium-Ion Batteries with Ultralong Cycle Lifespan. *Angew. Chem. Int. Ed.* **2023**, *62*, e202215865. [CrossRef] [PubMed]
5. Peng, J.; Gao, Y.; Zhang, H.; Liu, Z.; Zhang, W.; Li, L.; Qiao, Y.; Yang, W.; Wang, J.; Dou, S.; et al. Ball Milling Solid-State Synthesis of Highly Crystalline Prussian Blue Analogue Na_{2-x} $MnFe(CN)_6$ Cathodes for All-Climate Sodium-Ion Batteries. *Angew. Chem. Int. Ed.* **2022**, *61*, e202205867. [CrossRef]
6. Dutta, M.; Bhattacharjee, S.; De, S. Separation of reactive dyes from textile effluent by hydrolyzed polyacrylonitrile hollow fiber ultrafiltration quantifying the transport of multicomponent species through charged membrane pores. *Sep. Purif. Technol.* **2020**, *234*, 116063. [CrossRef]
7. Shao, L.; Liu, J.; Li, F.; Yang, J.; Zhang, X.; Li, S.; Li, Y.; Sun, P. Preparation and adsorption properties of TiO_2/MoS_2 nanocomposites. *Mater. Res. Express* **2019**, *6*, 055046. [CrossRef]
8. Huang, Z.-H.; Ji, Z.-Y.; Zhao, Y.-Y.; Liu, J.; Li, F.; Yuan, J.-S. Treatment of wastewater containing 2-methoxyphenol by persulfate with thermal and alkali synergistic activation: Kinetics and mechanism. *Chem. Eng. J.* **2020**, *380*, 122411. [CrossRef]
9. Yi, F.; Gan, H.; Jin, H.; Zhao, W.; Zhang, K.; Jin, H.; Zhang, H.; Qian, Y.; Ma, J. Sulfur- and chlorine-co-doped g-C_3N_4 nanosheets with enhanced active species generation for boosting visible-light photodegradation activity. *Sep. Purif. Technol.* **2020**, *233*, 115997. [CrossRef]
10. Yuan, Y.-J.; Shen, Z.-K.; Song, S.; Guan, J.; Bao, L.; Pei, L.; Su, Y.; Wu, S.; Bai, W.; Yu, Z.-T.; et al. Co-P Bonds as Atomic-Level Charge Transfer Channel To Boost Photocatalytic H_2 Production of Co_2P/Black Phosphorus Nanosheets Photocatalyst. *ACS Catal.* **2019**, *9*, 7801–7807. [CrossRef]
11. Wang, J.; Xin, S.; Xiao, Y.; Zhang, Z.; Li, Z.; Zhang, W.; Li, C.; Bao, R.; Peng, J.; Yi, J.; et al. Manipulating the Water Dissociation Electrocatalytic Sites of Bimetallic Nickel-Based Alloys for Highly Efficient Alkaline Hydrogen Evolution. *Angew. Chem. Int. Ed.* **2022**, *61*, e202202518.
12. Jabbar, Z.H.; Graimed, B.H. Recent developments in industrial organic degradation via semiconductor heterojunctions and the parameters affecting the photocatalytic process: A review study. *J. Water Process Eng.* **2022**, *47*, 102671. [CrossRef]
13. Yuan, Y.-J.; Shen, Z.-K.; Wang, P.; Li, Z.; Pei, L.; Zhong, J.; Ji, Z.; Yu, Z.-T.; Zou, Z. Metal-free broad-spectrum PTCDA/g-C_3N_4 Z-scheme photocatalysts for enhanced photocatalytic water oxidation. *Appl. Catal. B Environ.* **2020**, *260*, 118179. [CrossRef]
14. Isac, L.; Cazan, C.; Enesca, A.; Andronic, L. Copper Sulfide Based Heterojunctions as Photocatalysts for Dyes Photodegradation. *Front. Chem.* **2019**, *7*, 694. [CrossRef]
15. Miyoshi, A.; Nishioka, S.; Maeda, K. Water Splitting on Rutile TiO_2-Based Photocatalysts. *Chem. Eur. J.* **2018**, *24*, 18204–18219. [CrossRef]
16. Hussain, T.; Junaid, M.; Qayyum, H.A. Preparation of Ba-doped $SrTiO_3$ photocatalyst by sol-gel method for hydrogen generation. *Chem. Phy. Lett.* **2020**, *754*, 137741. [CrossRef]
17. Hitam, C.N.C.; Jalil, A.A. A review on exploration of Fe_2O_3 photocatalyst towards degradation of dyes and organic contaminants. *J. Environ. Manag.* **2020**, *258*, 110050. [CrossRef]
18. Sun, Q.; Wang, N.; Yu, J.; Yu, J.C. A Hollow Porous CdS Photocatalyst. *Adv. Mater.* **2018**, *30*, 1804368. [CrossRef] [PubMed]
19. Dutta, V.; Sharma, S.; Raizada, P.; Thakur, V.K.; Khan, A.A.P.; Saini, V.; Asiri, A.M.; Singh, P. An overview on WO_3 based photocatalyst for environmental remediation. *J. Environ. Chem. Eng.* **2021**, *9*, 105018. [CrossRef]
20. Rajabi, H.R.; Khani, O.; Shamsipur, M.; Vatanpour, V. High-performance pure and Fe^{3+}-ion doped ZnS quantum dots as green nanophotocatalysts for the removal of malachite green under UV-light irradiation. *J. Hazard. Mater.* **2013**, *250–251*, 370–378. [CrossRef]
21. Karimi, L.; Zohoori, S.; Yazdanshenas, M.E. Photocatalytic degradation of azo dyes in aqueous solutions under UV irradiation using nano-strontium titanate as the nanophotocatalyst. *J. Saudi Chem. Soc.* **2014**, *18*, 581–588. [CrossRef]
22. Wang, Y.; Yang, X.; Lou, J.; Huang, Y.; Peng, J.; Li, Y.; Liu, Y. Enhance ZnO Photocatalytic Performance via Radiation Modified g-C_3N_4. *Molecules* **2022**, *27*, 8476. [CrossRef] [PubMed]

23. Ismael, M. A review on graphitic carbon nitride (g-C_3N_4) based nanocomposites: Synthesis, categories, and their application in photocatalysis. *J. Alloys Compd.* **2020**, *846*, 156446. [CrossRef]
24. Devarayapalli, K.C.; Lee, K.; Do, H.B.; Dang, N.N.; Yoo, K.; Shim, J.; Prabhakar Vattikuti, S.V. Mesostructured g-C_3N_4 nanosheets interconnected with V_2O_5 nanobelts as electrode for coin-cell-type-asymmetric supercapacitor device. *Mater. Today Energy* **2021**, *21*, 100699. [CrossRef]
25. Vattikuti, S.V.P.; Nam, N.D.; Shim, J. Graphitic carbon nitride/$Na_2Ti_3O_7$/V_2O_5 nanocomposite as a visible light active photocatalyst. *Ceram. Int.* **2020**, *46*, 18287–18296. [CrossRef]
26. Mohamed, M.M.; Ghanem, M.A.; Khairy, M.; Naguib, E.; Alotaibi, N.H. Zinc oxide incorporated carbon nanotubes or graphene oxide nanohybrids for enhanced sonophotocatalytic degradation of methylene blue dye. *Appl. Surf. Sci.* **2019**, *487*, 539–549. [CrossRef]
27. Li, H.; Ding, J.; Cai, S.; Zhang, W.; Zhang, X.; Wu, T.; Wang, C.; Foss, M.; Yang, R. Plasmon-enhanced photocatalytic properties of Au/ZnO nanowires. *Appl. Surf. Sci.* **2022**, *583*, 152539. [CrossRef]
28. Habba, Y.G.; Capochichi-Gnambodoe, M.; Leprince-Wang, Y. Enhanced Photocatalytic Activity of Iron-Doped ZnO Nanowires for Water Purification. *Appl. Sci.* **2017**, *7*, 1185. [CrossRef]
29. Braiek, Z.; Roques-Carmes, T.; Assaker, I.B.; Gannouni, M.; Arnoux, P.; Corbel, S.; Chtourou, R. Enhanced solar and visible light photocatalytic activity of In_2S_3-decorated ZnO nanowires for water purification. *J. Photochem. Photobiol. A Chem.* **2019**, *368*, 307–316. [CrossRef]
30. Liu, S.; Zhu, H.; Yao, W.; Chen, K.; Chen, D. One step synthesis of P-doped g-C_3N_4 with the enhanced visible light photocatalytic activity. *Appl. Surf. Sci.* **2018**, *430*, 309–315. [CrossRef]
31. Yang, H.; Jin, Z.; Hu, H.; Lu, G.; Bi, Y. Fivefold Enhanced Photoelectrochemical Properties of ZnO Nanowire Arrays Modified with C_3N_4 Quantum Dots. *Catalysts* **2017**, *7*, 99. [CrossRef]
32. Dong, P.; Yang, B.; Liu, C.; Xu, F.; Xi, X.; Hou, G.; Shao, R. Highly enhanced photocatalytic activity of WO_3 thin films loaded with Pt-Ag bimetallic alloy nanoparticles. *RSC Adv.* **2017**, *7*, 947–956. [CrossRef]
33. Khaing, K.K.; Yin, D.; Ouyang, Y.; Xiao, S.; Liu, B.; Deng, L.; Li, L.; Guo, X.; Wang, J.; Liu, J.; et al. Fabrication of 2D–2D Heterojunction Catalyst with Covalent Organic Framework (COF) and MoS_2 for Highly Efficient Photocatalytic Degradation of Organic Pollutants. *Inorg. Chem.* **2020**, *59*, 6942–6952. [CrossRef] [PubMed]
34. Vijayakumar, T.P.; Benoy, M.D.; Duraimurugan, J.; Suresh Kumar, G.; Shkir, M.; Maadeswaran, P.; Senthil Kumar, A.; Ramesh Kumar, K.A. Hydrothermal synthesis of CuO/g-C_3N_4 nanosheets for visible-light driven photodegradation of methylene blue. *Diam. Relat. Mater.* **2022**, *121*, 108735. [CrossRef]

molecules

MDPI

Article

Outstanding Electrochemical Performance of Ni-Rich Concentration-Gradient Cathode Material $LiNi_{0.9}Co_{0.083}Mn_{0.017}O_2$ for Lithium-Ion Batteries

Hechen Li [1], Yiwen Guo [1], Yuanhua Chen [2,3,*], Nengshuang Gao [1], Ruicong Sun [1], Yachun Lu [1] and Quanqi Chen [1,3,*]

1 Guangxi Key Laboratory of Electrochemical and Magneto-Chemical Functional Materials, College of Chemistry and Bioengineering, Guilin University of Technology, Guilin 541004, China; 2120200819@glut.edu.cn (H.L.)
2 College of Mechanical and Vehicle Engineering, Hunan University, Changsha 410082, China
3 School of Automobile Engineering, Guilin University of Aerospace Technology, Guilin 541004, China
* Correspondence: cyh@guat.edu.cn (Y.C.); quanqi.chen@glut.edu.cn (Q.C.)

Abstract: The full-concentrationgradient $LiNi_{0.9}Co_{0.083}Mn_{0.017}O_2$ (CG-LNCM), consisting of core Ni-rich $LiNi_{0.93}Co_{0.07}O_2$, transition zone $LiNi_{1-x-y}Co_xMn_yO_2$, and outmost shell $LiNi_{1/3}Co_{1/3}Mn_{1/3}O_2$ was prepared by a facile co-precipitation method and high-temperature calcination. CG-LNCM was then investigated with an X-ray diffractometer, ascanning electron microscope, a transmission electron microscope, and electrochemical measurements. The results demonstrate that CG-LNCM has a lower cation mixing of Li^+ and Ni^{2+} and larger Li^+ diffusion coefficients than concentration-constant $LiNi_{0.9}Co_{0.083}Mn_{0.017}O_2$ (CC-LNCM). CG-LNCM presents a higher capacity and a better rate of capability and cyclability than CC-LNCM. CG-LNCM and CC-LNCM show initial discharge capacities of 221.2 and 212.5 mAh g^{-1} at 0.2C (40 mA g^{-1}) with corresponding residual discharge capacities of 177.3 and 156.1 mAh g^{-1} after 80 cycles, respectively. Even at high current rates of 2C and 5C, CG-LNCM exhibits high discharge capacities of 165.1 and 149.1 mAh g^{-1} after 100 cycles, respectively, while the residual discharge capacities of CC-LNCM are as low as 148.8 and 117.9 mAh g^{-1} at 2C and 5C after 100 cycles, respectively. The significantly improved electrochemical performance of CG-LNCM is attributed to its concentration-gradient microstructure and the composition distribution of concentration-gradient $LiNi_{0.9}Co_{0.083}Mn_{0.017}O_2$. The special concentration-gradient design and the facile synthesis are favorable for massive manufacturing of high-performance Ni-rich ternary cathode materials for lithium-ion batteries.

Keywords: concentration-gradient; Ni-rich ternary; cathode materials; lithium-ion batteries

1. Introduction

Compared with other secondary batteries such as lead acid, nickel–cadmium and nickel–metal hydride batteries, lithium-ion batteries (LIBs) have been widely used in portable electronic devices and electric vehicles (EVs) because of their higher energy density and longer life span [1–6]. With the increasing demand for LIBs with a high energy density and outstanding cyclability, it is urgent and imperative to develop high-energy density and cyclability of cathode materials for LIBs. Numerous efforts have been made to optimize cathode materials, and dozens of these have been well developed [7], but the Ni-based ternary cathode materials $LiNi_{1-x-y}Co_xMn_yO_2$ (NCM) and $LiNi_{1-x-y}Co_xAl_yO_2$ (NCA) have received extensive and intensive attention and have been extensively utilized [8–13]. Ni-rich ternary cathode materials supply a large capacity that rises proportionately with Ni content, even though they suffer from inferior cyclic stability during cycling and thermal instability [14–18]. These shortcomings mainly result from the microcracking of electroactive particles upon the charging/discharging processes, which tremendously enlarges the inner surface exposed to electrolyte attack [15,19].

To improve the performance of Ni-rich ternary cathode materials, diverse approaches such as doping heterogeneous ions [20–22], surface coating [23–25], the construction of a single crystal phase [26–28], core/shell microstructure [29,30] and particles with concentration-gradient composition [31–33] have been adopted to inhibit the formation of microcracking. Among the preceding approaches, building particles with concentration-gradient composition is one of the most effective ways to improve the cyclability of electroactive materials. Generally, core composition can provide high capacity, and shell composition can improve cyclability and thermal stability in the ideal concentration-gradient microstructure [34]. In the present investigation, full concentration-gradient $LiNi_{0.9}Co_{0.087}Mn_{0.013}O_2$, in which the composition varies gradiently from core composition $LiNi_{0.93}Co_{0.07}O_2$ to outmost shell composition $LiNi_{1/3}Co_{1/3}Mn_{1/3}O_2$, was proposed and prepared by a facile co-precipitation method combined with high-temperature calcination in an oxygen atmosphere. Benefiting from the high capacity provided by Ni-rich core composition and the excellent cyclability of shell composition $LiNi_{1/3}Co_{1/3}Mn_{1/3}O_2$, the concentration-gradient $LiNi_{0.9}Co_{0.087}Mn_{0.013}O_2$ exhibits significantly improved electrochemical performance in comparison with the concentration-constant $LiNi_{0.9}Co_{0.087}Mn_{0.013}O_2$ prepared by the same synthesis procedure.

2. Results and Discussions

Figure 1 shows the X-ray diffraction (XRD) patterns of concentration-gradient precursors (CG-NCMOH), concentration-constant precursors (CC-NCMOH), the concentration-gradient products (CG-LNCM), and the concentration-constant products (CC-LNCM). Figure 1a and the Rietveld refinement results (Figure S1a,b in Supplementary File) demonstrate that no manganese and cobalt hydroxides or other oxides exist in the XRD patterns of hydroxide precursors CC-NCMOH and CG-NCMOH, indicating that Mn^{2+} and Co^{2+} are successfully doped into Ni sites of $Ni(OH)_2$ to form Ni-Co-Mn ternary hydroxide. The diffraction peaks of CC-NCMOH and CG-NCMOH shift to a lower angle, which implies that doping of Mn^{2+} and Co^{2+} enlarges the layer distance of $Ni(OH)_2$ according to the Bragg equation $2d\sin\theta = n\lambda$. As observed in Figure 1b, the diffraction peaks of CC-LNCM and CG-LNCM are strong and sharp, suggesting that both CC-LNCM and CG-LNCM have high crystallinity, and all the diffractions can be well indexed by the R3m space group. The structural parameters for CC-LNCM and CG-LNCM are listed in Tables S1 and S2, and the results reveal that CG-LNCM has a larger cell volume than CC-LNCM, hinting that CG-LNCM may have better electrochemical performance in comparison with CC-LNCM because a larger cell volume favors the more rapid transport of Li^+ in electroactive particles and hence the better electrochemical performance. The layered structure of the material could be judged by the splitting of characteristic peaks. The more obvious the splits of (006)/(102) and (018)/(110) are, the more they indicate a higher degree of layered structure [35]. In addition, the ratio of c/a for both CC-LNCM and CG-LNCM is bigger than 4.9, showing that both samples have a well-developed layered crystal structure [36,37]. Furthermore, the quota of intensity ratio of (003) to (104), $I_{(003)}/I_{(104)}$ reflects the cation mixing of Li^+ and Ni^{2+}, and the larger value of $I_{(003)}/I_{(104)}$ means the lower cation mixing. It was reported that if the value of $I_{(003)}/I_{(104)}$ is bigger than 1.2, the materials will have lower cation mixing [38,39] and improved electrochemical performance. According to Tables S1 and S2, the comparisons of CC-LNCM and CG-LNCM in $I_{(003)}/I_{(104)}$ and the occupation of Li and Ni in Wyckoff sites 3a and 3b reveal that CG-LNCM has a lower cation mixing of Li^+ and Ni^{2+} than CC-LNCM, and the special concentration-gradient microstructure may be responsible for the lower cation mixing.

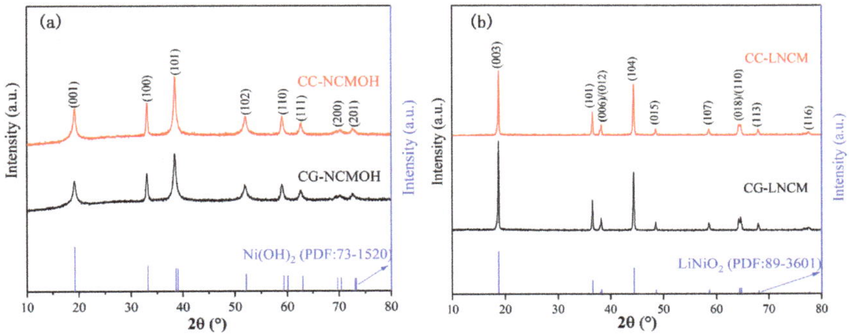

Figure 1. XRD patterns of (**a**) CG-NCMOH and CC-NCMOH; (**b**) CG-LNCM and CC-LNCM.

The scanning electron microscope (SEM) images of CC-NCMOH, CG-NCMOH, CC-LNCM and CG-LNCM are presented in Figure 2. As observed in Figure 2a,b, both two hydroxide precursors display sphere-like morphology and a rough surface and are composed of nanoplates agglomerated loosely together. The loose agglomeration of CC-NCMOH and CG-NCMOH is beneficial for rapid reaction with LiOH to form $LiNi_{0.9}Co_{0.083}Mn_{0.017}O_2$ with high crystallinity. Figure 2c,d demonstrate that the final products CC-LNCM and CG-LNCM present a similar sphere-like morphology to that of hydroxide precursors. However, unlike the precursors CC-NCMOH and CG-NCMOH, CC-LNCM and CG-LNCM consist of nanoparticles. Compared with the compact CC-LNCM, the CG-LNCM particles are loosely agglomerated, which is favorable for the permeation of the electrolyte and rapid diffusion of Li^+ in the electroactive particles. The high-resolution transmission electron microscope (HRTEM) images of CC-LNCM and CG-LNCM are shown in Figure 2e,f, respectively, and d-spacing of about 0.2040 and 0.2042 nm is observed in the lattice fringes of CC-LNCM and CG-LNCM, respectively, corresponding to the lattice plane (104). The minor difference of the d-space may result from the different surface composition of $LiNi_{0.9}Co_{0.083}Mn_{0.017}O_2$ and $LiNi_{1/3}Co_{1/3}Mn_{1/3}O_2$ for CC-LNCM and CG-LNCM, respectively.

The SEM image of cross-section of CG-LNCM presented in Figure 3a further clearly demonstrates that the sphere-like secondary particles are composed of primary nanoparticles, and the corresponding EDX elemental mappings in Figure 3b–d indicate that the elements Ni, Co and Mn are evenly dispersive, similar to the concentration-constant $LiNi_{0.8}Co_{0.1}Mn_{0.1}O_2$ [36], implying that it is difficult to separate concentration-gradient samples from concentration-constant samples by elemental mappings. Figure 3e,f display the SEM images of CC-LNCM and CG-LNCM primary particles, respectively. The element contents of the selected area of particles are determined by energy dispersive X-ray spectroscopy (EDXS), and EDXS mappings of the selected area are shown in Figure S2. It can be observed that contents of Ni, Co, and Mn in the edge and interior of the primary particle are almost identical and close to the molar ratio of Ni, Co and Mn of the concentration-constant $LiNi_{0.9}Co_{0.083}Mn_{0.017}O_2$. The compositions are different in the different areas of the CG-LNCM, further confirming that composition varies gradiently from interior to edge and CG-LNCM is truly a concentration-gradient oxide.

To learn the effects of the concentration-gradient and concentration-constant microstructure on the oxidate state of Ni in $LiNi_{0.9}Co_{0.083}Mn_{0.017}O_2$, an X-ray photoelectron spectroscopy (XPS) measurement was conducted and the corresponding Ni 2p XPS spectra were shown in Figure 4. As can be observed in Figure 4a,b, the XPS spectra of both CC-LNCM and CG-LNCM consist of two satellite peaks and two main peaks, and the two peaks centered at about 855.96 and 873.5 eV are deconvoluted into 856.1 and 854.9 eV, and 873.9 and 872.4 eV, respectively. The Binding energy of 854.9 and 872.4 eV correspond to Ni $2p_{3/2}$ and Ni $2p_{1/2}$ of Ni(II) [40,41], respectively, and the binding energy of 856.1 and 873.9 eV match Ni $2p_{3/2}$ and Ni $2p_{1/2}$ of Ni(III) [40,42], respectively. The molar ratio of

Ni^{2+}/Ni^{3+} is calculated according to the ratio of the closed area of Ni $2p_{3/2}$ XPS spectra of Ni^{2+} to the closed area of Ni $2p_{3/2}$ XPS spectra of Ni^{3+}. The molar ratios of Ni^{2+}/Ni^{3+} are 0.48/0.52 and 0.58/0.42 for CC-LNCM and CG-LNCM, respectively. The different content of Ni^{2+} results from the different surface compositions of CC-LNCM and CG-LNCM, similar to the previous report that the molar ratio of Ni^{2+}/Ni^{3+} is larger on the surface of $LiNi_{1-x-y}Co_xMn_yO_2$ with a lower content of Ni [43]. This finding further confirms the concentration-gradient microstructure of CG-LNCM, in which the composition varies from the core $LiNi_{0.93}Co_{0.07}O_2$ to the outmost shell $LiNi_{1/3}Co_{1/3}Mn_{1/3}O_2$.

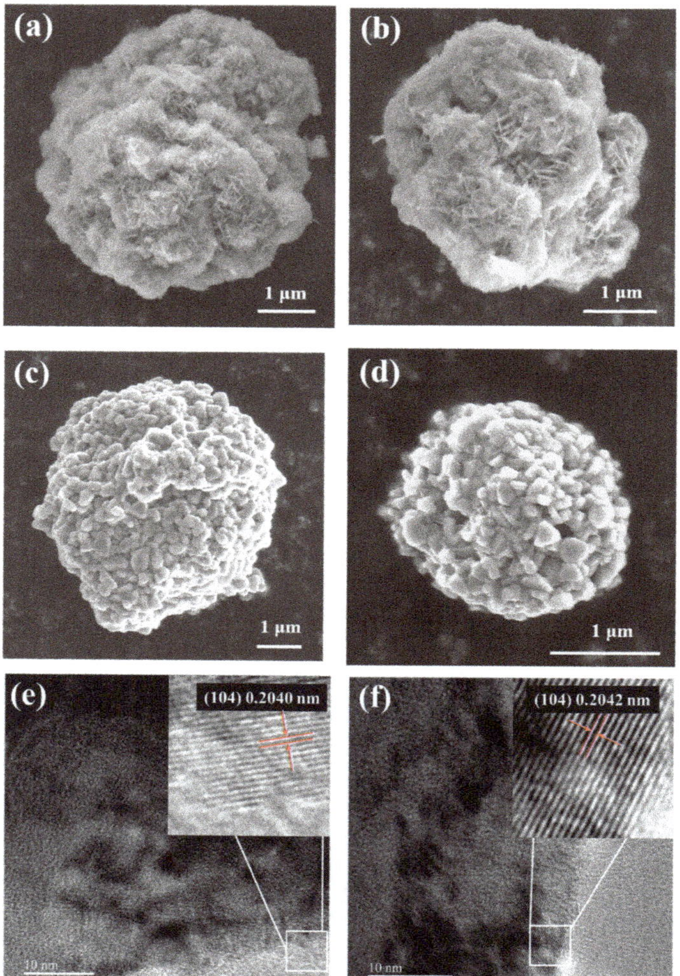

Figure 2. SEM images of (**a**) CC-NCMOH, (**b**) CG-NCMOH, (**c**) CC-LNCMO and (**d**) CG-LNCM; HRTEM images of (**e**) CC-NCMOH and (**f**) CG-NCMOH.

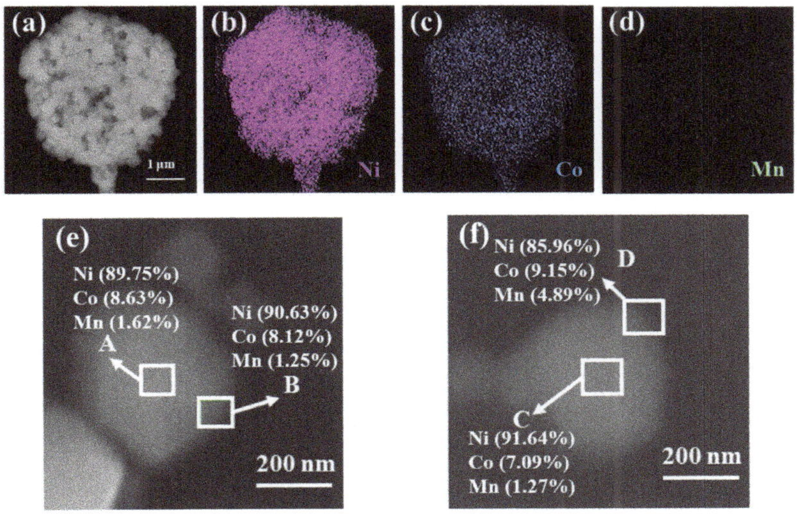

Figure 3. SEM image of cross-section of (**a**) CG-LNCM and the corresponding EDX elemental mappings of (**b**) Ni, (**c**) Co and (**d**) Mn; SEM images of primary particles and compositions of the selected area of (**e**) CC-LNCM and (**f**) CG-LNCM.

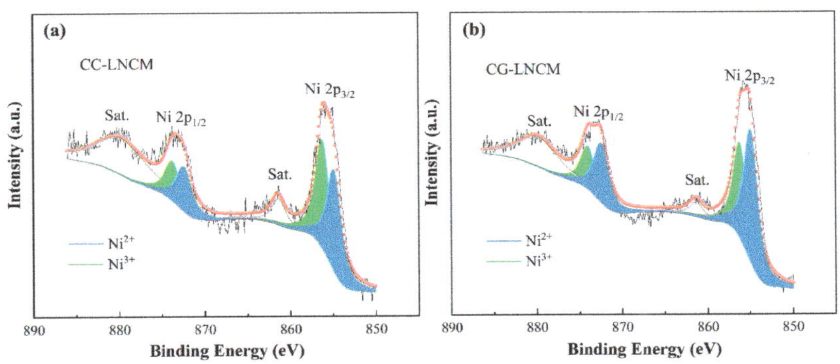

Figure 4. Ni 2p XPS spectra of (**a**) CC-LNCM and (**b**) CG-LNCM.

To investigate the effect of composition distribution of $LiNi_{0.9}Co_{0.083}Mn_{0.017}O_2$ on the electrochemical mechanism, cyclic voltammetry measurements were conducted on CC-LNCM and CG-LNCM at a scan rate of 0.1 mV s^{-1} in the potential scope of 3.0–4.3 V (vs. Li^+/Li) at room temperature, and the first three cycles of cyclic voltammograms (CVs) of CC-LNCM and CG-LNCM electrodes are shown in Figure 5. The shape and the closed area of the first three CVs are almost unchanged, suggesting that both CC-LNCM and CG-LNCM have the better cyclability. As seen in Figure 5a,b, two CVs exhibit a similar shape, and three couples of redox peaks have close potential, indicating that CC-LNCM and CG-LNCM possess the identical electrochemical mechanism upon the charging and discharging processes. The three redox peaks are associated with the interconversion of the hexagonal phase to the monoclinic phase (H1↔M), the monoclinic phase to the hexagonal phase (M↔H2), and the hexagonal phase to the hexagonal phase (H2↔H3), respectively [44,45]. The accurate comparison of redox peak potentials of CC-LNCM and CG-LNCM demonstrates that CG-LNCM has lower oxidation peak potentials and larger reduction peak potentials than CC-LNCM, implying that CG-LNCM exhibits lower electrochemical polarization and better electrochemical reaction reversibility than CC-LNCM.

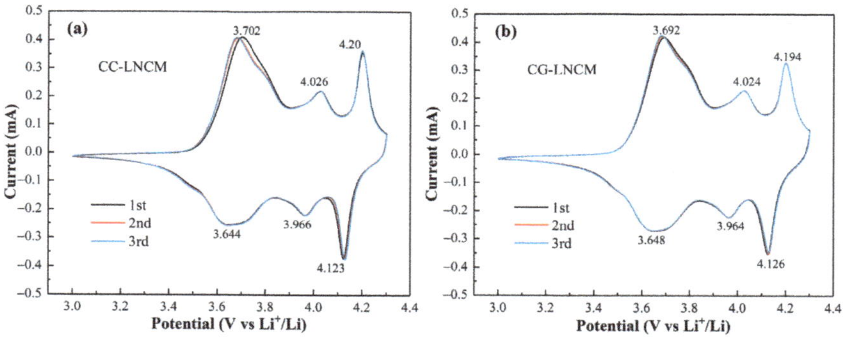

Figure 5. The first three cycles of cyclic voltammograms of (**a**) CC-LNCM and (**b**) CG-LNCM at 0.1 mV s^{-1} in the potential range of 3.0–4.3 V (vs. Li^+/Li).

The electrodes used to evaluate galvanostatic electrochemical performance at each given current rate were fresh CC-LNCM or CG-LNCM electrodes without an activation process. The initial charge/discharge profiles of CC-LNCM and CG-LNCM electrodes at various current rates (1C = 200 mA g^{-1}) in the voltage range of 3.0–4.3 V (vs. Li^+/Li) were shown in Figure 6a,b, and the big difference indicated between charge and discharge capacities is observed. The low initial Coulombic efficiency of fresh CC-LNCM and CG-LNCM electrodes, similar to that of $LiNi_{0.9}Co_{0.08}Al_{0.02}O_2$ [45], may be associated with properties of Ni-rich based oxide cathode materials. Low initial Coulombic efficiency is presented in rich Ni-based oxide cathode materials, especially nanoparticles. After several cycles, the Coulombic efficiency greatly increases, as observed in Figure S3, to about 97.8% in the fifth cycle, implying that the activation of electrodes can remarkably improve the Coulombic efficiency. It can be found that the discharge and charge capacities decrease with the increase of current rates due to a larger polarization at higher current rates. The initial discharge capacities of CC-LNCM are 212.5, 200.7, 183.9, 167.1 and 139.2 mAh g^{-1} at 0.2C, 0.5C, 1C, 2C and 5C, respectively, while CG-LNCM presents the somewhat higher initial discharge capacities of 221.2, 203.3, 185.0, 168.9 and 140.8 mAh g^{-1} at 0.2C, 0.5C, 1C, 2C and 5C, respectively. The initial discharge capacity of CC-LNCM at 0.2C is close to that of 210 mAh g^{-1} for $LiNi_{0.9}Co_{0.05}Mn_{0.05}O_2$ at 0.2C [37], higher than that of 203.8 mAh g^{-1} for $LiNi_{0.91}Co_{0.06}Mn_{0.03}O_2$ single crystal at 0.1C [28], 207 mAh g^{-1} for $LiNi_{0.9}Co_{0.08}Al_{0.02}O_2$ at 0.2C [46]. The initial discharge capacity of CG-LNCM at 0.2C is close to that of about 221 mAh g^{-1} for concentration-gradient $LiNi_{0.84}Co_{0.06}Mn_{0.09}Al_{0.01}O_2$ at 0.1C [32], and larger than that of 200 mAh g^{-1} for concentration-gradient $LiNi_{0.9}Mn_{0.1}O_2$ at 0.1C [47]. The initial discharge capacities of CG-LNCM at 0.5C, 1C and 2C are close to those of 200 mAh g^{-1} for $LiNi_{0.92}Co_{0.03}Mn_{0.03}Al_{0.02}O_2$ at 0.5C [48], 180 mAh g^{-1} for TiO_2-coated $LiNi_{0.9}Co_{0.08}Al_{0.02}O_2$ at 1C [47] and 165 mAh g^{-1} for TiO_2-coated $LiNi_{0.9}Co_{0.08}Al_{0.02}O_2$ at 2C [45], respectively. The above-mentioned results demonstrate that CG-LNCM is competitive in the first discharge capacity with other Ni-based ternary cathode materials.

The cycle performance of CC-LNCM and CG-LNCM at 0.2C, 2C and 5C are depicted in Figure 6c,d. As observed in Figure 6c,d, discharge capacities of CC-LNCM and CG-LNCM at all current rates increase before the first number of specific cycles, which is ascribed to the activation process that originates from the insufficient contact between electrolyte and electroactive particles. After activation, the discharge capacity of CC-LNCM reaches the highest value of 218.3 mAh g^{-1} at 0.2C and decreases to 156.1 mAh g^{-1} after 80 cycles. The ratio of the residual discharge capacity to the highest discharge capacity corresponds to 71.4%. While CG-LNCM exhibits the highest capacity of 223.4 mAh g^{-1} in the third cycle and higher residual capacity of 177.3 mAh g^{-1} after 80 cycles at 0.2C, the ratio of the residual discharge capacity to the highest discharge capacity corresponds to 79.4%. The residual capacity of CG-LNCM at 0.2C is close to that of about 174 mAh g^{-1} for $LiNi_{0.9}Co_{0.08}Al_{0.02}O_2$ [46], about 170 mAh g^{-1} for TiO_2-coated $LiNi_{0.9}Co_{0.08}Al_{0.02}O_2$ [45], and 169 mAh g^{-1} for $LiTaO_3$ modified $LiNi_{0.9}Co_{0.06}Mn_{0.04}O_2$ [49] after 80 cycles at 0.2C. The residual discharge capacities of CC-LNCM are 148.8 and 117.9 mAh g^{-1} at 2C and 5C after 100 cycles, respectively, which are much smaller than the corresponding residual discharge capacities of CG-LNCM. The special concentration-gradient microstructure of CG-LNCM may be responsible for the improved electrochemical performance. The residual discharge capacity of CG-LNCM at 2C is 165.1 mAh g^{-1} after 100 cycles, which is larger than that of 130 mAh g^{-1} for TiO_2-coated $LiNi_{0.9}Co_{0.08}Al_{0.02}O_2$ [45], 134.4 mAh g^{-1} for $LiNi_{0.9}Co_{0.05}Mn_{0.05}O_2$ [50], and 156.9 mAh g^{-1} for Li_2SiO_3 coated $LiNi_{0.9}Co_{0.05}Mn_{0.05}O_2$ [50] at 2C after 100 cycles. Even at the high current rate of 5C, CG-LNCM presents the high discharge capacity of 149.1 mAh g^{-1} after 100 cycles, significantly higher than that of 117.9 mAh g^{-1} for CC-LNCM. It is noted that both CC-LNCM and CG-LNCM show better cycle performance at higher current rates than the cycle performance of these two cathodes at 0.2C, and this phenomenon is similar to that of $LiNi_{0.8}Co_{0.1}Mn_{0.1}O_2$ [51]. This phenomenon may be ascribed to the lower tolerance of great deformation of CC-LNCM and CG-LNCM with loose microstructure and nanosized primary particles. Compared with cycles at higher current rates, extraction of more Li^+ from electroactive material $LiNi_{0.9}Co_{0.083}Mn_{0.017}O_2$ and insertion of more Li^+ into delithiated $LiNi_{0.9}Co_{0.083}Mn_{0.017}O_2$ occur at lower current rates during charging and discharging processes, resulting in huge deformation, structure instability and the resultant inferior cyclability at lower current rates. In summary, the aforementioned electrochemical tests demonstrate that CG-LNCM has a higher capacity and better rate capability and cyclability than CC-LNCM, which can be attributed to the advantages of the concentration-gradient microstructure that the Ni-rich core $LiNi_{0.93}Co_{0.07}O_2$ provides. Furthermore, the high capacity and shell composition of $LiNi_{1/3}Co_{1/3}Mn_{1/3}O_2$ supplies excellent structural stability, and the multiple core/shell structure of concentration-gradient composition is favorable for alleviation of microcracks upon cycling.

To further understand the difference in electrochemical properties of concentration-gradient and concentration-constant $LiNi_{0.9}Co_{0.083}Mn_{0.017}O_2$, electrochemical impedance spectroscopy (EIS) was carried out on the fresh CG-LNCM and CC-LNCM electrodes and other electrodes that had been cycled for 100 cycles at 5 C, and the corresponding Nyquist plots are shown in Figure 7. As observed in Figure 7a, the Nyquist plots of fresh electrodes are composed of a pressed semicircle and a sloped line, which is different from the Nyquist plots (Figure 7b) of cycled electrodes that consist of two compressed semicircles and an inclined line. The equivalent circuit models for the different Nyquist plots are depicted in the insert of Figure 7a,b, respectively. In equivalent circuit models, R_e, R_f and R_{ct} represent internal resistance of cell, charge transfer resistance, and resistance of the solid electrolyte interface (SEI) [52,53], respectively, while CPE and W_o stand for double layer capacitance and capacity of the surface layer, and Warburg impedance, respectively. The fitting results are listed in Table S3 and indicate that values of R_f and R_{ct} of CC-LNCM electrodes are larger than those of CG-LNCM electrodes at the corresponding states. The smaller R_{ct} of CG-LNCM electrodes suggests that CG-LNCM possesses better electrochemical kinetics and hence better electrochemical performance in comparison with CC-LNCM. In addition, the smaller R_f of cycled CG-LNCM electrodes demonstrates that the resistance of SEI of a cycled CG-LNCM electrode is smaller than that of a cycled CG-LNCM electrode, implying that CG-LNCM has a lower polarization than CC-LNCM during cycling, favorable for improvement of cyclability.

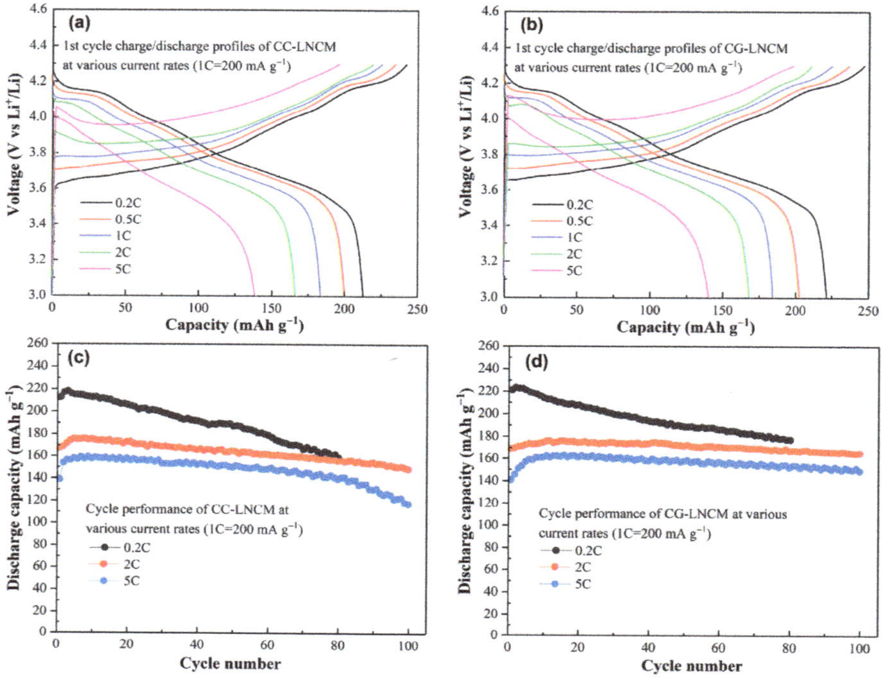

Figure 6. Initial charge/discharge profiles of (**a**) CC-LNCM and (**b**) CG-LNCM at various current rates, cycle performance of (**c**) CC-LNCM and (**d**) CG-LNCM at various current rates.

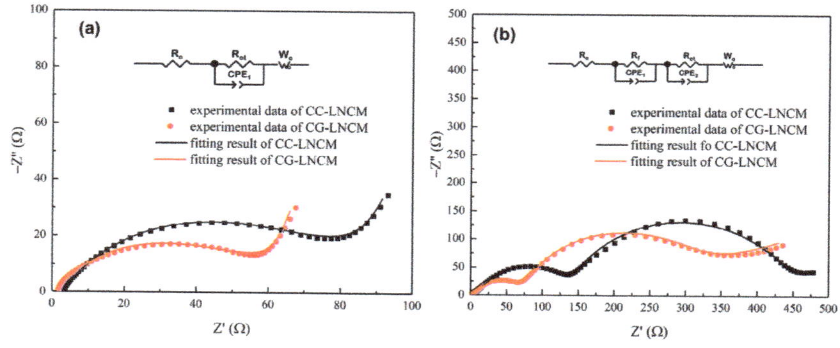

Figure 7. Nyquist plots of (**a**) fresh electrodes and (**b**) electrodes after cycling for 100 cycles at 5C.

It is well-known that the diffusion of Li^+ in electroactive particles is a rate-determining step of the electrochemical reaction of electroactive materials, and the Li^+ diffusion coefficient, D_{Li}, of electrode materials is a key parameter to evaluate the kinetics of electrochemical reaction. D_{Li} was calculated according to the following Equation (1) [54].

$$D_{Li} = \frac{4}{\pi\tau}\left(\frac{m_B V_m}{M_B S}\right)^2\left(\frac{\Delta E_s}{\Delta E_\tau}\right)^2 \tag{1}$$

The meanings of symbols in Equation (1) are the same as in our previous report, and the detailed calculation procedure is also akin to our previous report [55]. To compare D_{Li}

of CC-LNCM and CG-LNCM, the galvanostatic intermittent titration technique (GITT) measurement was carried out and the value of D_{Li} was estimated by Equation (1). The GITT curves of CC-LNCM and CG-LNCM are shown in Figure 8a,b. The calculated values of D_{Li} of CC-LNCM and CG-LNCM at charged and discharged states are depicted in Figure 8c,d, demonstrating that CG-LNCM has higher values of D_{Li} in both charged and discharged states than CC-LNCM. The differences of D_{Li} result from the discrepancies of CG-LNCM and CC-LNCM. In the charging process, the values of D_{Li} of CG-LNCM vary from 3.73×10^{-11} to 6.14×10^{-10} cm^2 s^{-1}, while D_{Li} of CC-LNCM lies in the scope of 1.00×10^{-11} to 1.96×10^{-10} cm^2 s^{-1}. During discharging, the CG-LNCM and CC-LNCM range in D_{Li} from 1.20×10^{-12} to 5.34×10^{-10} and 4.15×10^{-13} to 2.16×10^{-10} cm^2 s^{-1}, respectively. The comparison of D_{Li} in CG-NCM and CC-NCM suggests that the electrochemical reaction rate of CG-LNCM is faster than that of CC-LNCM, which can explain why CG-LNCM has a better electrochemical performance than CC-LNCM. The larger diffusion coefficients of Li^+ benefit from the concentration-gradient composition and microstructure, and result in the significantly improved electrochemical performance of $LiNi_{0.9}Co_{0.083}Mn_{0.017}O_2$.

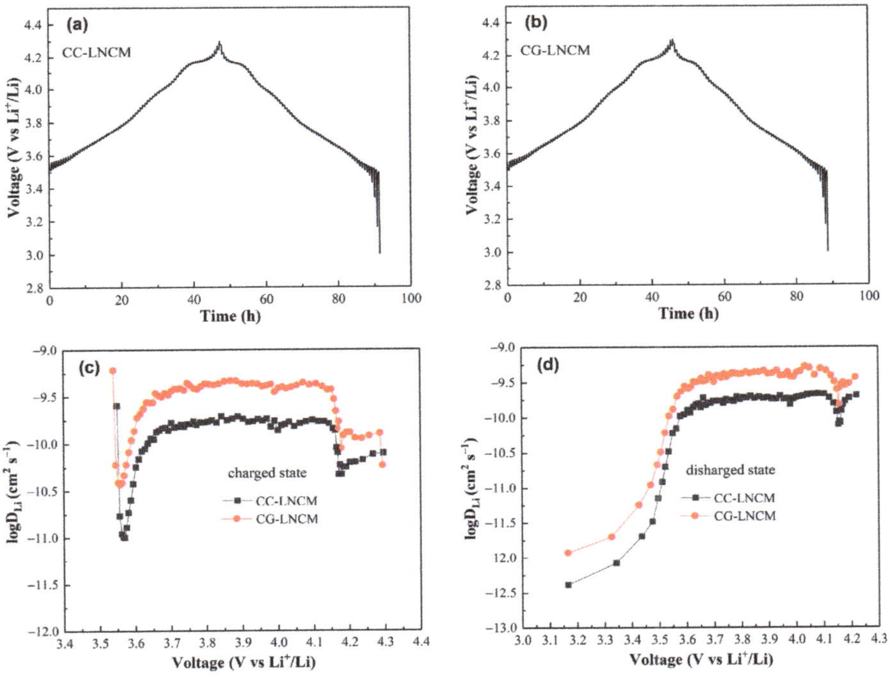

Figure 8. GITT curves of (**a**) CC-LNCM and (**b**) CG-LNCM, relationship between voltage and Li^+ diffusion coefficient of CC-LNCM and CG-LNCM at (**c**) charged state and (**d**) discharged state.

3. Materials and Methods

3.1. Synthesis of Materials

The concentration-gradient precursor $Ni_{0.9}Co_{0.083}Mn_{0.017}(OH)_2$ was synthesized by a facile co-precipitation method using $NiSO_4{\cdot}6H_2O$, $CoSO_4{\cdot}7H_2O$, $MnSO_4{\cdot}H_2O$, NaOH and $NH_3{\cdot}H_2O$ as raw materials, and the corresponding schematic illustration is depicted in Figure 9. To avoid the oxidation of Ni^{2+}, Co^{2+} and Mn^{2+} during the synthesis of precursor $Ni_xCo_yMn_{1-x-y}(OH)_2$ precipitate, both solution and reaction are in a nitrogen atmosphere. The total concentration of transition metal ions for both the aqueous solution A (tank 1) and

B (tank 2) is 0.075 and 1.425 mol L^{-1}, respectively, and the molar ratios of Mn^{2+}:Co^{2+}:Ni^{2+} for solution A and B are 1:1:1 and 0:0.07:0.93, respectively. Furthermore, solution A is equal to solution B in volume. At the start of the co-precipitation procedure, Ni-rich solution (tank 2) was firstly fed at a constant rate of 100 mL h^{-1} into the reactor containing a certain amount of distilled water and $NH_3 \cdot H_2O$ at 50 °C. Simultaneously, the solution A (tank 1) was pumped into tank 2 at a constant rate of 50 mL h^{-1}. At the same time, the mixture solution containing 1.66 mol L^{-1} NaOH and adequate quantity of $NH_3 \cdot H_2O$ was added to the reactor at a reasonable rate for adjusting pH of the reaction solution to 11.5~11.8, which is suitable for formation of $Ni_{1-x-y}Co_xMn_y(OH)_2$ precipitate. With the continuous addition of solution A to solution B, the molar ratio of Mn^{2+}:Co^{2+}:Ni^{2+} of solution B changes gradiently from 0:0.07:0.93 to 1:1:1; thus the concentration distribution of Ni, Co and Mn of the target precursor $Ni_{1-x-y}Co_xMn_y(OH)_2$ precipitate will be gradient. The final concentration-gradient precipitate $Ni_{1-x-y}Co_xMn_y(OH)_2$ consists of a $Ni_{0.93}Co_{0.07}(OH)_2$ core, a transitional zone containing a series of $Ni_{1-x-y}Co_xMn_y(OH)_2$ ($0.07 < x < 1/3$, $0 < y < 1/3$) precipitate and a $Ni_{1/3}Co_{1/3}Mn_{1/3}(OH)_2$ shell. After complete addition of solutions A and B, the precipitate was aged for 15 h. Subsequently, the suspension was filtered and washed three times to get the dark green precipitate. Finally, the dark green precipitate was dried for 24 h at 110 °C in a vacuum oven. The molar ratio of total transition metal ions for solution A and solution B was 5:95 (0.075:1.425), so the chemical formula of the concentration-gradient precipitate can be simply expressed as $Ni_{0.9}Co_{0.083}Mn_{0.017}(OH)_2$ (CG-NCMOH).

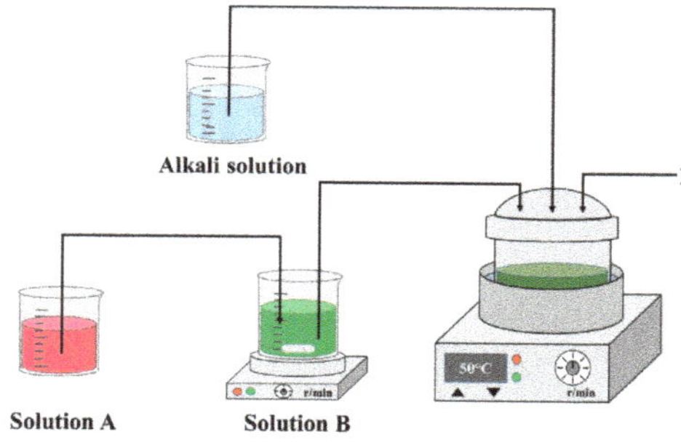

Figure 9. Schematic illustration of the preparation of concentration-gradient precursor $Ni_{0.9}Co_{0.083}Mn_{0.017}(OH)_2$.

The concentration-gradient $LiNi_{0.9}Co_{0.083}Mn_{0.017}O_2$ (CG-LNCM) was prepared by calcination of the mixture of $LiOH \cdot H_2O$ and the dried concentration-gradient precursor $Ni_{0.9}Co_{0.083}Mn_{0.017}(OH)_2$ with a molar ratio of 1.04:1 under an oxygen atmosphere in a tube furnace in which the mixed reactants were heated with an adequate rate of 5 °C per minute to 480 °C and maintained for 5 h, and then subsequently heated to 750 °C and held for 13.5 h, and finally cooled naturally to room temperature to get concentration-gradient $LiNi_{0.9}Co_{0.083}Mn_{0.017}O_2$. For comparison, the control concentration-constant $LiNi_{0.9}Co_{0.083}Mn_{0.017}O_2$ (CC-LNCM) was prepared by the same calcination of the stochiometric mixture of concentration-constant precursor $Ni_{0.9}Co_{0.083}Mn_{0.017}(OH)_2$ (CC-NCMOH) and $LiOH \cdot H_2O$ with a molar ratio of 1:1.04 under oxygen atmosphere, and the concentration-constant precursor $Ni_{0.9}Co_{0.083}Mn_{0.017}(OH)_2$ was prepared by the same co-precipitation method without tank 1, and the molar ratio of Ni^{2+}, Co^{2+} and Mn^{2+} is fixed to 0.9:0.083:0.017.

3.2. Material Characterizations

The phases of the concentration-gradient and concentration-constant samples were investigated by X-ray diffraction (XRD, PANalytical, X'Pert3 powder) using Cu k_α radiation in the 2θ range of 10–80°. The morphologies of the prepared materials were observed by scanning with an electron microscope (SEM, HITACHI, SU5000) and a high-resolution transmission electron microscope (HRTEM, JEOL, JEM-2100F). X-ray photoelectron spectroscopy (XPS, Ulvac-Phi, PHI 5000 VersaProbe III) with monochromatic Al k_α radiation was applied to determine the oxidation state of the elements in the prepared materials.

3.3. Electrochemical Properties Characterizations

The 2016-type coin cell was used to evaluate the electrochemical performance of concentration-gradient and concentration-constant $LiNi_{0.9}Co_{0.083}Mn_{0.017}O_2$ electrodes by electrochemical tests. The coin cell consists of a lithium foil-counter electrode, a working electrode, a Cellguard 2500 film separator and an electrolyte of 1 mol/L $LiPF_6$ in the mixed solvent of ethylene carbonate (EC), ethyl methyl carbonate (EMC) and diethyl carbonate (DEC) with a volume ratio of 4:2:4. The dried working electrodes are composed of 80 wt.% of electroactive material CG-LNCM or CC-LNCM, 10 wt.% of acetylene black and 10 wt.% of polyvinylidene fluoride, and the loading of electroactive materials is about 1.8 mg cm^{-2}. The galvanostatic charge/discharge tests were carried out in the voltage range of 3.0 to 4.3 V (vs. Li^+/Li) at 25 °C. The cyclic voltammetry (CV) was conducted on a CHI660E electrochemical workstation at 0.1 mV s^{-1} in the potential range of 3.0 to 4.3 V. To compare the diffusion coefficients of Li^+, D_{Li}, of CG-LNCM and CC-LNCM, the galvanostatic intermittent titration technique (GITT) was conducted on the electrodes cycled at 0.1C for three times. The GITT measurement was performed at a pulse current of 20 mA g^{-1} for 10 min, followed by a relaxation of 30 min.

4. Conclusions

The full concentration-gradient $LiNi_{0.9}Co_{0.083}Mn_{0.017}O_2$, in which composition varies from the Ni-rich core $LiNi_{0.93}Co_{0.07}O_2$ to the outmost shell $LiNi_{1/3}Co_{1/3}Mn_{1/3}O_2$, was prepared by a facile co-precipitation method combined with high-temperature calcination under an inert atmosphere. Benefiting from the special functions of a concentration-gradient microstructure for which the Ni-rich core provides a high capacity, the shell supplies the excellent structural stability of the surface, and concentration-gradient distribution of compositions alleviates the formation of microcrack. Moreover, concentration-gradient $LiNi_{0.9}Co_{0.083}Mn_{0.017}O_2$ possesses a higher capacity and a better rate capability and cyclability in comparison with the concentration-constant $LiNi_{0.9}Co_{0.083}Mn_{0.017}O_2$ prepared by the same method. The combination of the concentration-gradient design and the rapid co-precipitation synthesis may provide an effective strategy for large-scale production of Ni-based ternary cathode materials with high-performance for lithium-ion batteries.

Supplementary Materials: The following supporting information can be downloaded at: https://www.mdpi.com/article/10.3390/molecules28083347/s1, Figure S1: Refine XRD patterns of (a) CC-NCMOH, (b) CG-NCMOH, (c) CC-LNCM and (d) CG-LNCM; Figure S2: The EDX graph of (a) area A and (b) area B of CC-LNCM, (c) area C and (d) area D of CG-LNCM; Figure S3: Activated charge/discharge profiles of (a) CC-LNCM and (b) CG-LNCM at various current rates; Table S1: Cell parameters of as-prepared samples; Table S2: Atomic parameters of as-prepared samples; Table S3: The fitting results of fresh electrodes and electrodes after cycled for 100 cycles at 5C.

Author Contributions: H.L.: Conceptualization, methodology, validation, formal analysis, writing-original draft, and editing; Y.G.: Conceptualization, methodology, software, investigation.; Y.C.: Supervision, conceptualization, methodology, data curation, formal analysis, writing-review and editing; N.G.: Conceptualization, methodology, software, investigation, data curation; R.S.: Conceptualization, methodology, software, investigation; Y.L.: Conceptualization, methodology, software, investigation; Q.C.: Supervision, project administration, funding acquisition, conceptualization, review, and editing. All authors have read and agreed to the published version of the manuscript.

Funding: This research was funded by the National Natural Science Foundation of China (Grant No. 52164027) and the Natural Science Foundation of Guangxi Province (Grant No. 2021GXNSFAA220063).

Institutional Review Board Statement: Not applicable.

Informed Consent Statement: Not applicable.

Data Availability Statement: Not applicable.

Conflicts of Interest: The authors declare no conflict of interest.

Sample Availability: Not applicable.

References

1. Tarascon, J.M.; Armand, M. Issues and challenges facing rechargeable lithium batteries. *Nature* **2001**, *414*, 359. [CrossRef] [PubMed]
2. Palacin, M.R. Recent advances in rec hargeable battery materials: A chemist's perspective. *Chem. Soc. Rev.* **2009**, *38*, 2565–2575. [CrossRef]
3. Sun, Y.-K.; Myung, S.-T.; Park, B.-C.; Prakash, J.; Belharouak, I.; Amine, K. High-energy cathode material for long-life and safe lithium batteries. *Nat. Mater.* **2009**, *8*, 320. [CrossRef] [PubMed]
4. Scrosati, B.; Hassoun, J.; Sun, Y.-K. Lithium-ion batteries. A look into the future. *Energy Environ. Sci.* **2011**, *4*, 3287–3295. [CrossRef]
5. Zhao, J.; Hong, M.; Ju, Z.; Yan, X.; Gai, Y.; Liang, Z. Durable lithium metal anodes enabled by interfacial layers based on mechanically interlocked networks capable of energy dissipation. *Angew. Chem. Int. Ed. Engl.* **2022**, *61*, e202214386. [CrossRef] [PubMed]
6. Yang, S.J.; Yao, N.; Jiang, F.N.; Xie, J.; Sun, S.Y.; Chen, X.; Yuan, H.; Cheng, X.B.; Huang, J.Q.; Zhang, Q. Thermally stable polymer-rich solid electrolyte interphase for safe lithium metal pouch cells. *Angew. Chem. Int. Ed. Engl.* **2022**, *61*, e202214545. [CrossRef]
7. Miao, Y.; Hynan, P.; von Jouanne, A.; Yokochi, A. Current Li-ion battery technologies in electric vehicles and opportunities for advancements. *Energies* **2019**, *12*, 1074. [CrossRef]
8. Shao, Y.; Huang, B.; Liu, Q.; Liao, S. Preparation and modification of Ni-Co-Mn ternary cathode materials. *Prog. Chem.* **2018**, *30*, 410–419. [CrossRef]
9. Bai, X.; Ban, L.; Zhuang, W. Research progress on coating and doping modification of nickel rich ternary cathode materials. *J. Inorg. Mater.* **2020**, *35*, 972–986. [CrossRef]
10. Liu, L.; Li, M.; Chu, L.; Jiang, B.; Lin, R.; Zhu, X.; Cao, G. Layered ternary metal oxides: Performance degradation mechanisms as cathodes, and design strategies for high-performance batteries. *Prog. Mater. Sci.* **2020**, *111*, 100655. [CrossRef]
11. Song, L.; Du, J.; Xiao, Z.; Jiang, P.; Cao, Z.; Zhu, H. Research progress on the surface of high-nickel nickel-cobalt-manganese ternary cathode materials: A mini review. *Front. Chem.* **2020**, *8*, 761. [CrossRef]
12. Choi, J.U.; Voronina, N.; Sun, Y.-K.; Myung, S.-T. Recent progress and perspective of advanced high-energy Co-less Ni-rich cathodes for Li-ion batteries: Yesterday, today, and tomorrow. *Adv. Energy Mater.* **2020**, *10*, 2002027. [CrossRef]
13. Xu, R.; Xu, W.; Wang, J.; Liu, F.; Sun, W.; Yang, Y. A review on regenerating materials from spent lithium-ion batteries. *Molecules* **2022**, *27*, 2285. [CrossRef]
14. Wang, X.; Ding, Y.-L.; Deng, Y.-P.; Chen, Z. Ni-rich/Co-poor layered cathode for automotive Li-ion batteries: Promises and challenges. *Adv. Energy Mater.* **2020**, *10*, 1903864. [CrossRef]
15. Jiang, M.; Danilov, D.L.; Eichel, R.-A.; Notten, P.H.L. A review of degradation mechanisms and recent achievements for Ni-rich cathode-based Li-ion batteries. *Adv. Energy Mater.* **2021**, *11*, 2103005. [CrossRef]
16. Kim, J.; Lee, H.; Cha, H.; Yoon, M.; Park, M.; Cho, J. Prospect and reality of Ni-rich cathode for commercialization. *Adv. Energy Mater.* **2018**, *8*, 1702028. [CrossRef]
17. Butt, A.; Ali, G.; Kubra, K.T.; Sharif, R.; Salman, A.; Bashir, M.; Jamil, S. Recent advances in enhanced performance of Ni-rich cathode materials for Li-ion batteries: A review. *Energy Technol.* **2022**, *10*, 2100775. [CrossRef]
18. Liao, C.; Li, F.; Liu, J. Challenges and modification strategies of Ni-rich cathode materials operating at high-voltage. *Nanomaterials* **2022**, *12*, 1888. [CrossRef]
19. Yin, S.; Deng, W.; Chen, J.; Gao, X.; Zou, G.; Hou, H.; Ji, X. Fundamental and solutions of microcrack in Ni-rich layered oxide cathode materials of lithium-ion batteries. *Nano Energy* **2021**, *83*, 105854. [CrossRef]
20. Ahaliabadeh, Z.; Kong, X.; Fedorovskaya, E.; Kallio, T. Extensive comparison of doping and coating strategies for Ni-rich positive electrode materials. *J. Power Sources* **2022**, *540*, 231633. [CrossRef]
21. Tao, J.; Mu, A.; Geng, S.; Xiao, H.; Zhang, L.; Huang, Q. Influences of direction and magnitude of Mg^{2+} doping concentration gradient on the performance of full concentration gradient cathode material. *J. Solid State Electrochem.* **2021**, *25*, 1959–1974. [CrossRef]
22. Li, Q.; Li, Z.; Wu, S.; Wang, Z.; Liu, X.; Li, W.; Li, N.; Wang, J.; Zhuang, W. Utilizing diverse functions of zirconium to enhance the electrochemical performance of Ni-rich layered cathode materials. *ACS Appl. Energy Mater.* **2020**, *3*, 11741–11751. [CrossRef]
23. Wang, Y.; Liu, K.; Wang, B. Coating strategies of Ni-rich layered cathode in LIBs. *Chem. Res. Chin. Univ.* **2021**, *42*, 1514–1529. [CrossRef]
24. Wang, X.; Ruan, X.; Du, C.-F.; Yu, H. Developments in surface/interface engineering of Ni-rich layered cathode materials. *Chem. Rec.* **2022**, *22*, e202200119. [CrossRef] [PubMed]

25. Wang, H.; Ge, W.; Li, W.; Wang, F.; Liu, W.; Qu, M.Z.; Peng, G. Facile fabrication of ethoxy-functional polysiloxane wrapped $LiNi_{0.6}Co_{0.2}Mn_{0.2}O_2$ cathode with improved cycling performance for rechargeable Li-ion battery. *ACS Appl. Mater. Interfaces* **2016**, *8*, 18439–18449. [CrossRef] [PubMed]
26. Ni, L.; Zhang, S.; Di, A.; Deng, W.; Zou, G.; Hou, H.; Ji, X. Challenges and strategies towards single-crystalline Ni-rich layered cathodes. *Adv. Energy Mater.* **2022**, *12*, 2201510. [CrossRef]
27. Pang, P.; Tan, X.; Wang, Z.; Cai, Z.; Nan, J.; Xing, Z.; Li, H. Crack-free single-crystal $LiNi_{0.83}Co_{0.10}Mn_{0.07}O_2$ as cycling/thermal stable cathode materials for high-voltage lithium-ion batteries. *Electrochim. Acta* **2021**, *365*, 137380. [CrossRef]
28. Lee, S.-H.; Sim, S.-J.; Jin, B.-S.; Kim, H.-S. High performance well-developed single crystal $LiNi_{0.91}Co_{0.06}Mn_{0.03}O_2$ cathode via LiCl-NaCl flux method. *Mater. Lett.* **2020**, *270*, 127615. [CrossRef]
29. Ran, Q.; Zhao, H.; Hu, Y.; Hao, S.; Liu, J.; Li, H.; Liu, X. Enhancing surface stability of $LiNi_{0.8}Co_{0.1}Mn_{0.1}O_2$ cathode with hybrid core-shell nanostructure induced by high-valent titanium ions for Li-ion batteries at high cut-off voltage. *J. Alloys Compd.* **2020**, *834*, 155099. [CrossRef]
30. Sun, Y.K.; Myung, S.T.; Kim, M.H.; Prakash, J.; Amine, K. Synthesis and characterization of $Li[(Ni_{0.8}Co_{0.1}Mn_{0.1})_{(0.8)}(Ni_{0.5}Mn_{0.5})_{(0.2)}]O_2$ with the microscale core-shell structure as the positive electrode material for lithium batteries. *J. Am. Chem. Soc.* **2005**, *127*, 13411–13418. [CrossRef]
31. Lim, B.-B.; Myung, S.-T.; Yoon, C.S.; Sun, Y.-K. Comparative Study of Ni-rich layered cathodes for rechargeable lithium batteries: $Li[Ni_{0.85}Co_{0.11}Al_{0.04}]O_2$ and $Li[Ni_{0.84}Co_{0.06}Mn_{0.09}Al_{0.01}]O_2$ with two-step full concentration gradients. *ACS Energy Lett.* **2016**, *1*, 283–289. [CrossRef]
32. Park, G.-T.; Ryu, H.-H.; Noh, T.-C.; Kang, G.-C.; Sun, Y.-K. Microstructure-optimized concentration-gradient NCM cathode for long-life Li-ion batteries. *Mater. Today* **2022**, *52*, 9–18. [CrossRef]
33. Park, N.Y.; Ryu, H.H.; Park, G.T.; Noh, T.C.; Sun, Y.K. Optimized Ni-rich NCMA cathode for electric vehicle batteries. *Adv. Energy Mater.* **2021**, *11*, 200377. [CrossRef]
34. Hou, P.; Zhang, H.; Zi, Z.; Zhang, L.; Xu, X. Core-shell and concentration-gradient cathodes prepared via co-precipitation reaction for advanced lithium-ion batteries. *J. Mater. Chem. A* **2017**, *5*, 4254–4279. [CrossRef]
35. Tsai, S.-Y.; Fung, K.-Z. Synthesis routes on electrochemical behavior of Co-free layered $LiNi_{0.5}Mn_{0.5}O_2$ cathode for Li-ion batteries. *Molecules* **2023**, *28*, 794. [CrossRef] [PubMed]
36. Jiang, Y.; Liu, Z.; Zhang, Y.; Hu, H.; Teng, X.; Wang, D.; Gao, P.; Zhu, Y. Full-gradient structured $LiNi_{0.8}Co_{0.1}Mn_{0.1}O_2$ cathode material with improved rate and cycle performance for lithium ion batteries. *Electrochim. Acta* **2019**, *309*, 74–85. [CrossRef]
37. Tan, Z.; Li, Y.; Xi, X.; Yang, J.; Xu, Y.; Xiong, Y.; Wang, S.; Liu, S.; Zheng, J. Lattice engineering to alleviate microcrack of $LiNi_{0.9}Co_{0.05}Mn_{0.05}O_2$ cathode for optimization their Li^+ storage functionalities. *Electrochim. Acta* **2022**, *401*, 139482. [CrossRef]
38. Li, G.; Zhang, Z.; Wang, R.; Huang, Z.; Zuo, Z.; Zhou, H. Effect of trace Al surface doping on the structure, surface chemistry and low temperature performance of $LiNi_{0.5}Co_{0.2}Mn_{0.3}O_2$ cathode. *Electrochim. Acta* **2016**, *212*, 399–407. [CrossRef]
39. Zhou, H.; Zhou, F.; Shi, S.; Yang, W.; Song, Z. Influence of working temperature on the electrochemical characteristics of Al_2O_3-coated $LiNi_{0.8}Co_{0.1}Mn_{0.1}O_2$ cathode materials for Li-ion batteries. *J. Alloys Compd.* **2020**, *847*, 156412. [CrossRef]
40. Lian, K.K.; Kirk, D.W.; Thorpe, S.J. Investigation of a "two-state" tafel phenomenon for the oxygen evolution reaction on an amorphous Ni-Co alloy. *J. Electrochem. Soc.* **1995**, *142*, 3704. [CrossRef]
41. Mansour, A.N. Characterization of NiO by XPS. *Surf. Sci. Spectra* **1994**, *3*, 231–238. [CrossRef]
42. Mansour, A.N. Characterization of $LiNiO_2$ by XPS. *Surf. Sci. Spectra* **1996**, *3*, 279–286. [CrossRef]
43. Kosova, N.V.; Devyatkina, E.T.; Kaichev, V.V. Optimization of Ni^{2+}/Ni^{3+} ratio in layered $Li(Ni,Mn,Co)O_2$ cathodes for better electrochemistry. *J. Power Sources* **2007**, *174*, 965–969. [CrossRef]
44. Yang, J.; Xia, Y. Suppressing the phase transition of the layered Ni-rich oxide cathode during high-voltage cycling by introducing low-content Li_2MnO_3. *ACS Appl. Mater. Interfaces* **2016**, *8*, 1297–1308. [CrossRef] [PubMed]
45. Li, W.-W.; Zhang, X.-J.; Si, J.-J.; Yang, J.; Sun, X.-Y. TiO_2-coated $LiNi_{0.9}Co_{0.08}Al_{0.02}O_2$ cathode materials with enhanced cycle performance for Li-ion batteries. *Rare Met.* **2021**, *40*, 1719–1726. [CrossRef]
46. Huang, J.; Duan, J.; Du, K.; Cao, Y.; Peng, Z.; Hu, G. Enhanced cycling performance of $LiNi_{0.9}Co_{0.08}Al_{0.02}O_2$ via Co-rich surface. *JOM* **2020**, *72*, 738–744. [CrossRef]
47. Mo, Y.; Guo, L.; Jin, H.; Du, B.; Cao, B.; Chen, Y.; Li, D.; Chen, Y. Building nickel-rich cathodes with large concentration gradient for high performance lithium-ion batteries. *J. Power Sources* **2020**, *468*, 228405. [CrossRef]
48. Zhang, C.; Wan, J.; Li, Y.; Zheng, S.; Zhou, K.; Wang, D.; Wang, D.; Hong, C.; Gong, Z.; Yang, Y. Restraining the polarization increase of Ni-rich and low-Co cathodes upon cycling by Al-doping. *J. Mater. Chem. A* **2020**, *8*, 6893–6901. [CrossRef]
49. Su, Y.; Chen, G.; Chen, L.; Shi, Q.; Lv, Z.; Lu, Y.; Bao, L.; Li, N.; Chen, S.; Wu, F. Roles of fast-ion conductor $LiTaO_3$ modifying Ni-rich cathode material for Li-ion batteries. *Chemsuschem* **2021**, *14*, 1955–1961. [CrossRef]
50. Qian, R.; Liu, Y.; Cheng, T.; Li, P.; Chen, R.; Lyu, Y.; Guo, B. Enhanced surface chemical and structural stability of Ni-rich cathode materials by synchronous lithium-ion conductor coating for lithium-ion batteries. *ACS Appl. Mater. Interfaces* **2020**, *12*, 13813–13823. [CrossRef]
51. Hao, S.; Zhang, D.; Li, Y.; Xi, X.; Wang, S.; Li, X.; Shen, X.; Liu, S.; Zheng, J. Multifunctionality of cerium decoration in enhancing the cycling stability and rate capability of a nickel-rich layered oxide cathode. *Nanoscale* **2021**, *13*, 20213–20224. [CrossRef]

52. Park, J.-S.; Hong, Y.J.; Kim, J.H.; Kang, Y.C. Carbon-templated strategy toward the synthesis of dense and yolk-shell multi-component transition metal oxide cathode microspheres for high-performance Li ion batteries. *J. Power Sources* **2020**, *461*, 15. [CrossRef]
53. Yang, X.; Tang, Y.; Shang, G.; Wu, J.; Lai, Y.; Li, J.; Qu, Y.; Zhang, Z. Enhanced cyclability and high-rate capability of $LiNi_{0.88}Co_{0.095}Mn_{0.025}O_2$ cathodes by homogeneous Al^{3+} doping. *ACS Appl. Mater. Interfaces* **2019**, *11*, 32015–32024. [CrossRef]
54. Weppner, W.; Huggins, R.A. Determination of the kinetic parameters of mixed-conducting flectrodes and application to the system Li3Sb. *J. Electrochem. Soc.* **1977**, *124*, 1569. [CrossRef]
55. Chen, Q.; Qiao, X.; Wang, Y.; Zhang, T.; Peng, C.; Yin, W.; Liu, L. Electrochemical performance of Li3−xNaxV2(PO4)3/C composite cathode materials for lithium ion batteries. *J. Power Sources* **2012**, *201*, 267–273. [CrossRef]

molecules

MDPI

Article

Preparation of Heavily Doped P-Type PbSe with High Thermoelectric Performance by the NaCl Salt-Assisted Approach

Xinru Ma [1], Xuxia Shai [1,*], Yu Ding [2,3,*], Jie Zheng [4], Jinsong Wang [5], Jiale Sun [1], Xiaorui Li [1], Weitao Chen [1], Tingting Wei [1], Weina Ren [1], Lei Gao [1], Shukang Deng [4,*] and Chunhua Zeng [1,*]

1 Faculty of Science, Institute of Physical and Engineering Science, Kunming University of Science and Technology, Kunming 650500, China
2 College of Chemistry and Materials Science, Hubei Engineering University, Xiaogan 432000, China
3 Nuode New Materials Co., Ltd., Shenzhen 518048, China
4 Education Ministry Key Laboratory of Renewable Energy Advanced Materials and Manufacturing Technology, Yunnan Normal University, Kunming 650500, China
5 Faculty of Materials Science and Engineering, Kunming University of Science and Technology, Kunming 650093, China
* Correspondence: xuxiashai@kust.edu.cn (X.S.); dy9802@126.com (Y.D.); skdeng@126.com (S.D.); chzeng83@kust.edu.cn (C.Z.)

Abstract: Thermoelectric (TE) technology, which can convert scrap heat into electricity, has attracted considerable attention. However, broader applications of TE are hindered by lacking high-performance thermoelectric materials, which can be effectively progressed by regulating the carrier concentration. In this work, a series of $PbSe(NaCl)_x$ (x = 3, 3.5, 4, 4.5) samples were synthesized through the NaCl salt-assisted approach with Na^+ and Cl^- doped into their lattice. Both theoretical and experimental results demonstrate that manipulating the carrier concentration by adjusting the content of NaCl is conducive to upgrading the electrical transport properties of the materials. The carrier concentration elevated from 2.71×10^{19} cm^{-3} to 4.16×10^{19} cm^{-3}, and the materials demonstrated a maximum power factor of 2.9×10^{-3} W m^{-1} K^{-2}. Combined with an ultralow lattice thermal conductivity of 0.7 W m^{-1} K^{-1}, a high thermoelectric figure of merit (ZT) with 1.26 at 690 K was attained in $PbSe(NaCl)_{4.5}$. This study provides a guideline for chemical doping to improve the thermoelectric properties of PbSe further and promote its applications.

Keywords: chemical doping; salt-assisted; NaCl doping; PbSe; thermoelectric material; thermoelectric performance

1. Introduction

Thermoelectric (TE) materials capable of transforming scrap thermal energy into electric power have attracted considerable attention [1,2]. To promote the sustainable development of the environment and economy, it is crucial to explore suitable TE materials with high TE performance for their practical application. Currently, based on the existing thermoelectric materials such as Bi_2Te_3 and SiGe, the corresponding conversion efficiency of them are around the value of 10%, which is far away from the demanding value of 20% [3]. Among the advanced *p*-type thermoelectric materials, lead chalcogenide has generated widespread interest due to its low thermal conductivity, high carrier fluidity, and high symmetrical structure [4,5], in which the PbTe has been widely studied. Notably, regarding the high melting point and rich content of the element, PbSe (a close analog of PbTe), with a similar two-valley valence band structure to PbTe, has also captivated extensive research, but the unsatisfactory TE properties still hinder its practical applications [6]. To obtain higher conversion efficiency, it is necessary to improve the dimensionless of merit (ZT), which is described as ZT = $\alpha^2\sigma$ T/κ, where σ represents the electrical conductivity and α denotes the Seebeck coefficient. The product $\alpha^2\sigma$ is referred to as the power factor. The total thermal conductivity is composed of the lattice thermal conductivity (κ_L) and the electron thermal conductivity (κ_e), and T represents the absolute temperature [7]. Generally, there are two ways to enhance the ZT values, one is optimizing carrier concentration and changing the energy band to boost the PF, and another is reducing the lattice thermal conductivity through the initiation of atomic defects and formation of nanostructures [8–11]. Excitingly, chemical doping can not only modify the electronic structure but also optimize the carrier concentration, which has been illustrated as a successful control approach for tuning the TE material properties [12–14]. By properly introducing dopants, the energy distance from the light band to the heavy valence band of lead chalcogenide can be greatly decreased—such "band convergence" is beneficial to improve the TE performance and especially increasing the Seebeck coefficient [15].

To tune the band structure of *p*-type PbSe, many different dopants have been examined, such as by substituting Cu/Sr/Ba/K/Ag for Pb or Te/Cl for Se [16–19]. In addition, the doping method has also been explored [20,21]. Compared with other processes, the flux method is widely considered to obtain highly crystallized and phase-pure samples [22]. Due to the slow cooling process, the stable liquid environment (flux) provides sufficient space and ample time for the constituent atoms to nucleate and self-organize [23]; it is accessible for the experiment since the atmospheric pressure and the working temperature are below the melting point [24]. In the preliminary work, both Sn and Ga have been proven as the feasible effective flux for thermoelectric materials [25,26]. Moreover, given the cost, environmental friendliness, and sustainability, NaCl has become the best choice as the flux. Excitingly, Deng's research group has adopted NaCl as the flux to prepare the $Cu_2Se_{1+x}(NaCl)_{2.5}$ (x = 0, 0.01, 0.02, 0.03) and β-Zn_4Sb_3 successfully, thus presenting excellent thermoelectric properties [27,28]. Furthermore, theoretical calculations have been widely performed to understand the effect of chemical doping on band structure and thermal transport properties of TE materials due to the predictability and reliability, providing guidance for optimizing and designing TE materials [29–31].

Thus, chemical doping combined with the suitable preparation process, the PbSe-based materials with high TE performance could be achieved by structural optimization or electronic band regulation, associating with the theoretical calculation, contributing to understanding the nature of the doping effect, and providing a substantive improvement in TE materials. Aiming toward a fundamental understanding and practical development of PbSe-based thermoelectric materials, the pursuit of highly effective TE materials toward ultimate commercialization necessitates greater knowledge of the effects and regulators of chemical doping in light of the significance of chemical doping for TE materials. Herein, we present tuning TE performance via chemical doping and processing improvement by NaCl flux, employing the Na and Cl substitution as a prototype from theoretical simulation and experimental verification to explore the formation of high $PbSe(NaCl)_x$ (x = 3, 3.5, 4, 4.5) crystal quality and its inherent influence on TE properties. By the NaCl salt-assisted approach, the Na and Cl are involved in PbSe successfully, and the Na doping exerts a dominant influence on the valence band convergence. The charge carrier density of $PbSe(NaCl)_x$ was gradually optimized from 2.71×10^{19} cm^{-3} to 4.16×10^{19} cm^{-3} as the increasing content of NaCl, concurrently with the appropriate Seebeck coefficient and conductivity value, a high power factor of 2.9×10^{-3} Wm^{-1}K^{-2} and an appreciable ZT value of 1.26 at 690 K were caught by $PbSe(NaCl)_{4.5}$. This demonstration indicates the positive role of Na^+ and Cl^- on the thermoelectric performance of PbSe compounds, suggesting an effective strategy for upgrading the thermoelectric characteristics of thermoelectric materials by tuning the carrier concentration through NaCl salt-assisted approach, thus hoping to promote the development in PbSe-based TE technology for addressing global energy and environmental challenges.

2. Results and Discussion

2.1. Structure and Composition

The morphological image of $PbSe(NaCl)_3$ with metallic luster and about 12 mm in size, shown in Figure 1a. Figure 1b demonstrates the X-ray diffraction (XRD) patterns of all samples; the standard XRD spectrum of PbSe is presented for a comparison at the bottom. It can be observed that the diffraction peaks of the doped sample were slightly shifted towards a high diffraction angle along with the introduction of NaCl, especially the peak at 60.381° in the inset. Generally, the shift of XRD peaks reflects residual stress or variations in chemical constituents [32]. In this context, the primary cause of the peak shift is the relatively reduced lattice constant, which can be attributed to the comparatively smaller ionic radius of Na^+ (102 picometer) and Cl^- (181 picometer) in comparison to Pb^{2+} (119 picometer) and Se^{2-} (198 picometer), respectively. It demonstrates that the doping content has not reached the solid solubility limit of PbSe and the successful presence of NaCl in the lattice of PbSe. The intensity of the (200) diffraction peaks of $PbSe(NaCl)_x$ enhanced and then decreased with the addition of more NaCl. At the same time, there is no discernible variation in other peaks, indicating that the Na^+ and Cl^- facilitate the crystallinity, specifically the growth of (200) crystal planes.

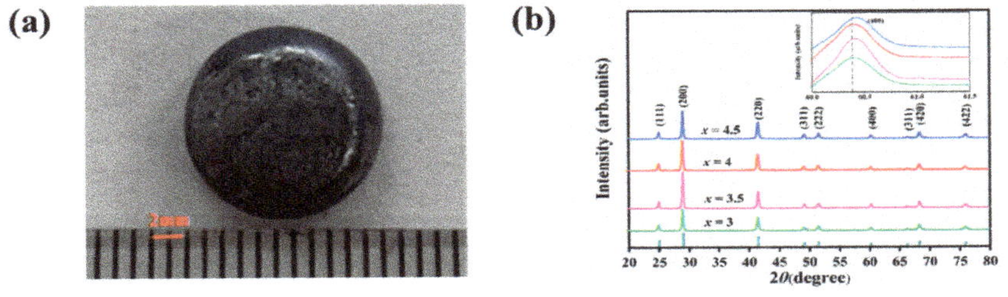

Figure 1. External appearance picture and X-ray diffraction patterns. (**a**) The photograph of $PbSe(NaCl)_3$ synthesized using the sodium chloride fusion approach. (**b**) XRD patterns of $PbSe(NaCl)_x$ (x = 3, 3.5, 4, 4.5) with the standard XRD spectrum of PbSe for comparison.

Figure 2a exhibits the FESEM pictures of the $PbSe(NaCl)_3$ sample, without any cracks or voids, even with a magnification of 100,000 times. And the SEM analysis with different magnifications and multiple locations emerged in Figure S1, indicating the feasibility of the NaCl salt-assisted approach to synthesize the dense crystal structure, which agrees with the high density obtained from the Archimedes method. The range of this (8.47–8.74 g cm^{-3}) is in close agreement with the theoretical density of 8.27 g cm^{-3} [33]. To further determine the elemental composition, an EDS map was performed on the $PbSe(NaCl)_3$ in Figure 2b. Additionally, the results suggested that only Pb, Se, Na, and Cl elements were present in the sample without any other impurity elements. The electron probe microscope (EPMA) was used to investigate the content of every component, and the pertinent statistics are presented in Table 1. By increasing the proportion of NaCl, the content of Cl rose significantly relative to the content of Na, from 0.61% to 1.91%. The content difference of Se is similar, and the actual content of Pb in the sample is slightly reduced, indicating the substitute doping of NaCl inter to the lattice of PbSe. HRTEM was implemented to characterize the crystals at x = 3 (Figure 2c). The completed lattice stripes of $PbSe(NaCl)_{3.0}$ display interplanar spacings of 0.306 nm in the particle, which matched well with the (200) planes of PbSe. Figure 2d exhibits the image of SAED acquired along the [002] zone axis of $PbSe(NaCl)_{3.0}$, further confirming the high crystal quality of the specimen synthesized via the NaCl salt-assisted approach [34].

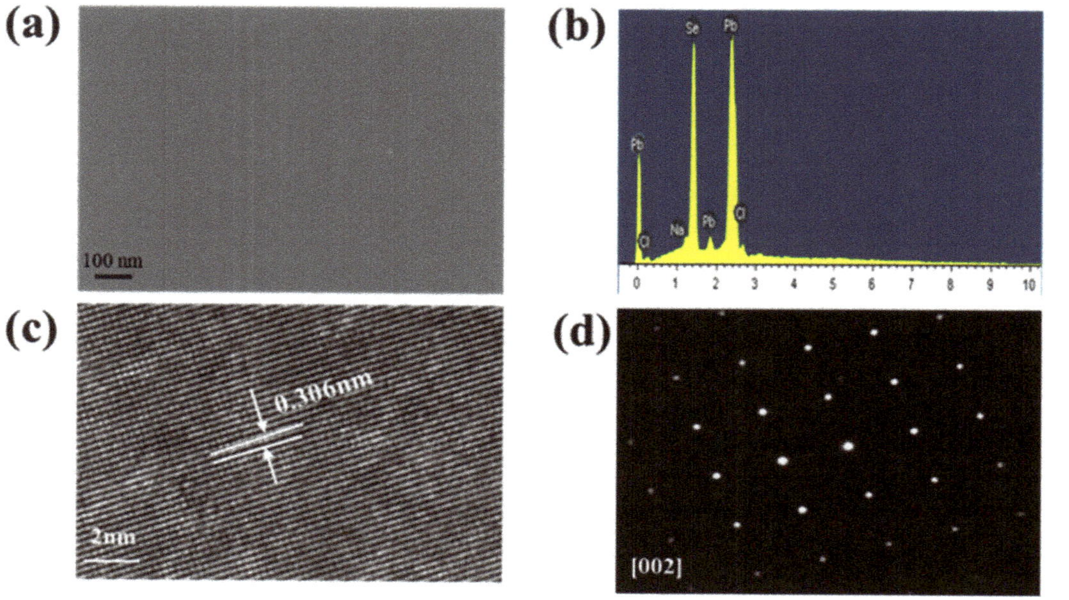

Figure 2. The micro characterization of $PbSe(NaCl)_3$ with high crystal quality. (**a**) FESEM image, (**b**) EDS image, (**c**) HRTEM image, (**d**) SAED image.

Table 1. A list of samples and some of their temperature-dependent properties at 300 K.

Samples	**Crystal Compositions PbSe-(NaCl)x**				R_H **cm^3/C**	n_H **10^{19} cm^{-3}**	μ_H **cm^2/Vs**	ρ **g·cm^3**
	Pb	Se	Na	Cl				
$x = 3$	48.52	50.7	0.17	0.61	0.23	2.717	276	8.47
$x = 3.5$	48.17	50.6	0.2	1.03	0.21	2.976	292	8.59
$x = 4$	47.96	50.1	0.3	1.64	0.18	3.472	295	8.52
$x = 4.5$	47.58	50.1	0.41	1.91	0.15	4.167	270	8.74

The XPS technology was employed to explore the chemical oxidation state of constituents in the sample, as displayed in Figure 3. The Pb $4f_{7/2}$ and Pb $4f_{5/2}$ orbital peaks of PbSe samples can be observed at 137.6 and 142.4 eV in Figure 3a, which corresponds to Pb^{2+} valence, indicating that the Pb element has a valence of +2. In Figure 3b, the XPS spectrum of Se 3d is presented, revealing the chemical binding energy peak of Se $3d_{5/2}$ at 53.4 eV and the chemical binding energy peak of Se $3d_{3/2}$ at 54.2 eV. These peaks correspond to the characteristics of Se^{-2}, indicating that the Se element is −2 valence [34]. In Figure 3c, the XPS spectrum of Na 1s is displayed, and the chemical binding energy peak of Na 1s is observed at 1071.46 eV, suggesting that Na has a valence state of +1, thus promising the increase in hole concentration. Figure 3d exhibits the XPS spectrum of Cl 2p, where the chemical binding energy peak of Cl 2p is located within the range of 198.5–199 eV, implying that Cl has a valence state of −1.

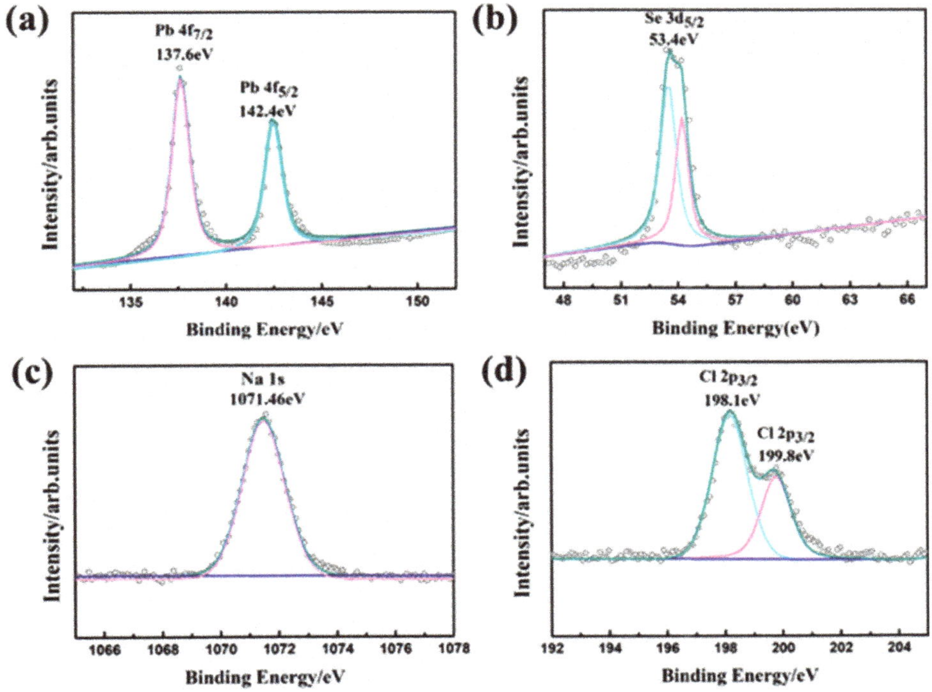

Figure 3. XPS fittings for Pb 4f (**a**), Se 3d (**b**), Na 1s (**c**), and Cl 2p (**d**).

The electronic structure of TE material can be commendably optimized by chemical doping, determining the carrier concentration and ultimately affecting the TE performance of material [35,36]. Therefore, it is crucial to explore the electronic structure by theoretical calculations. Based on the XRD and XPS results, it is not double that the Na^+ and Cl^- ions are inserted in the PbSe's lattice by substitutional doping. In order to show the way of NaCl doping intuitively and the influence of doping content on electronic structure of PbSe$(NaCl)_x$ compounds, supercells with standards of $3 \times 3 \times 3$ and $4 \times 4 \times 4$ were built and further electronic structure information was explored. The construction of the PbSe $(NaCl)_x$ model involves the substitution of Pb with Na and Se with Cl. We replaced Pb and Se with different stations, according to the principle of minimum energy to determine the optimal structure, the most stable configuration of the PbSe$(NaCl)_x$ system is shown in Figure S2, which is utilized to further calculate the density of states (DOS) and band structure (Figure 4). Although these models cannot accurately correspond to the experimental doping concentration, they are sufficient to explain the effect of doping on the electronic structure and even properties of PbSe $(NaCl)_x$ compounds. Figure 4a displays the energy band structure of the primary PbSe with a band gap of 0.419 eV, which corresponds well with the value of 0.439 eV calculated by J.P. Perdew et al. [37], and is also in good agreement with the bandgap range of 0.4–0.475 eV for PbSe thin films [38]. As shown in Figure 4b, it can be concluded that the L bands and the second valence band (Σ bands) can be found at the valence band regardless of doping NaCl, which is determined by the intrinsic properties of PbSe. In addition, the energy difference between the L and Σ bands is reduced from 0.28 eV in undoped PbSe to 0.18 eV in PbSe-NaCl, indicating a convergence of the two bands upon NaCl doping, of which is beneficial to enhance the TE performance of NaCl-doped PbSe. Similarly, this "band convergence" is also observed with the supercell of $4 \times 4 \times 4$ (Figure S4), and the tendency of band structure is consistent with that of 3 × 3 × 3. Furthermore, in order to comprehensively explore the function of Na and Cl dopants in $PbSe(NaCl)_x$ compounds, additional calculations were performed for Na and Cl doping in PbSe, as presented in Figure S3a,c, respectively. The results indicate that both Na and Cl doping exhibit a significantly similar band convergence effect. Nonetheless, the ΔEv of Na-doped PbSe is 0.18 eV, which is significantly smaller than that of Cl-doped PbSe (0.22 eV). This suggests that Na doping exerts a dominant influence on the valence band convergence, which is related to the report by Wu et al. [39], thereby leading ultimately to the improvement in electron transport and thermoelectric performance [17]. Compared with the calculated projected density-of-states of pure PbSe (Figure 4c), Na, Cl (Figure S3b,d), or NaCl-doped PbSe (Figure 4d) systems, the conduction band minimum stems mainly from Pb p states, while the valence Σ band comes from p orbitals of Se, Pb, and Cl for PbSe-NaCl.

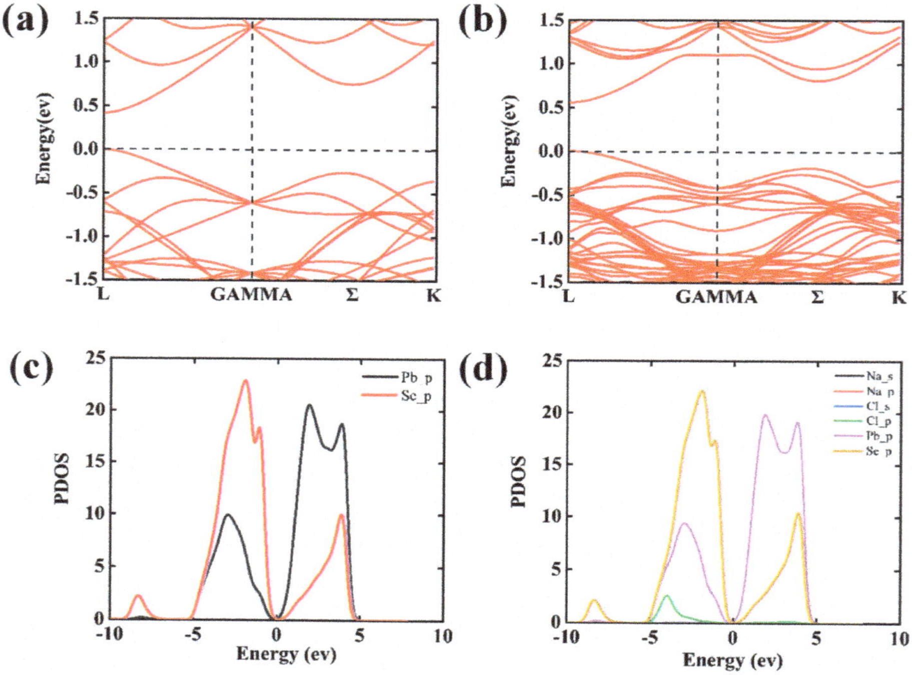

Figure 4. Band structure of pure PbSe with ΔEv~0.28 eV (**a**); NaCl-doped PbSe with ΔEv~0.18 eV (**b**); the corresponding DOS for pure PbSe (**c**) and PbSe with NaCl doping (**d**), where the corresponding states are presented in a clear manner to demonstrate their respective contributions.

2.2. *Thermoelectric Properties*

In Figure 5a–c, temperature-dependent electrical conductivity, the Seebeck coefficient, and a power factor of $PbSe(NaCl)_x$ (x = 3, 3.5, 4, 4.5) are presented, respectively. The charge transport ability of thermoelectric materials can be judged by a conductivity test, and the result is shown in Figure 5a. With the temperature rising from 300 to 700 K, the electrical conductivity of $PbSe(NaCl)_x$ compounds has a downward tendency, suggesting typical, degenerated semiconductor behavior. As the temperatures rise, the atoms and molecules in the material vibrate more violently, leading to an increase in the scattering of electrons in the material, thereby decreasing the conductivity. Notably, for temperatures beyond 600 K, PbSe-NaCl has a slower rate of decrease in conductivity since the suppression of bipolar conduction. The electrical conductivity of all samples is in the range from 12×10^4 S m^{-1} to 18×10^4 S m^{-1} at 300 K, with the most significant conductivity achieved for the sample defined as $PbSe(NaCl)_{4.5}$. In combination with the Hall tests, further electrical properties of the samples were identified according to the calculated carrier concentration n_H ($n_H = 1/(eR_H)$) and carrier mobility μ_H ($\mu_H = \sigma R_H$) at room temperature (Table 1). As a result, the Hall coefficients R_H of all samples are positive, indicating that hole conduction plays a dominant role in the samples. With an increase in the content of NaCl, the hole concentration increased gradually, resulting in a higher carrier concentration, as previously reported [40]. The carrier concentration increased from 2.71×10^{19} cm^{-3} to 4.16×10^{19} cm^{-3} because of the substitution of Na^+ for Pb^{2+} and Cl^- for Se^{2-}. Simultaneously, the carrier mobility decreased to 270 cm^2 $V^{-1}s^{-1}$ in S4, which may be induced by the higher defect density. The carrier mobility and concentration together determine the conductivity of the semiconductor according to $\sigma = ne\mu$. The results reveal that the higher the content of NaCl, the better the conductivity; the maximum electrical conductivity of 18×10^4 S m^{-1} was achieved in $PbSe(NaCl)_{4.5}$, indicating that NaCl can be used as an excellent dopant for transferring electronics, in agreement with the conclusion reported by Jing Li et al. [41].

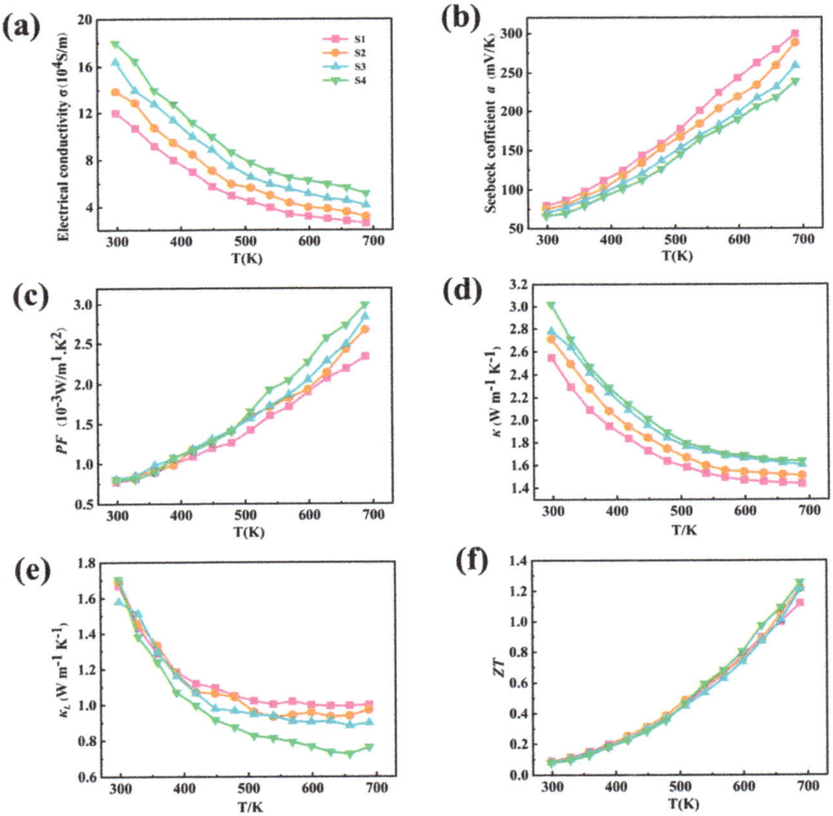

Figure 5. Temperature-dependent (**a**) electrical conductivity, (**b**) the Seebeck coefficient, (**c**) power factor, (**d**) thermal conductivity, (**e**) lattice thermal conductivity and (**f**) figureof merit ZT for $PbSe(NaCl)_x$ (x = 3, 3.5, 4, and 4.5).

The Seebeck coefficient refers to the ratio between the voltage generated by a thermoelectric material under a temperature gradient and the temperature gradient, which is one of the key parameters to measure the performance of TE materials. It is obvious that the Seebeck coefficients of samples are positive (Figure 5b), indicating that the

$PbSe(NaCl)_x$ samples are p-type semiconductors, and the values of S increase with the rising temperature. It is noteworthy that the Seebeck coefficient value of $PbSe(NaCl)_x$ is about 66~80 mV/K, which is two quantities larger than that of pure PbSe compounds prepared by the conventional melting technique or the microwave-assisted chemical deposition (MA-CBD) technique [42]. This improvement corresponds well to the energy band convergence of $PbSe(NaCl)_x$. The relationship between power factor (PF), calculated according to PF = $a^2\sigma$, and temperature is shown in Figure 5c. At the outset, the PF increases slightly from 7×10^{-4} W $m^{-1}K^{-2}$ to 8×10^{-4} W $m^{-1}K^{-2}$ with the increasing content of NaCl at room temperature, and the PF increases with the increasing temperature for all samples. As a consequence of the higher σ combined with the excellent a, the maximum power factor of 2.9×10^{-3} W m^{-1} K^{-2} was achieved in $PbSe(NaCl)_{4.5}$ at 700 K, which exceeds the value obtained for the sample prepared by the melt rotation method (1.25×10^{-3} W $m^{-1}K^{-2}$) [43].

In addition, thermal conductivity is also one of the important factors affecting the properties of thermoelectric materials, which refers to the ability of heat transport in a material. Figure 5d,e show the temperature-dependent thermal conductivity (total (κ) and lattice (κ_L)) of samples with varying content. In thermoelectric materials, thermal conductivity is influenced by the lattice structure and the transport model charge. The total thermal conductivity at room temperature of $PbSe(NaCl)_{3.0}$ is ~2.54 W m^{-1} K^{-1}, rising to ~3.0 W m^{-1} K^{-1} for $PbSe(NaCl)_{4.5}$, which may be due to the increases in electronic thermal conductivity caused by the higher doping concentration. Over the temperature span of 300 K to 700 K, the total thermal conductivity of all specimens decreased monotonically with the escalating temperature. The main reason for this is the increase in the intensity of phonons vibration with the rising temperature, leading to more frequent scattering and a decrease in phonon thermal conductivity, thus resulting in a decrease in the total thermal conductivity. Additionally, after 500 K, the thermal conductivity decreases more slowly with temperature due to the intensification of lattice vibration and the appearance of a bipolar effect [34].

The lattice thermal conductivity (κ_L) was estimated by subtracting the electronic thermal conductivity from the total thermal conductivity ($\kappa_L = \kappa - \kappa_e$). The electronic thermal conductivity was calculated according to the Wiedemann–Franz law, $\kappa_e = L\sigma T$, where L is the Lorenz number ($L = 2.45 \times 10^{-8}$ W Ω K^{-2}) [44]. At room temperature, the lattice thermal conductivity κ_L of all PbSe-$(NaCl)_x$ samples are in the range of 1.7 ± 0.2 W m^{-1} K^{-1}, consistent with the previous report [31,45], and that of $PbSe(NaCl)_{3.0}$ is around 1.67 W m^{-1} k^{-1}. With the increasing temperature, the lattice thermal conductivity of $PbSe(NaCl)_{3.0}$ shows a downward trend to the lowest value of ~0.99 W m^{-1} K^{-1} at 630 K contributed by the lattice vibration. Surprisingly, it further decreased to ~0.72 W $m^{-1}K^{-1}$ at ~660 K as the stoichiometric ratio of NaCl increased to 4.5, which was primarily caused by the enhancement of the point defect scattering from the high doping level [43,46]. The point defect scattering in solid solution systems originates from the differences in atomic mass (mass fluctuations) and size, as well as the interatomic coupling force differences (strain fluctuations) between the impurity atoms and the host lattice, as described by the Callaway model [47,48]. Here, the substantial atomic mass fluctuation is mainly manifested between Na (atomic mass is 23) and Pb (atomic mass is 207), which predominately contributes to the reduction in lattice thermal conductivity. In particular, the lattice thermal conductivity of $PbSe(NaCl)_{4.5}$ is slightly increased at 690 K, which may be due to the redistribution of doping elements at this temperature, resulting in changes in the lattice structure, and then changing the propagation mode of phonons, engendering a slight increase in the κ_L.

Importantly, In Figure 5f, the temperature-dependent ZT is presented, which is the most convincing, and highlights the impressive thermoelectric performance of heavily doped p-type PbSe. At a temperature of 690 K, a ZT value of ~1.26 was observed when the carrier concentration, n_H, was 4.2×10^{19} cm^{-3}. The figure is superior to that of PbSe-based materials processed using the hot-pressing technique (~1.12 at 690 K) and spark plasma sintering (~1.0 at 690 K), indicating that the low thermal conductivity and significantly improved power factor (PF) in heavily doped samples contribute to the promising thermoelectric performance in p-type PbSe. The reduction in thermal conductivity can be attributed to the introduction of dopants, which scatter phonons and decrease their mean free path. Meanwhile, the high carrier concentration provokes a remarkable amelioration in electrical conductivity and power factor. These two factors, in combination, result in a high ZT value for the heavily doped p-type PbSe material. The promising thermoelectric performance of heavily doped p-type PbSe suggests its potential use in various applications, such as waste heat recovery, thermoelectric cooling, and power generation.

3. Materials and Methods

3.1. Synthesis Method

Stoichiometric quantities of high-quality raw materials, namely, 99.99% lead (Pb) shots, 99.999% selenium (Se) shots, and 99.999% sodium chloride (NaCl) granules, all purchased from Aladdin, were precisely measured in the proportion of 1:1:x (where x = 3, 3.5, 4, or 4.5) and labeled as S1, S2, S3, and S4 in sequence. The raw materials were mixed thoroughly using a high-energy ball mill (model XQM-4L, Tencan Powder Technology Co., Ltd. Changsha, China) for 4 h under an argon atmosphere. The mixture was then loaded into a vitreous silica tube (diameter = 30 mm, length = 500 mm) with a carbon layer, which was subsequently evacuated and sealed under a vacuum. The sealed tube was placed in a horizontal tube furnace (model STF-1200, Kejing Electric Furnace Co., Ltd., Foshan, China) and heated to 932 K at a rate of 178 K/h, held at this temperature for 3 h, and then heated to 1373 K for 1 h, and kept at this temperature for 20 h. After that, the temperature was decreased to 723 K within 60 h and then cooled to room temperature. Distilled water was used to extract the lustrous crystals from the solvent, and the samples were tested with a series of representations.

3.2. Characterization and Measurement

The crystal structure of the sample was determined through powder X-ray diffraction (XRD) analysis using Cu Kα radiation at an ambient temperature, employing an Ultima IV instrument. Electron probe microanalysis (EPMA) (JXA-8230) was utilized to examine the actual contents of each element in the samples. Field emission scanning electron microscopy (FESEM) (SUPRA 55VP) was utilized to analyze the microstructures of the samples, and qualitative elemental analysis of the samples was conducted using electron energy dispersive X-ray spectroscopy (EDS) with an Oxford instrument. Chemical analysis of Pb, Se, Na, and Cl chemical states was performed using X-ray photoelectron spectroscopy (XPS) on a PHI5000 Versaprobe-II system. Furthermore, the transmission electron microscope (TEM) was used to obtain images, along with corresponding selected area electron diffraction (SAED) patterns, using a Tecnai G20 instrument. The Seebeck coefficient (α) and electrical conductivity (σ) were assessed from 300 K to 690 K in the vacuum condition. The common four-point probe approach measured the σ with a direct current of 20 mA. The α were attained, employing a comparison approach with Konstantin (Ni:40%) as a guide and a temperature gradient of approximately 1.5 K. We usually use the formula $\alpha = \Delta V/\Delta T$ to calculate the Seebeck coefficient, where ΔV is the thermoelectromotive force, and ΔT is the temperature difference between the two sides of the material. The thermal conductivity ($\kappa = DC_P\rho$) was directly obtained by the transient plate heat source method, in which the density ρ was measured by the Archimedes method. Electronic thermal conductivity κ_e and lattice thermal conductivity κ_L were calculated according to $\kappa_e = L\sigma T$ and $\kappa_L = \kappa_{total} - \kappa_e$, respectively. The Hall coefficient (R_H) was determined using the van der Pauw method and the HL5500 Hall system from Nanometrics, under a magnetic field of 0.55 T and a direct current of 1 mA at room temperature. The samples, which had thicknesses ranging from 0.3 to 0.5 mm, were subjected to this analysis. The Hall carrier concentration n_H and Hall mobility μ_H were determined using the formulas $n = 1/(eR_H)$ and $\mu_H = \sigma R_H$, respectively.

3.3. Details of Computational Methods

The projector augmented wave (PAW) method within the density functional theory (DFT) was employed using the Vienna ab initio simulation package (VASP) for all computations. The Perdew–Burke–Ernzerhof (PBE) functional and an energy cutoff of 400 eV were utilized for the density-of-states and band-structure calculations based on a $3 \times 3 \times 3$ supercell. The energy and relaxation force convergence criteria were set to 10^{-5} eV and 0.01 eV $Å^{-1}$, respectively. For the crystal structure relaxation, a k-grid of $3 \times 3 \times 3$ was used, while a k-grid of $5 \times 5 \times 5$ was employed for the static calculation in the intrinsic supercell and $PbSe(NaCl)_x$. The density-of-states calculation utilized a k-grid of $7 \times 7 \times 7$. In addition, the band structure of pristine PbSe and PbSe-NaCl was calculated by considering spin–orbit coupling (SOC).

4. Conclusions

In summary, the influence of different levels of NaCl doping on the microstructure, energy band structure, and thermoelectric properties of $PbSe\text{-}(NaCl)_x$ are systematically investigated by the first-principles calculations combined with practical experiments. All samples exhibited the PbSe-type crystal structure—without any other phase. Based on the NaCl salt-assisted approach, the carrier concentration of $PbSe(NaCl)_x$ is regulatable by the doping content of NaCl, which was adjusted up to 4.2×10^{19} cm^{-3} in $PbSe(NaCl)_{4.5}$, leading to the appreciable α and σ simultaneously, thereby resulting in a super ZT value (1.26 at 690 K) eventually. The feasibility of the NaCl salt-assisted approach to modify the TE performance of the PbSe compound promotes the development of PbSe-based TE technology for addressing global energy and environmental challenges and providing reference for other TE materials.

Supplementary Materials: The following supporting information can be downloaded at: https://www.mdpi.com/article/10.3390/molecules28062629/s1, Figure S1: FESEM images of $PbSe(NaCl)_3$, which with different magnifications (10,000–100,000 magnification) (a–d) and from different positions at 100,000 magnification (e–h).; Figure S2: The crystal structure of PbSe doped with only one atom in each species.; Figure S3: Band structure and DOS calculated using the $3 \times 3 \times 3$ supercell. Band structure of Na-doped PbSe with $\Delta Ev \sim 0.18$ eV (a); Cl-doped PbSe with $\Delta Ev \sim 0.22$ eV (c); and corresponding density of states with only one atom substitution for Na-doped PbSe (b); and Cl-doped PbSe (d), where the corresponding states are displayed to clearly show their contribution.; Figure S4: Band structure calculated using the $4 \times 4 \times 4$ supercell. Band structure of PbSe with $\Delta Ev \sim 0.42$ eV (a); NaCl-doped PbSe with $\Delta Ev \sim 0.33$ eV (b).

Author Contributions: X.M., X.S., Y.D., J.W., S.D. and C.Z. devised the research plan and proofread the manuscript. X.M., J.Z., J.S., W.C. and X.L. took part in developing the research design, executed the experimental work, evaluated the results, and prepared the initial manuscript. T.W., W.R. and L.G. performed the experiments and analyzed the data. All authors have read and agreed to the published version of the manuscript.

Funding: Financial support from the National Nature Science Foundation of China (Grant No. 61904072 and 12265017), the Yunnan Fundamental Research Projects (Nos. 202001AU070025, 202101BE070001-049, 202101AS070018, 202101AV070015), and the Creative Research Group Scheme (in Science and Technology) at the University of Yunnan Province.

Institutional Review Board Statement: Not applicable.

Informed Consent Statement: Not applicable.

Data Availability Statement: Request for the data presented in the study can be made to the corresponding author.

Conflicts of Interest: The authors declare no conflict of interest.

Sample Availability: The compounds' samples can be provided by the authors of the study.

References

1. Channegowda, M.; Mulla, R.; Nagaraj, Y.; Lokesh, S.; Nayak, S.; Mudhulu, S.; Rastogi, C.K.; Dunnill, C.W.; Rajan, H.K.; Khosla, A. Comprehensive Insights into Synthesis, Structural Features, and Thermoelectric Properties of High-Performance Inorganic Chalcogenide Nanomaterials for Conversion of Waste Heat to Electricity. *ACS Appl. Energy Mater.* **2022**, *5*, 7913–7943. [CrossRef]
2. Zhao, K.; Qiu, P.; Shi, X.; Chen, L. Recent Advances in Liquid-Like Thermoelectric Materials. *Adv. Funct. Mater.* **2019**, *30*, 1903867. [CrossRef]
3. Pei, J.; Cai, B.; Zhuang, H.L.; Li, J.F. Bi_2Te_3-based applied thermoelectric materials: Research advances and new challenges. *Natl. Sci. Rev.* **2020**, *7*, 1856–1858. [CrossRef]
4. Ul Haq, B.; AlFaify, S.; Alshahrani, T.; Ahmed, R.; Mahmood, Q.; Tahir, S.A.; Alhashim, H.H.; Laref, A. Exploring the potential of lead-chalcogenide monolayers for room-temperature thermoelectric applications. *Ceram. Int.* **2021**, *47*, 3380–3388. [CrossRef]
5. Byrnes, J.; Mitchell, D.R.G.; Aminorroaya Yamini, S. Thermoelectric performance of thermally aged nanostructured bulk materials—A case study of lead chalcogenides. *Mater. Today Phys.* **2020**, *13*, 100190. [CrossRef]
6. Chasapis, T.C.; Lee, Y.; Hatzikraniotis, E.; Paraskevopoulos, K.M.; Chi, H.; Uher, C.; Kanatzidis, M.G. Understanding the role and interplay of heavy-hole and light- hole valence bands in the thermoelectric properties of PbSe. *Phys. Rev. B* **2015**, *91*, 085207. [CrossRef]
7. Fiedler, C.; Kleinhanns, T.; Garcia, M.; Lee, S.; Calcabrini, M.; Ibanez, M. Solution-Processed Inorganic Thermoelectric Materials: Opportunities and Challenges. *Chem. Mater.* **2022**, *34*, 8471–8489. [CrossRef]
8. Shuai, J.; Sun, Y.; Tan, X.; Mori, T. Manipulating the Ge Vacancies and Ge Precipitates through Cr Doping for Realizing the High-Performance GeTe Thermoelectric Material. *Small* **2020**, *16*, 1906921. [CrossRef]
9. Li, C.; Yuan, H.; Wang, Y.; Liu, H. Enhancement of the power factor of SnSe by adjusting the crystal and energy band structures. *Phys. Chem. Chem. Phys.* **2022**, *24*, 24130–24136. [CrossRef]
10. Hussain, T.; Li, X.; Danish, M.H.; Rehman, M.U.; Zhang, J.; Li, D.; Chen, G.; Tang, G. Realizing high thermoelectric performance in eco-friendly SnTe via synergistic resonance levels, band convergence and endotaxial nanostructuring with Cu_2Te. *Nano Energy* **2020**, *73*, 104832. [CrossRef]
11. Gayner, C.; Amouyal, Y. Energy Filtering of Charge Carriers: Current Trends, Challenges, and Prospects for Thermoelectric Materials. *Adv. Funct. Mater.* **2019**, *30*, 1901789. [CrossRef]
12. Li, X.; Liu, J.; Li, S.; Zhang, J.; Li, D.; Xu, R.; Zhang, Q.; Zhang, X.; Xu, B.; Zhang, Y.; et al. Synergistic band convergence and endotaxial nanostructuring: Achieving ultralow lattice thermal conductivity and high figure of merit in eco-friendly SnTe. *Nano Energy* **2020**, *67*, 104261–104271. [CrossRef]
13. Liu, S.; Yu, Y.; Wu, D.; Xu, X.; Xie, L.; Chao, X.; Bosman, M.; Pennycook, S.J.; Yang, Z.; He, J. Coherent Sb/CuTe Core/Shell Nanostructure with Large Strain Contrast Boosting the Thermoelectric Performance of n-Type PbTe. *Adv. Funct. Mater.* **2021**, *31*, 2007340. [CrossRef]
14. Lee, K.H.; Kim, Y.M.; Park, C.O.; Shin, W.H.; Kim, S.W.; Kim, H.S.; Kim, S.i. Cumulative defect structures for experimentally attainable low thermal conductivity in thermoelectric $(Bi,Sb)_2Te_3$ alloys. *Mater. Today Energy* **2021**, *21*, 100795. [CrossRef]
15. Hodges, J.M.; Hao, S.Q.; Grovogui, J.A.; Zhang, X.M.; Bailey, T.P.; Li, X.; Gan, Z.H.; Hu, Y.Y.; Uher, C.; Dravid, V.P.; et al. Chemical Insights into PbSe-x%HgSe: High Power Factor and Improved Thermoelectric Performance by Alloying with Discordant Atoms. *J. Am. Chem. Soc.* **2018**, *140*, 18115–18123. [CrossRef] [PubMed]
16. Zhou, C.; Yu, Y.; Lee, Y.-L.; Ge, B.; Lu, W.; Cojocaru-Mirédin, O.; Im, J.; Cho, S.-P.; Wuttig, M.; Shi, Z. Exceptionally high average power factor and thermoelectric figure of merit in n-type PbSe by the dual incorporation of Cu and Te. *J. Am. Chem. Soc.* **2020**, *142*, 15172–15186. [CrossRef]
17. Luo, Z.Z.; Cai, S.; Hao, S.; Bailey, T.P.; Spanopoulos, I.; Luo, Y.; Xu, J.; Uher, C.; Wolverton, C.; Dravid, V.P.; et al. Strong valence band convergence to enhance thermoelectric performance in PbSe with two chemically independent controls. *Angew. Chem. Int. Ed.* **2021**, *60*, 268–273. [CrossRef]
18. Cai, S.; Hao, S.; Luo, Z.-Z.; Li, X.; Hadar, I.; Bailey, T.; Hu, X.; Uher, C.; Hu, Y.-Y.; Wolverton, C.; et al. Discordant nature of Cd in PbSe: Off-centering and core-shell nanoscale CdSe precipitates lead to high thermoelectric performance. *Energy Environ. Sci.* **2020**, *13*, 200–211. [CrossRef]
19. Tan, G.; Hao, S.; Cai, S.; Bailey, T.P.; Luo, Z.; Hadar, I.; Uher, C.; Dravid, V.P.; Wolverton, C.; Kanatzidis, M.G. All-Scale Hierarchically Structured p-Type PbSe Alloys with High Thermoelectric Performance Enabled by Improved Band Degeneracy. *J. Am. Chem. Soc.* **2019**, *141*, 4480–4486. [CrossRef] [PubMed]
20. Zhou, N.N.; Hu, Z.Q.; Nasser, R.; Song, J.-M. Thermoelectric Properties of PbSe Nanocomposites from Solution-Processed Building Blocks. *ACS Appl. Energy Mater.* **2021**, *4*, 2014–2019. [CrossRef]
21. Raphel, A.; Vivekanandhan, P.; Kumaran, S. High entropy phenomena induced low thermal conductivity in $BiSbTe_{1.5}Se_{1.5}$ thermoelectric alloy through mechanical alloying and spark plasma sintering. *Mater. Lett.* **2020**, *269*, 127672. [CrossRef]
22. Liu, M.; Zhang, J.; Xu, J.; Hu, B.; Sun, K.; Yang, Y.; Wang, J.; Du, B.; Zhang, H. The crystallization, thermodynamic and thermoelectric properties of vast off-stoichiometric Sn–Se crystals. *J. Mater. Chem. C* **2020**, *8*, 6422–6434. [CrossRef]
23. Canfield, P.C.; Fisher, I.R. High-temperature solution growth of intermetallic single crystals and quasicrystals. *J. Cryst. Growth* **2001**, *225*, 155–161. [CrossRef]

24. Teshima, K.; Lee, S.; Sakurai, M.; Kameno, Y.; Yubuta, K.; Suzuki, T.; Shishido, T.; Endo, M.; Oishi, S. Well-Formed One-Dimensional Hydroxyapatite Crystals Grown by an Environmentally Friendly Flux Method. *Cryst. Growth Des.* **2009**, *9*, 2937–2940. [CrossRef]
25. Rakshit, M.; Jana, D.; Banerjee, D. General strategies to improve thermoelectric performance with an emphasis on tin and germanium chalcogenides as thermoelectric materials. *J. Mater. Chem. A* **2022**, *10*, 6872–6926. [CrossRef]
26. Giraldo, S.; Jehl, Z.; Placidi, M.; Izquierdo-Roca, V.; Perez-Rodriguez, A.; Saucedo, E. Progress and Perspectives of Thin Film Kesterite Photovoltaic Technology: A Critical Review. *Adv. Mater.* **2019**, *31*, e1806692. [CrossRef]
27. Chen, J.; Shen, L.; Liu, W.; Tang, Y.; Lu, J.; Sun, L.; Ge, W.; Yang, P.; Deng, S. High thermoelectric performance of Na-doped beta-Zn_4Sb_3 prepared by NaCl salt-assisted approach. *J. Alloys Compd.* **2021**, *856*, 157345. [CrossRef]
28. Lu, J.; Li, D.; Liu, W.; Shen, L.; Chen, J.; Ge, W.; Deng, S. Thermal stability and thermoelectric properties of Cd-doped nano-layered Cu_2Se prepared using NaCl flux method. *Chin. Phys. B* **2020**, *29*, 127403. [CrossRef]
29. Qian, X.; Wu, H.J.; Wang, D.Y.; Zhang, Y.; Wang, J.F.; Wang, G.T.; Zheng, L.; Pennycook, S.J.; Zhao, L.D. Synergistically optimizing interdependent thermoelectric parameters of n-type PbSe through alloying CdSe. *Energy Environ. Sci.* **2019**, *12*, 1969–1978. [CrossRef]
30. Zhu, Y.C.; Wang, D.Y.; Hong, T.; Hu, L.; Ina, T.; Zhan, S.P.; Qin, B.C.; Shi, H.N.; Su, L.Z.; Gao, X.; et al. Multiple valence bands convergence and strong phonon scattering lead to high thermoelectric performance in p-type PbSe. *Nat. Commun.* **2022**, *13*, 4179. [CrossRef]
31. Jood, P.; Male, J.P.; Anand, S.; Matsushita, Y.; Takagiwa, Y.; Kanatzidis, M.G.; Snyder, G.J.; Ohta, M. Na Doping in PbTe: Solubility, Band Convergence, Phase Boundary Mapping, and Thermoelectric Properties. *J. Am. Chem. Soc.* **2020**, *142*, 15464–15475. [CrossRef] [PubMed]
32. Jiang, B.; Yu, Y.; Cui, J.; Liu, X.; Xie, L.; Liao, J.; Zhang, Q.; Huang, Y.; Ning, S.; Jia, B.; et al. High-entropy-stabilized chalcogenides with high thermoelectric performance. *Science* **2021**, *371*, 830–834. [CrossRef] [PubMed]
33. Wang, H.; Pei, Y.; Lalonde, A.D.; Snyder, G.J. Heavily Doped p-Type PbSe with High Thermoelectric Performance: An Alternative for PbTe. *Adv. Mater.* **2011**, *23*, 1366–1370. [CrossRef] [PubMed]
34. Sun, L.; Shai, X.; Chen, Z.; Tang, Y.; Liu, W.; Shen, L.; Ge, W.; Deng, S. High performance in room temperature and thermoelectric properties of PbSe single crystal prepared by Pb-flux method. *Phys. B Condens. Matter* **2020**, *578*, 411797. [CrossRef]
35. Song, X.F.; Shai, X.X.; Deng, S.K.; Wang, J.S.; Li, J.; Ma, X.R.; Li, X.R.; Wei, T.T.; Ren, W.A.; Gao, L.; et al. Anisotropic Chalcogenide Perovskite $CaZrS_3$: A Promising Thermoelectric Material. *J. Phys. Chem. C* **2022**, *126*, 11751–11760. [CrossRef]
36. Kim, H.S.; Choi, G.; Ha, M.Y.; Kim, D.H.; Park, S.H.; Chung, I.; Lee, W.B. Thermoelectric transport properties of Pb doped SnSe alloys ($Pb_xSn_{1-x}Se$): DFT-BTE simulations. *J. Solid State Chem.* **2019**, *270*, 413–418. [CrossRef]
37. Perdew, J.; Burke, K.; Wang, Y. Generalized gradient approximation for the exchange-correlation hole of a many-electron system. *Phys. Rev. B* **1996**, *54*, 16533–16539. [CrossRef]
38. Perez, M.; Templeman, T.; Shandalov, M.; Ezersky, V.; Yahel, E.; Golan, Y. Citrate-controlled chemical solution deposition of PbSe thin films. *Crystengcomm* **2019**, *21*, 1818–1825. [CrossRef]
39. Wu, Y.X.; Nan, P.F.; Chen, Z.W.; Zeng, Z.Z.; Liu, R.H.; Dong, H.L.; Xie, L.; Xiao, Y.W.; Chen, Z.Q.; Gu, H.K.; et al. Thermoelectric Enhancements in PbTe Alloys Due to Dislocation-Induced Strains and Converged Bands. *Adv. Sci.* **2020**, *7*, 9. [CrossRef]
40. Huang, Z.; Zhang, Y.; Wu, H.; Pennycook, S.J.; Zhao, L.-D. Enhancing Thermoelectric Performance of p-Type PbSe through Suppressing Electronic Thermal Transports. *ACS Appl. Energy Mater.* **2019**, *2*, 8236–8243. [CrossRef]
41. Li, J.; Sui, J.; Pei, Y.; Meng, X.; Berardan, D.; Dragoe, N.; Cai, W.; Zhao, L.D. The roles of Na doping in BiCuSeO oxyselenides as a thermoelectric material. *J. Mater. Chem. A* **2014**, *2*, 4903–4906. [CrossRef]
42. Rodríguez-Lazcano, Y.; Barrios-Salgado, E.; Pérez-Orozco, J.P.; Campos, J.; Altuzar, P.; Regla, E.L.; Quesada-Saliba, D. Microwave-assisted chemical bath deposition of PbSe thermoelectric thin films. *Appl. Phys. A* **2021**, *127*, 537. [CrossRef]
43. Zhang, Q.; Cao, F.; Liu, W.; Lukas, K.; Ren, Z. Heavy doping and band engineering by potassium to improve the thermoelectric figure of merit in p-type PbTe, PbSe, and $PbTe_{(1-y)}Se_{(y)}$. *J. Am. Chem. Soc.* **2012**, *134*, 10031–10038. [CrossRef] [PubMed]
44. Biswas, K.; He, J.; Blum, I.; Wu, C.I.; Hogan, T.P.; Seidman, D.N.; Dravid, V.P.; Kanatzidis, M.G. High-performance bulk thermoelectrics with all-scale hierarchical architectures. *Nature* **2012**, *489*, 414–418. [CrossRef] [PubMed]
45. Parker, D.S.; Singh, D.J. High-temperature thermoelectric performance of heavily doped PbSe. *Phys. Rev. B* **2010**, *82*, 035204. [CrossRef]
46. Wang, S.; Zheng, G.; Luo, T.; She, X.; Li, H.; Tang, X. Exploring the doping effects of Ag in p-type PbSe compounds with enhanced thermoelectric performance. *J. Phys. D Appl. Phys.* **2011**, *44*, 475304–475312. [CrossRef]
47. Pei, Y.L.; He, J.; Li, J.F.; Li, F.; Liu, Q.; Pan, W.; Barreteau, C.; Berardan, D.; Dragoe, N.; Zhao, L.D. High thermoelectric performance of oxyselenides: Intrinsically low thermal conductivity of Ca-doped BiCuSeO. *NPG Asia Mater.* **2013**, *5*, e47. [CrossRef]
48. Wan, C.L.; Pan, W.; Xu, Q.; Qin, Y.X.; Wang, J.D.; Qu, Z.X.; Fang, M.H. Effect of point defects on the thermal transport properties of $(La_xGd_{1-x})_2Zr_2O_7$: Experiment and theoretical model. *Phys. Rev. B* **2006**, *74*, 144109. [CrossRef]

molecules

MDPI

Article

The Effect of Copper Sulfide Stoichiometric Coefficient and Morphology on Electrochemical Performance

Yanhong Ding [1,*], Rongpeng Lin [1], Shuicheng Xiong [1], Yirong Zhu [1,*], Meng Yu [1] and Xiaobo Duan [2]

[1] College of Materials and Advanced Manufacturing, Hunan University of Technology, Zhuzhou 412008, China
[2] GEM Co., Ltd., Shenzhen 518101, China
* Correspondence: dingyanhong@hut.edu.cn (Y.D.); zhuyirong@hut.edu.cn (Y.Z.)

Abstract: In this work, CuS, Cu_7S_4, Cu_9S_5, $Cu_{7.2}S_4$, and Cu_2S with the same morphology were successfully synthesized by the hydrothermal method. According to the calculation, their galvanostatic charge-discharge (GCD) curves were 43.29 (CuS), 86.3 (Cu_7S_4), 154 (Cu_9S_5), 185.4 ($Cu_{7.2}S_4$), and 206.9 F/g (Cu_2S) at the current density of 1 A/g. The results showed that the energy storage capacity of copper sulfide with the same morphology increased with the increase of the copper sulfide stoichiometric coefficient. At the second part of this work, the agglomerated cuprous sulfide and the microporous cuprous sulfide were successfully prepared, respectively. In addition, the porous spherical cuprous sulfide was annealed to get nano cuprous sulfide. It is found that the specific capacity of the agglomerated structure is the highest, which had reached 206.9 F/g at the current density of 1 A/g, and 547.9 F/g at the current density of 10 A/g after activation.

Keywords: copper sulfide; supercapacitor; specific capacity; hydrothermal method; morphology

1. Introduction

With the rapid development of energy technology, electric energy has become the protagonist of the energy era [1]. To meet the needs of rapid development, more and more high-efficient and environmentally friendly new energy sources and energy-storage devices are needed. Supercapacitors with high-power density, a long cycle life, and a fast-charging capacity have attracted the attention of electrochemical researchers in recent years [2,3]. Element copper ranks in the fourth period and the IB group in the periodic table with atomic No. 29. One electron in the outermost layer makes copper have more valence commonly. Rich raw materials, low price, excellent conductivity and easy machining make copper metal play an important role in industrial production and application [4]. Due to the more complex valence state of the sulfide phase and the higher electrochemical performance, transition metal sulfides as electrode materials for supercapacitors have attracted more attention. A high theoretical capacitance, high power density, abundant valence state, REDOX sites, and a low price make copper metal sulfide a widely concerned energy storage [5,6]. As for copper metal sulfide as electrode materials for supercapacitors, many researchers have reported their work. Wang Xiang et al. [7] used Cu_2O, synthesized by chemical precipitation, and Cu_2O as the template via a hydrothermal ion-exchange method to prepare nano Cu_7S_4, which provided a specific capacitance of 275 F/g at a current density of 1 A/g. Zhou et al. [8] used an ion exchange reaction to regulate the morphologies of $Cu_{7.2}S_4$ and synthesized submicron and micron porous spheres of $Cu_{7.2}S_4$. The material has a specific capacitance of 491.5 F/g at a current density of 1 A/g. Ravindra N prepared Cu_2S of flower-like nanotubes and integrated nanotubes by a continuous ion-adsorption reaction with a specific capacitance of 761 F/g and 470 F/g at a scanning rate of 5 mV/s [9]. Therefore, via different synthesis methods and a post-treatment of materials, adjusting the structure morphology of copper sulfide and so on contributed to the study of the structure with larger specific capacitance. Copper metal sulfide materials have a huge application potential in the field of supercapacitor materials.

In this work, to better explore the stoichiometric coefficient affect energy storage mechanism of different copper sulfides, CuS, Cu_7S_4, Cu_9S_5, $Cu_{7.2}S_4$, and Cu_2S with the same morphology were synthesized by hydrothermal method and adjusting morphology variables. The electrochemical properties of different copper sulfides were compared, and the energy storage performance of copper sulfides was discussed. Cuprous sulfide has good thermal stability, which will not decompose into cuprous sulfide and sulfur at high temperatures. Therefore, cuprous sulfide with good thermal stability was also selected as the research object. In order to further explore the effect of morphology and structure on the energy storage capacity of copper sulfide, the morphology of cuprous sulfide was regulated by different synthesis methods so as to carry out an electrochemical test and an exploration on different morphology of cuprous sulfides.

2. Results and Discussion

2.1. Energy Storage of Different Copper Sulfide Materials

2.1.1. XRD Pattern Analysis

Figure 1a shows the XRD map of the precursor of copper sulfide prepared by $Cu(CH_3COO)_2{\cdot}H_2O$ and TAA. The lower part is a PDF card of $Cu_4(SO_4)(OH)_6$ material. Through the comparative analysis of the map, it is found that the 2θ angles located at 11.3°, 13.8°, 16.5°, 22.8°, 27.9°, 33.5°, 35.6°, and 48.3° corresponded to the (110), (200), (210), (220), (400), (420), (22-2), and (250) crystal faces of $Cu_4(SO_4)(OH)_6$, respectively. However, in addition to the diffraction peak of $Cu_4(SO_4)(OH)_6$, the pattern has other diffraction peaks, indicating that the substance is a mixture. It may contain copper, sulfur, hydrogen, and oxygen based on the reactants. The diffraction peaks of the other 2θ of 16.2°, 18.2°, 18.8°, 22.3°, 24.0°, and 31.5° are exactly corresponding to the (−110), (020), (011), (021), (020), and (−221) crystal faces of copper sulfate pentahydrate. By XRD analysis of the precursor, it is proved that the precursor is a mixture of $Cu_4(SO_4)(OH)_6$ and $CuSO_4{\cdot}5H_2O$. The specific reaction equation is shown in Equation (1):

$$Cu(CH_3COO)_2{\cdot}H_2O + CH_3CSNH_2 \rightarrow Cu_4(SO_4)(OH)_6 + CuSO_4{\cdot}5H_2O \quad (1)$$

Cu_xS_y (I, II, III, IV, V; x and y are stoichiometric coefficients) was successfully prepared by the reaction of copper acetate monohydrate with TAA under different conditions. XRD images of the sample are shown in Figure 1b–f. The upper part of the figures with the same color are the test curves of the sample, and the lower part is the PDF standard card of the corresponding material. The intensity peaks of 27°, 29.1°, 31.6°, 32.7°, 47.8°, 52.6°, 59.2°, and 73.9° in Figure 1b were analyzed. It is found that the crystal faces of (101), (102), (103), (006), (110), (108), (116), and (208) of CuS coincided with the standard PDF cards of CuS (PDF#06-0464). Compared with the standard card of Cu_7S_4 (PDF#33-0489), the diffraction peaks in Figure 1c are found in 26.4°, 29.7°, 31°, 33.9°, 35.3°, 37.8°, 38.9°, 46.7°, and 48.8° diffraction peaks corresponding to the crystal plane (16,0,0), (804), (18,2,1), (20,0,1), (20,4,0), (155), (10,10,3), (0,16,0), and (886). In Figure 1d, the diffraction peaks at 2θ of 27.8°, 29.2°, 32.2°, 41.5°, 46.2°, and 54.7° are consistent with the crystal faces of Cu_9S_5 (PDF#47-1748) at (0,0,15), (107), (1,0,10), (0,1,17), (0,1,20), and (0,1,15). In addition, an analysis of the map in Figure 1e shows that the substance is $Cu_{7.2}S_4$ (PDF#24-0061) because its strong peak 2θ equals 27.8°, 32.1°, 46.2°, 54.8°, and 67.1°, and corresponds to the (111), (200), (220), (311), and (400) crystal faces of $Cu_{7.2}S_4$. The (100), (102), (110), (103), (112), and (114) crystal faces of Cu_2S at 26°, 37.3°, 46°, 48.5°, 54°, and 73.9° of 2θ in Figure 1f correspond to the crystal faces of Cu_2S.

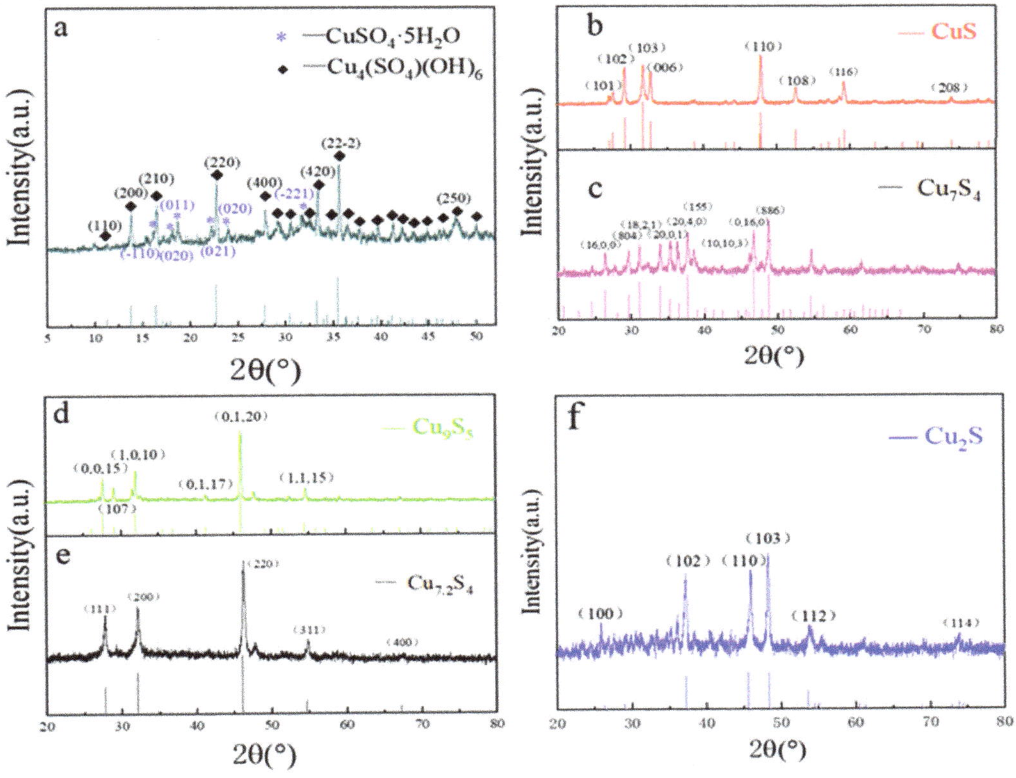

Figure 1. (**a**) is the XRD patterns figure of intermediates prepared by $Cu(CH_3COO)_2{\cdot}H_2O$ and TAA; (**b–f**) are the XRD patterns figures of CuS, Cu_7S_4, Cu_9S_5, $Cu_{7.2}S_4$, and Cu_2S.

To analyze XRD figures of five kinds of samples, it is proved that CuS, Cu_7S_4, Cu_9S_5, $Cu_{7.2}S_4$, and Cu_2S have been successfully prepared by controlling the conditions of hydrothermal synthesis with $Cu(CH_3COO)_2{\cdot}H_2O$ and TAA as raw materials.

2.1.2. Morphological Analysis

Figure 2a,c,e,g,i are the scanning images of CuS, Cu_7S_4, Cu_9S_5, $Cu_{7.2}S_4$, and Cu_2S. As shown in Figure 2, the morphology of all these five kinds of copper sulfide is agglomerated particles with particle diameters ranging from 10 to 55 µm. The purpose of adjusting the five samples to the same morphology is to avoid the influence of morphology on the following electrochemical test. When only the ratio of copper to sulfur atoms is univariate, the comparability of data from subsequent electrochemical tests can be better guaranteed. It can be seen from Figure 2b,d,f,h,j that the large granularity of copper sulfide is mainly due to the agglomeration of small particles. The main reason is that the viscosity of the solvent is small [10].

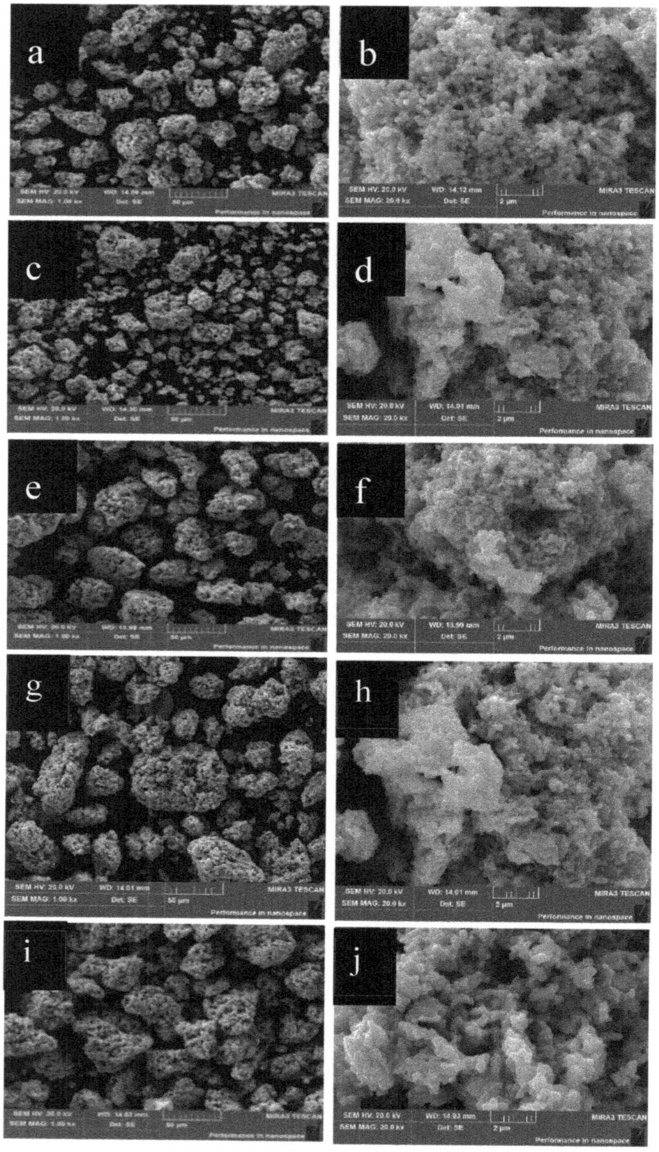

Figure 2. The morphology SEM images of different copper sulfides (**a**,**c**,**e**,**g**,**i**) (CuS, Cu_7S_4, Cu_9S_5, $Cu_{7.2}S_4$, and Cu_2S), whose particle diameter is 55 μm; Figures (**b**,**d**,**f**,**h**,**j**) are enlarged image part of figures (**a**,**c**,**e**,**g**,**i**).

At room temperature, TAA was hydrolyzed and combined with Cu^{2+} to form flocculent precursor $Cu_4(SO_4)(OH)_6$ and copper sulfate pentahydrate. As shown in Equation (2), with the rise of hydrothermal temperature, TAA continues to hydrolyze to form S^{2-}. Then S^{2-} reacts with $Cu_4(SO_4)(OH)_6$ to produce a CuS precipitate. When the ratio of Cu^{2+} to S^{2-} ion is 1:1, CuS is precipitated in the solution, and the reaction ends. When the ratio of Cu^{2+} to S^{2-} is greater than 1:1 ($Cu(CH_3COO)_2 \cdot H_2O$:TAA = 1.75:1, 1.8:1, 2:1), S^{2-} in the solution will reduce Cu^{2+}, according to the amount of Cu^{2+}, and will eventually produce different copper sulfides Cu_7S_4, Cu_9S_5, $Cu_{7.2}S_4$, and Cu_2S. The specific equations (Equations (3–7)) are as follows:

$$CH_3CSNH_2 + 3OH^- \overset{heating}{\rightarrow} CH_3COO^- + NH_3 + S^{2-} + H_2O \quad (2)$$

$$S^{2-} + Cu^{2+} \overset{heating}{\rightarrow} CuS \downarrow \quad (3)$$

$$4S^{2-} + 7Cu^{2+} \overset{heating}{\rightarrow} Cu_7S_4 \downarrow \quad (4)$$

$$5S^{2-} + 9Cu^{2+} \overset{heating}{\rightarrow} Cu_9S_5 \downarrow \quad (5)$$

$$4S^{2-} + 7.2Cu^{2+} \overset{heating}{\rightarrow} Cu_{7.2}S_4 \downarrow \quad (6)$$

$$S^{2-} + 2Cu^{2+} \overset{heating}{\rightarrow} Cu_2S \downarrow \quad (7)$$

In addition, as can be seen from the enlarged figure in Figure 2a, with the hydrothermal reaction, the crystal nuclei of copper sulfide grow into uniform nanoparticles. Then, due to the low viscosity of the solvent, the nanoparticles become close to agglomeration [11]. They come together, fuse at the boundary, and then eventually form aggregated particles of copper sulfide.

2.1.3. Electrochemical Test Analysis

Figure 3a shows cyclic voltammetry (CV) curves of CuS, Cu_7S_4, Cu_9S_5, $Cu_{7.2}S_4$, and Cu_2S. (A three-electrode test method was adopted using a self-made electrode working electrode, a calomel electrode as a reference electrode, and a platinum electrode as a reverse electrode, a 2 mol/L KOH solution as an electrolyte solution, a scanning rate 5 mV/s, and a scanning potential range 0–0.6 V.) It can be seen from Figure 3a that the CV curves of the five copper sulfides all have obvious REDOX peaks, indicating that the electrochemical reaction of copper sulfides in circulation has a pseudocapacitive reaction. In addition, the oxidation and reduction peaks of the four copper sulfides are relatively symmetrical, indicating that the reaction reversibility of the four copper sulfides is good. However, the oxidation peak of cuprous sulfide is significantly higher than its reduction peak, indicating that the reversibility of cuprous sulfide is poor in the process of charge and discharge. By comparing the area enclosed by the CV curve, cuprous sulfide is significantly larger than other copper sulfides, which indicates that its energy storage performance is superior to other copper sulfides.

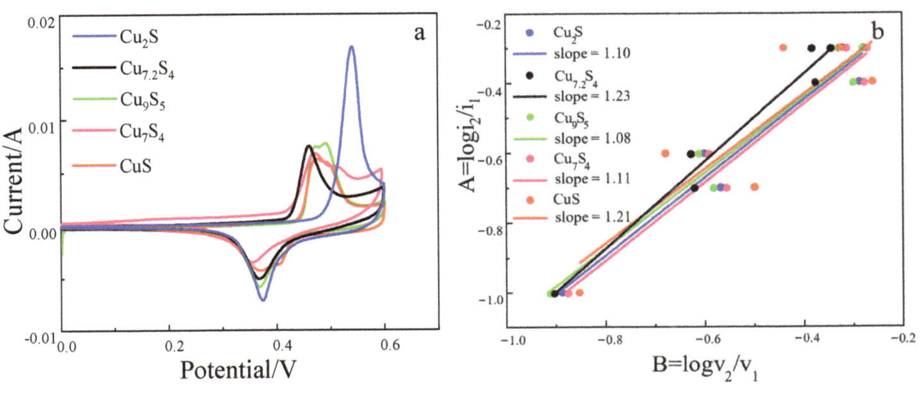

Figure 3. (**a**) is the CV curves of a CuS, Cu_7S_4, Cu_9S_5, $Cu_{7.2}S_4$, and Cu_2S at scanning rate of 5 mV/s; (**b**) is the capacitance contribution b of different copper sulfides.

The material capacitance contribution rate b can be calculated by Equation (8) (b = 0.5, the electrode material behaves as battery property; b∈(0.5–1), and the electrode material shows the properties of battery and pseudocapacitance (b $\geq$ 1, and the electrode material exhibits pseudocapacitance properties). In Equation (8), where i is the peak current in the CV figure, v is the scanning rate, and a is a constant. Equations (8) and (9) is transformed by formula transformation. Then, a binary system of first order equations was established through multiple sets of data to eliminate the constant a; Equation (10) was then obtained. According to Equation (10), six different $\log(i_2/i_1)$ and $\log(v_2/v_1)$ groups were calculated by combining the data with the peak current i of scanning rate v of 5, 10, 20, and 50 mV/s, and b was obtained by fitting the data of six groups. According to the fitting data in Figure 3b, the capacitance contribution ratio b of CuS, Cu_7S_4, Cu_9S_5, and Cu_2S is 1.21, 1.08, 1.11, and 1.1, respectively. The values are all greater than 1, so copper sulfide as a supercapacitor material shows a pseudocapacitance property. This result corresponds to the CV curve in Figure 3a.

$$i = av^{\,b} \tag{8}$$

$$\mathrm{log}i = b\mathrm{log}v + \mathrm{log}a \tag{9}$$

$$b = \log\left(\frac{i_2}{i_1}\right) / \log\left(\frac{v_2}{v_1}\right) \tag{10}$$

According to the CV figure in Figure 3a, the CV curves of Cu_2S has the largest area at the scanning rate of 5 mV/s, indicating that Cu_2S has the best energy storage capacity. In order to better explore the electrochemical properties of the material, the constant current charge-discharge curve of the material was tested, and the specific capacitance of the material was calculated by Equation (11) [12]. (I is the discharge current; Δt is the charging and discharging time; Δv is put the potential range of filling; m is the quality of the active substance).

$$Cm = I\Delta t/(m\ \Delta v) \tag{11}$$

Figure 4b is the GCD diagram of the material at a current density of 1 A/g. The specific capacitances of CuS, Cu_7S_4, $Cu_{7.2}S_4$, and Cu_2S were calculated as 43.2, 86.3, 185.4, and 206.9 F/g. In addition, Cu_9S_5 has the same Cu–S content ratio as $Cu_{7.2}S_4$, both of which are 1.8:1. Like $Cu_{7.2}S_4$, Cu_9S_5 has a specific capacitance of 154 F/g, which is between Cu_7S_4 and Cu_2S. It is found that with the increase of copper content in copper sulfide, the specific capacitance of the material increases, which also means that the energy storage capacity of the material is better. The results are in good agreement with the CV test. By observing the symmetry of GCD curves of five kinds of copper sulfides, different copper sulfides have excellent symmetry, which indicates that the coulomb efficiency of the materials is high. Equations (12) and (13) are the energy storage mechanisms of copper sulfide. According to the equations, the main reason for the energy storage mechanism of copper sulfide is the reversible pseudocapacitance reaction between copper sulfide and OH^- in the electrolyte, which can be explained as the continuous increase of copper valence. From this, the energy storage mechanism of the other copper sulfide can be deduced, as shown in Equation (14). According to the equations, with the increase of Cu element in CuS, Cu_7S_4, Cu_9S_5, $Cu_{7.2}S_4$, and Cu_2S, as well as the increase of x in Cu_xS, more hydroxide can be reacted in the material, and the specific capacity of the material increases accordingly. In addition, according to the capacitance contribution rate b, copper sulfide exhibits pseudocapacitive properties when storing energy, which means that its ability to store charge completely comes from the reversible REDOX reaction of the electrode material. Therefore, with the increase of the Cu element in copper sulfide in Figure 4a, the REDOX pair provided by the material will increase accordingly, thus enabling the material to store more capacity, and its energy storage capacity will improve.

$$CuS + OH^- \leftrightarrow CuSOH + e^- \tag{12}$$

$$CuSOH + OH^- \leftrightarrow CuSO + H_2O + e^- \tag{13}$$

$$Cu_xS + (4x-2)OH^- \leftrightarrow Cu_xSO_{(2x-1)} + (2x-1)H_2O + (4x-2)e^- \quad (14)$$

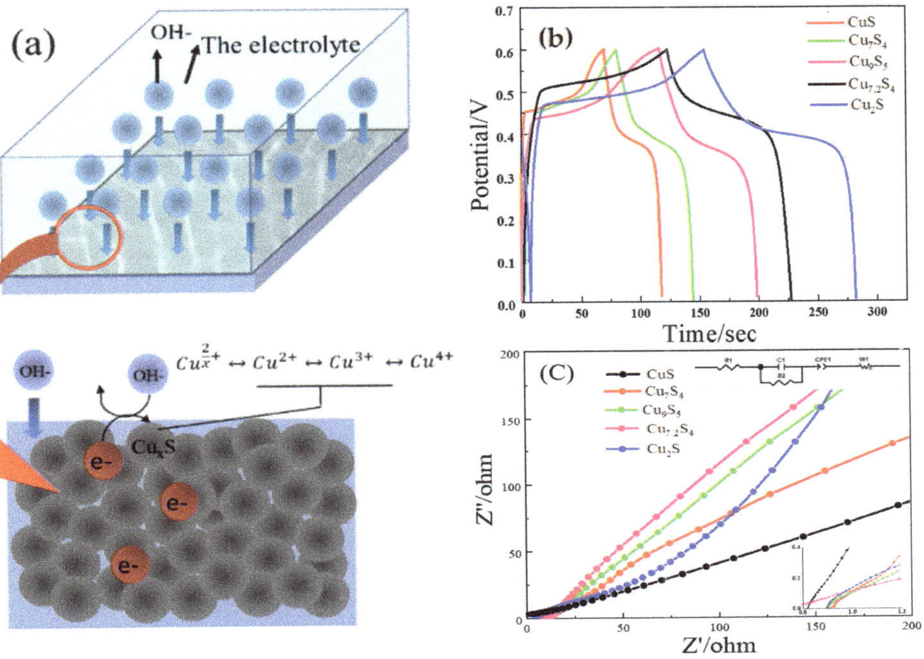

Figure 4. The schematic diagram of (**a**) energy storage of copper sulfide; the GCD diagrams of (**b**) different copper sulfides (CuS, Cu_9S_5, Cu_7S_4, $Cu_{7.2}S_4$, and Cu_2S) at 1 A/g current density; the electrochemical impedance spectroscopy (EIS) diagrams and equivalent circuit of (**c**) different copper sulfides.

The impedance of the material and strength of the charge-electron transport capacity can be analyzed by the AC impedance test. The test curve can be shown in the semi-circular diameter of the high-frequency region, which represents the charge-transfer resistance (Rct), and the low-frequency oblique line represents the diffusion resistance (Rw) of the electrode material, respectively. The electron transport speed in the reaction process is related with the Rct, and the slope of the oblique line is related with the Rw. The higher the slope of the oblique line, the lower the Rw. The ionic impedance of the electrolyte, the internal resistance of the electrode itself, and the interfacial impedance between the active material and the nickel foam collector can be shown as the intercept of the curve on the real axis, which represents the resistance (Rs).

Figure 4c shows the electrochemical impedance spectroscopy (EIS) of different copper sulfides. It can be seen from the figure that the high-frequency curves of the EIS diagrams of the five materials are almost ignored, indicating that the internal resistance of copper sulfide is very small when it is used as the material of supercapacitor, which proved that it has good energy storage performance. The low-frequency EIS curves of CuS, Cu_7S_4, Cu_9S_5, $Cu_{7.2}S_4$, and Cu_2S were analyzed. In this section, with the increase of the copper stoichiometry in the copper sulfide, the slope of the straight line in the low-frequency region gradually increases, indicating that the resistance of the interface charge transfer of the material is smaller and the diffusion resistance of the electrolyte of the material is lower. This also proves that its electrochemical ability is better.

Table 1 shows the specific capacitance and capacitance retention of different copper sulfides. According to the table, the retention rate of Cu_9S_5 reached 98.5%, which was higher than that of the other four materials. In addition, the capacitance retention of Cu_7S_4 and Cu_2S is 96.3% and 93.7%, which also show excellent capacitance retention. The retention rate of $Cu_{7.2}S_4$ was lower than that of the first three copper sulfides, which maintained at 80.3%. However, the capacitance retention of CuS is the lowest, only 63.4%. This may be related to the stability of the material. According to the CRC Handbook of Chemistry and Physics of 2014–2015 edition, the standard molar entropy of copper sulfide is 66.5 J/mol K, and that of cuprous sulfide is 120.9 J/mol K. This indicates that the structural stability of Cu_2S is better than CuS. Therefore, as a pseudocapacitive electrode, when the pseudocapacitive reaction occurs with the increase of current density, the resistance to a large current is stronger due to improved stability, so the capacitance retention rate is higher [13–17].

Table 1. The capacitance capability of different materials with the same morphology (CuS, Cu_9S_5, Cu_7S_4, $Cu_{7.2}S_4$, and Cu_2S) Cu–S ratio at 1 A/g and 10 A/g current density.

Material	Cu–S Ratio	C(1 A/g)	C(10 A/g)	Retention
CuS	1:1	43.2 F/g	27.4 F/g	63.4%
Cu_7S_4	1.75:1	86.3 F/g	83.1 F/g	96.3%
Cu_9S_5	1.8:1	154 F/g	151.7 F/g	98.5%
$Cu_{7.2}S_4$	1.8:1	185.4 F/g	148.9 F/g	80.3%
Cu_2S	2:1	206.9 F/g	193.8 F/g	93.7%

Figure 5 is the GCD cycle diagram of different copper sulfides at 10 A/g current density. It can be observed from Figure 5a–e that the cycle curve keeps an upward trend from 0 to 200 cycles. It shows that the specific capacitance of the material is increasing, and that the energy storage is increasing. The main reason is that the morphology of copper sulfide is agglomerate. With the continuous charge and discharge, the material continuously generates pseudocapacitive reactions. Then, the piles of copper sulfide spread out, increasing the specific surface area. Finally, the number of copper elements involved in the reaction continues to increase, thus increasing the charge stored in the material [18].

It can be seen from the analysis of Figure 5 that Cu_2S has the best cycle stability, and its capacity increases the most. It increased by 255.8% to 547.9 F/g. Secondly, the cyclic stability of Cu_9S_5 was excellent and remained stable until 2000 cycles, and the specific capacity increased by 130.3%. The cyclic stability of Cu_7S_4 is general. After rising, it stabilizes for a period of time, and then decreases with a small increase. The cycle stability of CuS is poor, and it begins to decline after 200 cycles without stable interval.

Finally, the cycle stability of $Cu_{7.2}S_4$ is the worst, with no rising period and no stable region. After 1000 cycles, the specific capacity is only 44.5%. The cyclic stability of materials is related to the structural stability of materials. As the material is recycled, the original structure will continue to collapse. As a result, the number of materials involved in the reaction will continue to decrease, resulting in a decrease in the specific capacity of the material. With a high and non-integral Cu–S ratio, the crystal structure of $Cu_{7.2}S_4$ may be more unstable, which resulted it in being infected the deepest and collapsing the most easily while the material was recycled.

2.2. *Effect of Morphology of Cuprous Sulfide on Electrochemical Properties of Materials*

2.2.1. XRD Analysis

Figure 6 is the XRD analysis of Cu_2S(I), Cu_2S(II), and Cu_2S(III). It can be seen from the figure that the strength peaks of cuprous sulfide with three morphs are completely corresponding to the standard card of cuprous sulfide. The crystal faces of (412), (204), (630), (106), and (112) of cuprous sulfide at the 2θ angles are 36.1°, 37.3°, 45.8°, 48.3°, and 53.5°. Therefore, it can be determined that the synthesized substances Cu_2S(I), Cu_2S(II), and Cu_2S(III) are pure cuprous sulfide by adjusting the solvent of the solvothermal reaction and the subsequent annealing treatment of the hydrothermal products.

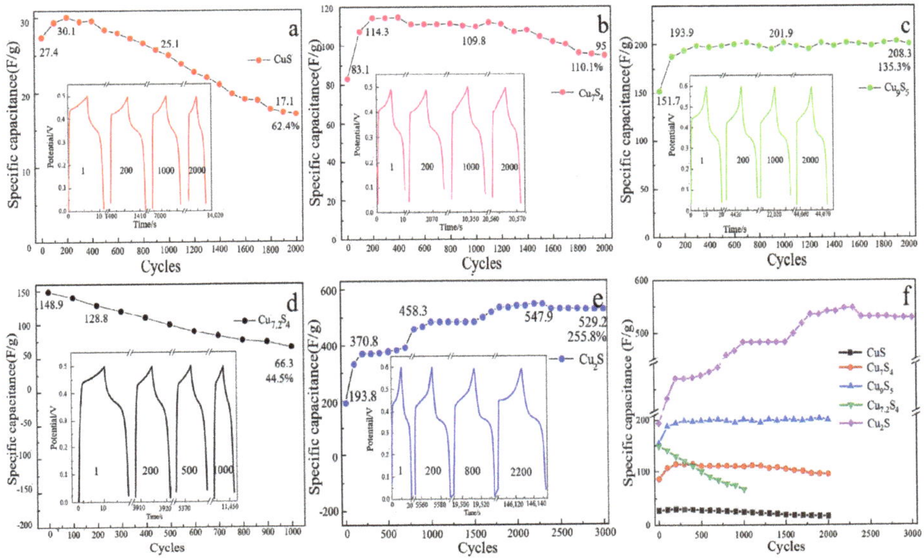

Figure 5. (**a**) to (**e**) are different cyclic properties of copper sulfides at 10 A/g current density: CuS (**a**), Cu_7S_4 (**b**), Cu_9S_5 (**c**), $Cu_{7.2}S_4$ (**d**), Cu_2S (**e**); a synthetic contrast of cyclic properties of (**f**) five kinds of copper sulfides.

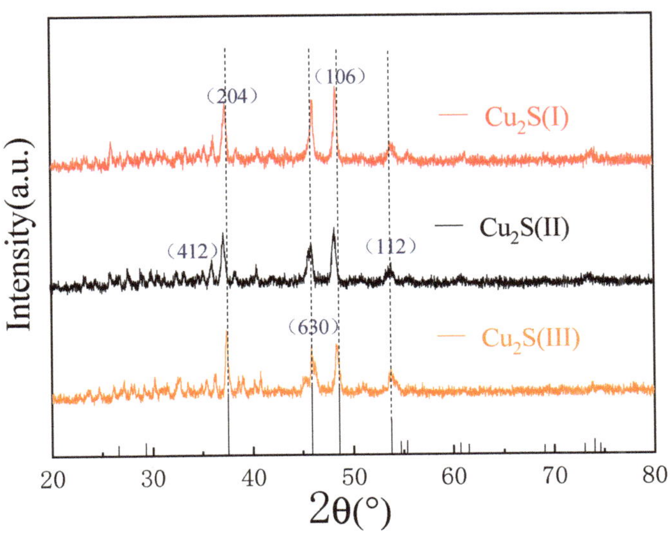

Figure 6. XRD patterns of cuprous sulfide with three different morphologies: Cu_2S(I), Cu_2S(II), and Cu_2S(III).

2.2.2. SEM Analysis

Figure 7 shows the SEM images of Cu_2S(I), Cu_2S(II), and Cu_2S(III). Through SEM images, we can observe that the cuprous sulfide prepared by the three synthetic methods has different morphologies. The materials shown in Figure 7a,b are also prepared by hydrothermal synthesis. The aggregate size of the synthesized large particles of cuprous sulfide is concentrated between 20 μm and 55 μm. Cuprous sulfide with a porous spherical structure, as shown in Figure 7c,d, were successfully prepared by changing the reaction solvent to 99% purity ethanol solution. The main reason is that the ethanol is weak acidic, which makes the S^{2-} and H^+ generated by the decomposition of thioacetamide generate H_2S, thus reducing the nucleation rate of cuprous sulfide and affecting the final morphology. In addition, compared with distilled water, ethanol as a solvent has lower viscosity and lower surface tension. Therefore, cuprous sulfide crystals are subjected to less external force during the growth process so that the crystals can expand more outwardly. In turn, the agglomeration is reduced and larger pores are formed, resulting in a porous morphology.

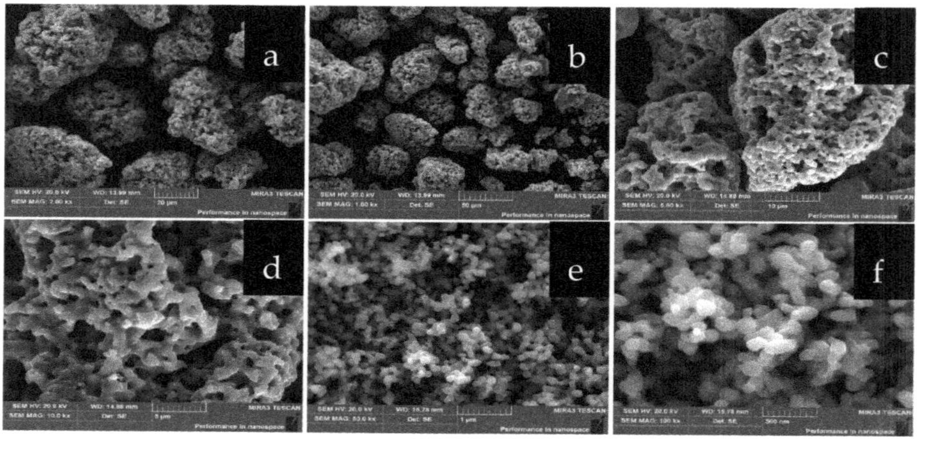

Figure 7. SEM images of cuprous sulfide with different morphologies: the scanning images of (**a**,**b**) Cu_2S(I); the scanning images of (**c**,**d**) Cu_2S(II); the scanning images of (**e**,**f**) Cu_2S(III).

Figure 7e,f are the scanning images of Cu_2S(III) prepared by annealing at 600 °C for 3 h with micropore structure cuprous sulfide. As can be seen from Figure 7d, the porous cuprous sulfide is formed by the fusion and combination of many nanoparticles. Since the subsequent annealing temperature of 600 °C is higher than the recrystallization temperature of cuprous sulfide, the junction of porous structure grains gradually melts under the action of high temperature, and the cuprous sulfide nanoparticles are separated. As the temperature increases, the particles gradually homogenize and eventually form nanoparticles cuprous sulfide. As can be seen from Figure 7f, the size of cuprous sulfide nanoparticles is concentrated at 500 nm.

2.2.3. Electrochemical Test Analysis

It can be seen from Figure 8 that cuprous sulfide with three morphologies all have obvious REDOX peaks. According to the calculation of the capacitance contribution rate in the last work, there is no double electric layer effect in cuprous sulfide energy storage, indicating that the Faraday reaction occurs in the process of energy storage. In addition, by comparing the symmetry of the REDOX peak of the three materials, it can be seen that the oxidation peak height of cuprous sulfide with the three morphologies is obviously higher than its reduction peak height. It shows that the reversibility of materials is related to the properties of materials and has nothing to do with the morphology of materials. By comparing the areas enclosed by the CV curves of cuprous sulfide in a different morphology, it can be seen that the area enclosed by the agglomerated cuprous sulfide is the largest, followed by the CV area in the microporous condition, and the smallest is the nanoparticle. It is proved that when the morphology is agglomerate, cuprous sulfide has more excellent energy storage capacity, while nanoparticles have the worst energy storage capacity. In addition, the agglomerated cuprous sulfide has a higher current intensity, indicating that it has a faster REDOX kinetics during charge and discharge [19,20].

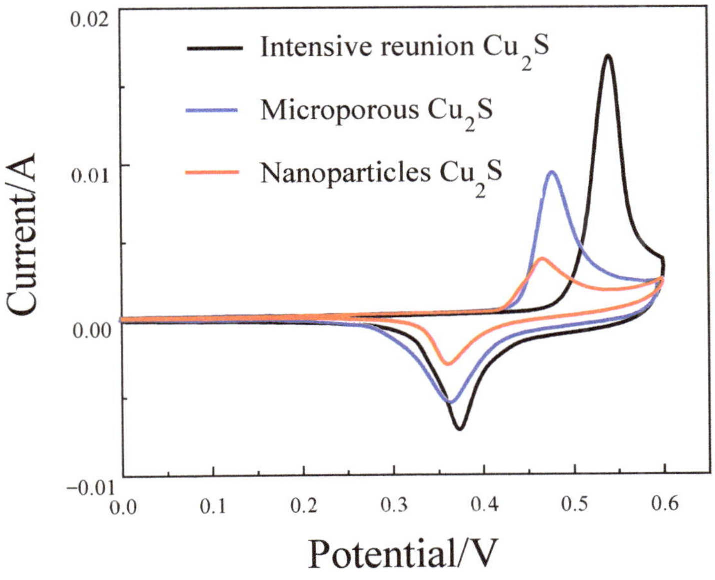

Figure 8. The cyclic voltammetry diagram curves of intensive reunion Cu_2S, microporous Cu_2S, and nanoparticles Cu_2S.

As shown in Figure 9, the influence of morphology on the energy storage performance of materials was explored by conducting a constant current charge and discharge at different current densities for cuprous sulfide with different morphologies. Figure 9a,b show the charge-discharge curves of agglomerated and microporous cuprous sulfide at the current density of 1, 3, 5, 10 A/g. It can be seen from the figure that the symmetry of the curve is improved, indicating that the coulomb efficiency of both is improved. Figure 9c shows the charge–discharge curve of cuprous sulfide nanoparticles. It can be seen from the figure that at the current density of 1 A/g, the charging time of the material is significantly longer than the discharging time, indicating that the coulomb efficiency of the material is poor at this current density. The reason is that the surface of the nanoparticles is flat, which makes it difficult for OH^- to attach during pseudocapacitance reaction. The separation and embedding of OH^- are affected, which makes the coulombic efficiency of the material poor. However, by increasing the current density, the symmetry of the material becomes better, indicating that increasing the current is conducive to the oxidation process of nano cuprous sulfide, thus obtaining excellent coulomb efficiency [21,22].

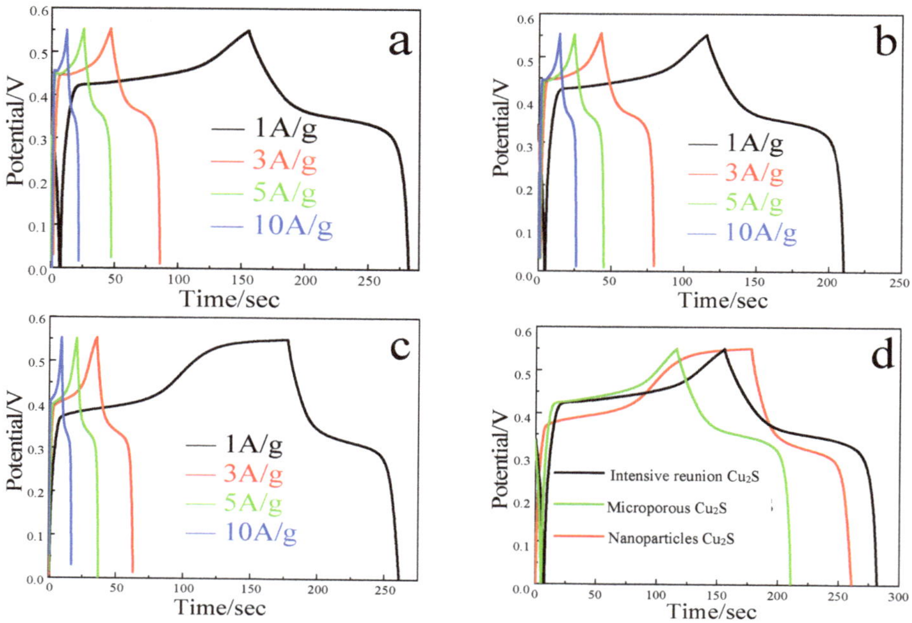

Figure 9. The GCD curves of (**a**–**c**) intensive reunion, microporous and nanoparticles three morphologies of Cu_2S; capacitance retention of (**d**) three different morphologies.

The specific capacitance of the material can be calculated by the GCD curve of the material. Table 2 shows that the specific capacitance of cuprous sulfide with agglomerated, microporous, and nanoparticle morphology at the current density of 1 A/g is 206.9, 169.2 and 147.2 F/g. The specific capacitance at the current density of 10 A/g is 193.8, 159.2, and 140.6 F/g, respectively. The capacitance retention rates of the three materials were 93.7%, 94.1%, and 97%, respectively, and the best capacitance retention rate was obtained under the morphology of nano particles. The important reason is that compared with the other two kinds of morphologies, the nanoparticle morphology has better structural stability and stronger resistance when facing large current, so the retention rate is the best.

Table 2. Specific capacitance and capacitance capability of cuprous sulfide with different morphologies.

Material	C(1 A/g)	C(10 A/g)	Retention
Intensive reunion Cu_2S	206.9 F/g	193.8 F/g	93.7%
Microporous Cu_2S	169.2 F/g	159.2 F/g	94.1%
Nanoparticles Cu_2S	147.2 F/g	140.6 F/g	97%

2.2.4. Specific Surface Analysis

According to Figure 10, it can be known that intensive reunion Cu_2S and nanoparticles Cu_2S has no obvious hysteresis ring, indicating that there is basically no pore structure. The microporous structure has a hysteretic ring, but the ring is not regular, indicating that the pore size distribution is uneven. The specific surface area of the agglomerated cuprous sulfide was 11 cm^2/g. The specific surface area of the microporous cuprous sulfide is 3.22 cm^2/g. The cuprous sulfide of nanoparticles was only 0.11 cm^2/g. It is found that the specific surface area corresponds to the specific capacitance of the material. It shows that the energy storage capacity of the material increases with the increase of the specific surface area. The main reason is that in the process of pseudocapacitive reaction, a large specific surface area can provide more active sites so that more materials can participate in the reaction and, thus, more energy can be stored [23].

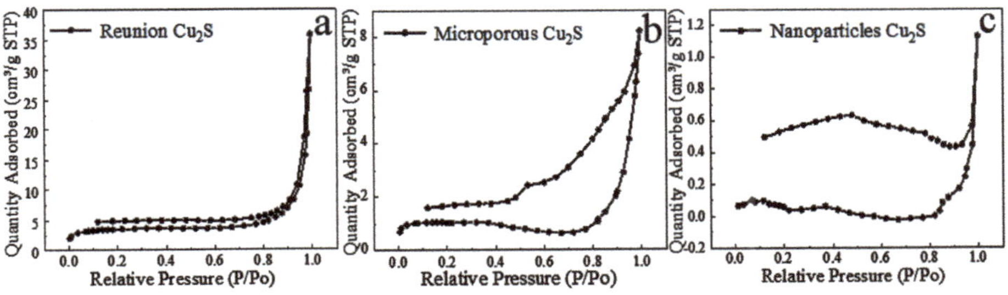

Figure 10. Nitrogen adsorption and desorption isotherms of Cu_2S with different morphologies.

Figure 11a shows the AC impedance curves of agglomerated, microporous, and nanoparticular cuprous sulfide. It can be seen from the figure that the semicircular radius of the high-frequency zone curve of the agglomerated cuprous sulfide is the smallest, indicating that the electrical conductivity of the agglomerated cuprous sulfide is the best under this topography. The cuprous sulfide of nanoparticles has the largest radius, indicating that its electrical conductivity is worse than that of other morphology [24]. On the whole, although the electrical conductivity of materials with different morphologies is different, the curves in the high-frequency region of cuprous sulfide materials with three morphologies are almost ignored. It shows that the internal resistance of cuprous sulfide with different morphology is very small when used as supercapacitor materials.

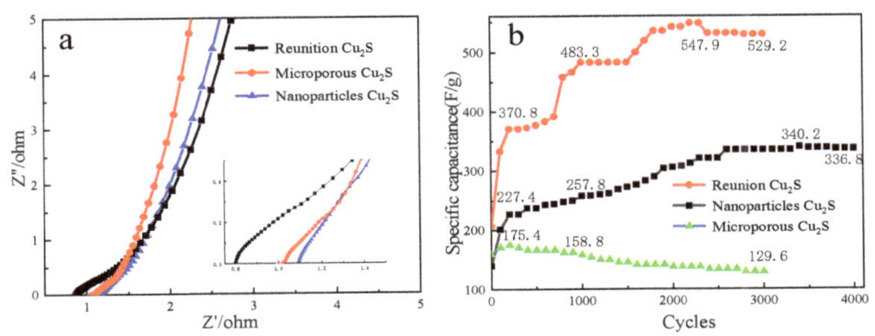

Figure 11. Different impedance curves of (**a**) three different morphologies of Cu_2S; different cyclic stability of (**b**) three different morphologies of Cu_2S.

The EIS curves of agglomerated, microporous, and nanoparticle Cu_2S were analyzed at a low frequency, and it was found that the slope of the line in the microporous Cu_2S low-frequency region was the highest. The results indicate that the resistance of interfacial charge transfer and the diffusion resistance of the electrolyte are lower when the material is used as the electrode. The reason is that the microporous Cu_2S has larger pores inside, which makes the diffusion channel of electrolyte wider than the other two morphologies. As a result, the diffusion resistance of ions is smaller, which is more conducive to the diffusion of ions, resulting in a smaller interfacial charge transfer resistance and a lower diffusion resistance of the electrolyte [25].

Figure 11b shows the influence of morphology on the cyclic stability of cuprous sulfide materials. As shown in Figure 11, the specific capacitance of agglomerated particle morphology at the beginning of the cycle is 193.8 F/g. Then, the capacity continues to rise to the maximum capacity of 547.9 F/g to 2300 laps. By Lap 3000, the remaining specific capacity was only 529.2 F/g. The square dot curve is the cycle graph of cuprous sulfide nanoparticles. As can be seen from the figure, the specific capacity of the material continues to rise from 140.6 F/g at the beginning to 227 F/g after 200 cycles during 0–3000 cycles. Then, it grows steadily to 3000 laps and stays steady. It briefly increases to a maximum of 340.2 F/g at 3400 laps. By 4000 cycles, the specific capacity was only 336.8 F/g. The triangulation curve shows the specific capacitance of the microporous cuprous sulfide during 3000 charge-discharge cycles. As shown in the figure, the microporous cuprous sulfide increases briefly from 159.2 F/g to 175.4 F/g. Then, it keeps going down. After 3000 cycles, only 129.6 F/g is left.

From the above analysis, it can be seen that both agglomerated cuprous sulfide and nanoparticle cuprous sulfide have good cycling stability, while the cycling stability of microporous cuprous sulfide is poor. The reason is that the stability of microporous cuprous sulfide is worse than the other structure. After repeated circulation, the microporous structure collapses, eventually forming a blockage. As a result, the specific surface area of the material decreases, and the active substances involved in the reaction decrease [26]. Finally, the specific capacity of the material continues to decline. The stability of cuprous sulfide particles is higher. Through a continuous reversible reaction, the active substances on the surface are constantly separated, thus making the specific surface area increase, making the specific capacity of the material increase. When the specific surface area increases to the limit, the specific capacity of the material begins to decline due to the cyclic loss of the material.

3. Materials and Methods

3.1. Chemicals and Materials

The experimental materials of thioacetamide (TAA), monohydrate copper acetate, sodium hydroxide, polyvinylidene fluoride (PVDF), *N*-methylpyrrolidone (NMP), acetylene black, distilled water, and 99.9% alcohol are analytical pure from Aladdin.

3.2. Synthesis of Different Copper Sulfides

By hydrothermal synthesis [27,28], 70 mL of distilled water was added into the reactor. Then, 1 mmol of copper acetate monohydrate was added and stirred until it was completely dissolved. In addition, 1 mmol of TAA was added, and the solution immediately generated flocculent precipitation. Then, the pH value of the solution was adjusted to alkaline by adding 3.5 mmol of sodium hydroxide. Then, the solution was transferred to a drying oven and kept at 180 °C for 24 h to get black precipitation. Finally, Cu_xS_y (I, CuS) was obtained by freeze-drying after cleaning it to neutral. Under the same method, Cu_xS_y (III, Cu_9S_5) and Cu_xS_y (V, Cu_2S) were obtained by adjusting the amount of monohydrate copper acetate to 1.8 mmol and 2 mmol. Cu_xS_y (IV, $Cu_{7.2}S_4$) was obtained by changing the pH value of 1.8 mol solution with potassium hydroxide. The ratio of copper acetate monohydrate to TAA was 1.75:1, and other reaction conditions remained the same, so Cu_xS_y (II, Cu_7S_4) was obtained.

3.3. Morphology Control of Cuprous Sulfide

Repeat 2.2 to obtain Cu_2S (I). We changed 70 mL of distilled water in the reactor into 70 mL of alcohol, while 2 mmol of monohydrate copper acetate and 1 mmol of TAA were successively added. Then, Cu_2S (II) was prepared by vacuum drying oven with 180 °C of heat preservation for 24 h. The Cu_2S (II) was transferred to the sintering furnace and annealed in argon atmosphere at 600 °C for 3 h to obtain Cu_2S (III).

3.4. Preparation of Test Electrodes

Common electrochemical workstation was used for electrode test and electrochemical test. The reference electrode was calomel electrode, the reverse electrode was platinum electrode, and the electrolyte solution was 2 mol/L of KOH in the test. We added 2 mg of PVDF and an appropriate amount of NMP into a 5 mL flask and electromagnetically stirred for 1.5 h. Then, 2 mg of acetylene black was added and stirred for 0.5 h. Active substances (Cu_xS_y (I, II, III, IV and V) and Cu_2S (I, II and III)) were added for 16 mg, sealed, and stirred for 12 h to get the solution. We evenly dropped 0.3 mL of solution on the round sheet nickel foam wafer with a diameter of 10 mm, and then ventilated and dried it for 8 h. It was then vacuum dried for 12 h. Finally, the small wafer was connected with the pole ear at a pressure of 10 Mpa, and the test electrode was prepared by vacuum drying for 12 h.

4. Conclusions

Different stoichiometric coefficient copper sulfides, such as CuS, Cu_7S_4, Cu_9S_5, $Cu_{7.2}S_4$, and Cu_2S with uniform morphology, were prepared by a hydrothermal method. Electrochemical tests show that different copper sulfides exhibit pseudocapacitance properties when used as electrode materials for supercapacitors. In addition, the specific capacitance of copper sulfide increases with the increase of copper stoichiometric coefficient to sulfur. The main reason is that the increase of active sites makes more OH^- participate in the pseudocapacitive reaction. It can be seen from the cyclic test at the current density of 10 A/g that cuprous sulfide has the best circulability. After activation, the specific capacitance can reach 547.9 F/g. Cu_9S_5 was followed by Cu_7S_4, CuS was average, and $Cu_{7.2}S_4$ was the worst. The agglomerated cuprous sulfide was successfully synthesized by the hydrothermal method. The microporous cuprous sulfide was successfully prepared by changing the reaction solvent. The nanometer cuprous sulfide was obtained by annealing microporous cuprous sulfide. Electrochemical tests show that the morphology has little effect on the pseudocapacity and impedance of cuprous sulfide. However, the specific capacitance is proportional to the specific surface area of different morphologies. In addition, the circulation of microporous cuprous sulfide is the worst. The reason is that its structure stability is poor; with the progress of the cycle, the structure of the material is destroyed, resulting in the constant decline of specific capacitance.

Author Contributions: Conceptualization, Y.D. and R.L.; methodology, R.L.; software, X.D.; validation, Y.Z., R.L. and S.X.; formal analysis, Y.D.; investigation, R.L.; resources, M.Y.; data curation, X.D.; writing—original draft preparation, R.L.; writing—review and editing, R.L.; visualization, R.L.; supervision, Y.Z.; project administration, Y.D.; funding acquisition, Y.D. All authors have read and agreed to the published version of the manuscript.

Funding: This work is financially supported by Hunan Provincial Natural Science Foundation of China (2022JJ50059).

Institutional Review Board Statement: Not applicable.

Informed Consent Statement: Not applicable.

Data Availability Statement: The data that support the findings of this work are available from the corresponding author upon reasonable request.

Conflicts of Interest: The authors declare no conflict of interest.

Sample Availability: Samples of the compounds different copper sulfides in this work are available from the authors.

References

1. Bhat, T.S.; Patil, P.S.; Rakhi, R.B. Recent trends in electrolytes for supercapacitors. *J. Energy Storage* **2022**, *50*, 104222. [CrossRef]
2. Cao, X.J.; Zeng, H.Y.; Cao, X.; Xu, S.; Alain, G.B.F.C.; Zou, K.M.; Liu, L. 3D-hierarchical mesoporous $CuCo_2S_4$@NiCoAl hydrotalcite/Ni foam material for high-performance supercapacitors. *Appl. Clay Sci.* **2020**, *199*, 105864. [CrossRef]
3. Li, M.; Yang, S.; Wang, M. $NiCo_2O_4$ and theirs composite materials for supercapacitors. *Trans. Nonferrous Met. Soc. China* **2017**, *27*, 2283–2292.
4. Ma, R.; Peng, C.; Wang, R.; Wang, X.; Zhang, C.; Tan, S. Preparation of Cu−diamond composite coating by composite electroplating. *Trans. Nonferrous Met. Soc. China* **2015**, *25*, 3414–3421.
5. Li, H.-B.; Xiao, G.-F.; Zeng, H.-Y.; Cao, X.-J.; Zou, K.-M.; Xu, S. Supercapacitor based on the $CuCo_2S_4$@NiCoAl hydrotalcite array on Ni foam with high-performance. *Electrochim. Acta* **2020**, *352*, 136500. [CrossRef]

6. Howarth, A.J.; Wang, T.C.; Al-Juaid, S.S.; Aziz, S.G.; Hupp, J.T.; Farha, O.K. One-Pot Synthesis of ZnS Nanowires_Cu_7S_4 Nanoparticles_ Reduce Graphene Oxide Nanocompositites for Supercapacitor and Photocatalytic Applications. *Dalton Trans.* **2016**, *45*, 93–97. [CrossRef]
7. Wang, X.; Chen, X.; Zhou, W.; Li, S. Cubic Nano Cu_7S_4 with Hollow Structures: Their Preparation and Electrochemical Capacitance Performance. *J. Mater. Sci. Eng.* **2017**, *35*, 542–547.
8. Zhou, K.; Liang, J.; Liu, J.; Sun, P.; Bu, J.; Zhang, W.; Chen, G. Synthesis of porous $Cu_{7.2}S_4$ sub-microspheres by an ion exchange method for high-performance supercapacitors. *RSC Adv.* **2016**, *6*, 16832–16837. [CrossRef]
9. Bulakhe, R.N.; Sahoo, S.; Nguyen, T.T.; Lokhande, C.D.; Roh, C.; Lee, Y.R.; Shim, J.J. Chemical synthesis of 3D copper sulfide with different morphologies for high performance supercapacitors application. *RSC Adv.* **2016**, *6*, 14844–14851. [CrossRef]
10. Zou, Y.; Jiang, L.; Zhai, T.; You, T.; Jing, X.; Liu, R.; Li, F.; Zhou, W.; Jin, S. Surface-enhanced Raman scattering by hierarchical CuS microflowers: Charge transfer and electromagnetic enhancement. *J. Alloy. Compd.* **2021**, *865*, 158919. [CrossRef]
11. Zhu, C.; Liu, T.; Qian, F.; Han, T.Y.J.; Duoss, E.B.; Kuntz, J.D.; Spadaccini, C.M.; Worsley, M.A.; Li, Y. Supercapacitors Based on Three-Dimensional Hierarchical Graphene Aerogels with Periodic Macropores. *Nano Lett.* **2016**, *16*, 3448–3456. [CrossRef]
12. Rajpurohit, A.S.; Punde, N.S.; Rawool, C.R.; Srivastava, A.K. Fabrication of high energy density symmetric supercapacitor based on cobalt-nickel bimetallic tungstate nanoparticles decorated phosphorus-sulphur co-doped graphene nanosheets with extended voltage. *Chem. Eng. J.* **2019**, *371*, 679–692. [CrossRef]
13. Lu, Y.; Li, B.; Zheng, S.; Xu, Y.; Xue, H.; Pang, H. Syntheses and Energy Storage Applications of M_xS_y(M = Cu, Ag, Au) and Their Composites: Rechargeable Batteries and Supercapacitors. *Adv. Funct. Mater.* **2017**, *27*, 1703949. [CrossRef]
14. Masarapu, C.; Zeng, H.F.; Hung, K.H.; Wei, B. Effect of Temperature on the Capacitance of Carbon Nanotube Supercapacitors. *ACS Nano.* **2009**, *3*, 2199–2206. [CrossRef]
15. Cheng, L.; Li, D.; Yu, W.; Ma, Q.; Dong, X.; Wang, X.; Wang, J.; Liu, G. A Novel Strategy to Fabricate CuS, $Cu_{7.2}S_4$, and $Cu_{2-x}Se$ Nanofibers via Inheriting the Morphology of Electrospun CuO Nanofibers. *Russ. J. Phys. Chem.* **2019**, *93*, 730–735. [CrossRef]
16. Gao, H.; Chen, Y.; Li, H.; Zhang, F.; Tian, G. Hierarchical Cu_7S_4-Cu_9S_8 heterostructure hollow cubes for photothermal aerobic oxidation of amines. *Chem. Eng. J.* **2019**, *363*, 247–258. [CrossRef]
17. Jia, C.; Wang, W.; Wang, Y.; Cui, J. Preparation and supercapacitive performance of VO_2/C-TiO_2 nanotube composite arrays. *Trans. Nonferrous Met. Soc. China* **2017**, *27*, 1632–1642.
18. Zhu, T.; Xia, B.; Zhou, L.; Lou, X.W.D. Arrays of ultrafine CuS nanoneedles supported on a CNT backbone for application in supercapacitors. *J. Mater. Chem.* **2012**, *22*, 7851. [CrossRef]
19. Huang, K.-J.; Zhang, J.-Z.; Fan, Y. One-step solvothermal synthesis of different morphologies CuS nanosheets compared as supercapacitor electrode materials. *J. Alloy. Compd.* **2015**, *625*, 158–163. [CrossRef]
20. Peng, H.; Ma, G.; Sun, K.; Mu, J.; Wang, H.; Lei, Z. High-performance supercapacitor based on multi-structural CuS@polypyrrole composites prepared by in situ oxidative polymerization. *J. Mater. Chem. A.* **2014**, *2*, 3303. [CrossRef]
21. Li, C.; Zhang, X.; Wang, K.; Su, F.; Chen, C.M.; Liu, F.; Wu, Z.S.; Ma, Y. Recent advances in carbon nanostructures prepared from carbon dioxide for high-performance supercapacitors. *J. Energy Chem.* **2021**, *54*, 352–367. [CrossRef]
22. Chen, Q.C.; Yan, G.J.; Luo, L.M.; Fei, C.H.E.N.; Xie, T.F.; Dai, S.C.; Yuan, M.L. Enhanced cycling stability of Mg–F co-modified $LiNi_{0.6}Co_{0.2}Mn_{0.2-y}Mg_yO_{2-z}F_z$ for lithium-ion batteries. *Trans. Nonferrous Met. Soc. China* **2018**, *28*, 1397–1403. [CrossRef]
23. Azami, M.; Jalil, A.; Hassan, N.; Hussain, I.; Fauzi, A.; Aziz, M. Green carbonaceous material–fibrous silica-titania composite photocatalysts for enhanced degradation of toxic 2-chlorophenol. *J. Hazard. Mater.* **2021**, *414*, 125524. [CrossRef] [PubMed]
24. Hu, W.; Chen, R.; Xie, W.; Zou, L.; Qin, N.; Bao, D. $CoNi_2S_4$ Nanosheet Arrays Supported on Nickel Foams with Ultrahigh Capacitance for Aqueous Asymmetric Supercapacitor Applications. *ACS Appl. Mater. Interfaces* **2014**, *6*, 19318–19326. [CrossRef]
25. Liu, L.; Annamalai, K.P.; Tao, Y. A hierarchically porous $CuCo_2S_4$/graphene composite as an electrode material for supercapacitors. *New Carbon Mater.* **2016**, *31*, 336–342. [CrossRef]
26. Xu, K.; Merlet, C.; Lin, Z.; Shao, H.; Taberna, P.L.; Miao, L.; Jiang, J.; Zhu, J.; Simon, P. Effects of functional groups and anion size on the charging mechanisms in layered electrode materials. *Energy Storage Mater.* **2020**, *33*, 460–469. [CrossRef]
27. Xia, X.; Zhu, C.; Luo, J.; Zeng, Z.; Guan, C.; Ng, C.F.; Zhang, H.; Fan, H.J. Synthesis of Free-Standing Metal Sulfide Nanoarrays via Anion Exchange Reaction and Their Electrochemical Energy Storage Application. *Small* **2014**, *10*, 766–773. [CrossRef]
28. Wang, Z.; Peng, C.; Wang, R.; Wang, X.; Liu, B. Typical properties and research progress of Al-doped-ZnO. *Trans. Nonferrous Met. Soc. China* **2012**, *22*, 416–426.

molecules

MDPI

Article

Facile One-Step Heat Treatment of Cu Foil for Stable Anode-Free Li Metal Batteries

Jie Chen [1], Linna Dai [1], Pei Hu [1,*] and Zhen Li [2]

[1] School of Science, Hubei University of Technology, Wuhan 430068, China
[2] State Key Laboratory of Material Processing and Die and Mold Technology, School of Materials Science and Engineering, Huazhong University of Science and Technology, Wuhan 430074, China
* Correspondence: hupei@hbut.edu.cn

Abstract: The anode-free lithium metal battery (AFLMB) is attractive for its ultimate high energy density. However, the poor cycling lifespan caused by the unstable anode interphase and the continuous Li consumption severely limits its practical application. Here, facile one-step heat treatment of the Cu foil current collectors before the cell assembly is proposed to improve the anode interphase during the cycling. After heat treatment of the Cu foil, homogeneous Li deposition is achieved during cycling because of the smoother surface morphology and enhanced lithiophilicity of the heat-treated Cu foil. In addition, Li_2O-riched SEI is obtained after the Li deposition due to the generated Cu_2O on the heat-treated Cu foil. The stable anode SEI can be successfully established and the Li consumption can be slowed down. Therefore, the cycling stability of the heat-treated Cu foil electrode is greatly improved in the Li | Cu half-cell and the symmetric cell. Moreover, the corresponding LFP | Cu anode-free full cell shows a much-improved capacity retention of 62% after 100 cycles, compared to that of 43% in the cell with the commercial Cu foil. This kind of facile but effective modification of current collectors can be directly applied in the anode-free batteries, which are assembled without Li pre-deposition on the anode.

Keywords: anode-free Li metal batteries; current collector modification; anode interface; solid electrolyte interphase

1. Introduction

Lithium metal batteries (LMBs) have received board interest mainly due to the highest theoretical specific capacity (3860 mAh g^{-1}) and the lowest redox potential (−3.04 V vs. the standard hydrogen electrode) of Li metal anodes [1–4]. AFLMB, whose Li source is provided only by the cathode, can be the ultimate choice among all the lithium metal batteries due to its high energy density and easy fabrication [5–7]. When the cathode is determined, the anode-free system achieves the highest gravimetric and volumetric densities among the combination options that use graphite, Sn, Si or Li metal anodes [8]. Avoiding the direct use of Li metal simplifies the battery fabrication and improves the security.

However, the AFLMB suffers from the predicaments caused by the extremely high reactivity of Li metal and colossal volume change during the repeated Li cycling, including the growth of Li dendrites, the generation and accumulation of dead Li, and the side reactions between metallic Li and the electrolyte [9–11]. Stable anode SEI is hard to establish and the active lithium is consumed continuously, which accelerates the battery failure [12]. The AFLMB without modification usually falls below 50% capacity retention in fewer than 50 cycles, which is far from the commercial application [5].

To improve the performance of AFLMB, many strategies have been proposed in recent years, such as the modification of current collectors [13–16], the electrolyte optimization [17–20], the lithium compensation in the cathode [21–23], and the optimization of external conditions (the mechanical pressure [24,25], the temperature [26,27], the cycling conditions [28], and so on). In AFLMB, it is worth noting that the initial lithium deposition morphology and formed solid electrolyte interphase (SEI) formed on the current collectors can influence the long-term evolution of the Li metal anode [29–32], which determines the cycle life of the batteries to a large extent. Thus, the properties of current collectors can play a crucial part in anode stabilization design. The "charge-induced growth model" reveals that the charge accumulation is beneficial to early Li dendrite growth [33]. Considering the "tip effect" that a protrusion on the electrode exhibits a stronger electrical field and gathers more charges [34,35], the surface structure of the current collectors should be smooth and flat. Moreover, whether the current collectors are lithiophilic or lithiophobic influences the Li nucleation pattern, which has an important effect on the lithium deposition morphology [29], and the current collectors should better be lithiophilic. The surface composition of some modified current collectors can also regulate the SEI through the reaction with the deposited Li, such as the content of LiF or Li_2O in SEI, which can be improved by modifying the current collectors with active fluoric or oxygenic species, while LiF and Li_2O are widely known to regulate the Li^+ transfer and enhance the structural modulus [36–39]. The rational design of the reaction can contribute to building high-quality SEI.

In this work, a facile one-step heat treatment method is proposed to modify the surface structure, lithiophilicity, and even surface composition of the commercial Cu current collectors, and the optimal heat treatment time is identified to be 10 min. The heat-treated Cu foil shows a flatter surface morphology with fewer fluctuations than the commercial Cu foil, which avoids the charge accumulation and controls the lithium-ion flux. Cu_2O is found to generate on the surface and the lithiophilicity is enhanced after the heat treatment. The Cu_2O is demonstrated to transform into Li_2O after Li deposition, constructing Li_2O-riched SEI and improving the anode stability. As a result, when the 3.5 mAh cm^{-2} areal capacity $LiFePO_4$ (LFP) cathode is used, the anode-free full cell with the heat-treated Cu foil maintains high capacity retentions of 62% and 47% after 100 and 200 cycles, compared to that of 43% and 31% in the cell with the commercial Cu foil.

2. Results and Discussion

The commercial Cu foils were heat-treated by a facile one-step method. They were heated at 320 °C in the atmosphere for 5, 10 and 15 min for comparison. The color of the Cu foils was changed from purplish red (Cu foil) to dark blue (Cu–5) and dark brown (Cu–10 and Cu–15) after the heat treatment, as shown in the insets of Figure 1a–d. The SEM image of the Cu foil shows rough morphology with distinct grains (Figure 1a), while the grains become blurred and the surface morphology becomes smooth gradually with the extension of heat treatment time (Figure 1b–d). The cross-sectional SEM images of the Cu foils in Figure 1e–h also confirm the trend of the morphology change.

To identify the changes in surface composition of the Cu foils, the Raman spectra and the XRD tests were performed. Raman spectrum is a kind of molecule scattering spectroscopy, which is characterized by the frequency excursion caused by the interactions of the molecule and photon to show the information of molecule. As shown in Figure 1i, the commercial Cu foil shows no signal because there is no molecular structure in the metal elementary substance, while the heat-treated Cu foils reveal obvious characteristic peaks of Cu_2O at 156, 222, 654 cm^{-1} and a weak characteristic peak of CuO at 305 cm^{-1} [40,41]. The XRD patterns of different samples in Figure 1j both show three strong characteristic peaks of Cu at 43°, 51° and 74° (PDF#04-0836), which means the main ingredient of the current collectors is still Cu. However, the intensity of the characteristic peaks of Cu_2O at 36° and 42° (PDF#05-0667) increases gradually, as shown in the partially enlarged drawing of Figure 1k, and no characteristic peak of CuO is found. Taking into account the results of SEM images, the Raman spectra and the XRD patterns, it is believed that a thin layer of Cu_2O with little CuO is generated on the surface of the Cu foils after the heat treatment and the amount of Cu_2O increases with the extension of the heat treatment time.

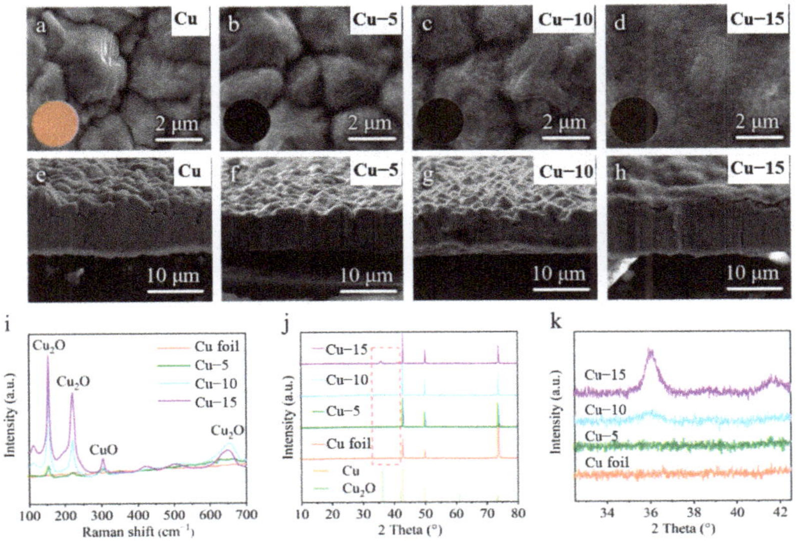

Figure 1. (**a–d**) Surface and (**e–h**) cross-sectional SEM images of the (**a**,**e**) commercial Cu foil, (**b**,**f**) Cu–5, (**c**,**g**) Cu–10 and (**d**,**h**) Cu–15. The inset is the corresponding optical photos. (**i**) Raman spectra of different Cu foils. (**j**) XRD patterns and (**k**) the partially enlarged drawing at 33–42° of different Cu foils.

CE tests of different Cu foils in Li | Cu half-cells were performed to reveal the optimum heat treatment time in Figure 2a. With the current density of 1 mA cm^{-2} and the areal capacity of 2 mAh cm^{-2}, the half-cell with the Cu–10 shows the best cycle life of 160 cycles. A total of 137 and 127 cycles are achieved in the cells with the Cu–5 and Cu–15, while only 56 cycles are achieved when the commercial Cu foil is used. It is worth noting that there is a certain amount of discharge capacity above 0 V in the first cycle, as shown in Figure 2b, which may mean the reaction between Cu_2O and the deposited Li. However, the reaction is almost irreversible in the following charge process and causes the loss of the active Li, which is a severe issue in the lithium-limited AFLMB. Thus, the discharge capacity above 0 V should be as little as possible. The amount of Cu_2O is critical. The effect is limited with little Cu_2O, and active Li is lost severely with much Cu_2O. Overall speaking, Cu–10 samples are better than Cu–5 and Cu–15. The optimum heat treatment is chosen to be 10 min through the above analysis. Only the characterization and electrochemical tests of the commercial Cu foil and Cu–10 are compared in the following part.

The atomic force microscope was applied to further characterize the change in the surface morphology of the Cu foils, as shown in Figure 2c,d. The AFM images of the commercial Cu foil show a longitudinal fluctuation of 943 nm in a 5 μm × 5 μm area, which is much larger than that of 521 nm in the Cu–10. In addition, the grains seem to be smaller after the heat treatment. Smaller longitudinal fluctuation and grains mean a smoother surface morphology which is beneficial to the homogenization of charges. Figure 2e shows the lithiophilicity tests of the Cu foil and the Cu–10. Two of the same Li foils were put on different Cu foils and heated at 200 °C simultaneously on the hot plate. The molten Li on the commercial Cu foil turns into a sphere so as to decrease the contact area with the lithiophobic surface, while the molten Li completely spreads on the Cu–10. The obvious contrast reveals the improvement of the lithiophilicity, which is beneficial to the Li nucleation on the current collectors.

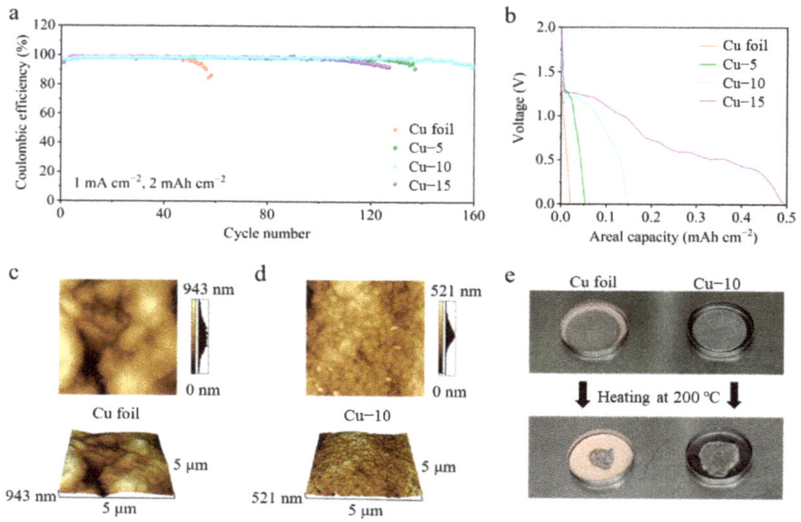

Figure 2. (**a**) Coulombic efficiency and (**b**) the first discharge profiles of the cells using the commercial Cu foil, Cu–5, Cu–10 or Cu–15. AFM images of the (**c**) commercial Cu foil and (**d**) Cu–10. (**e**) Optical photos of Li foil on the commercial Cu foil or Cu–10 before and after heating at 200 °C.

To demonstrate the effect of modified Cu foils on the Li deposition morphology, the Li | Cu half-cells were cycled at 1 mA cm^{-2} for 5 cycles and disassembled for SEM tests. Figure 3a,b shows the surface and cross-section SEM images when the 2 mAh cm^{-2} areal capacity is applied. Due to the rough and lithiophobic surface of the commercial Cu foil, the Li deposition morphology is uneven with many protrusions (Figure 3a,b), which will further cause the growth of Li dendrites and the instability of the anode interphase. As shown in Figure 3c,d, when the areal capacity is increased to 4 mAh cm^{-2}, the Li dendrites are evident on the surface. It can be speculated that there must have been many side reactions occurring at the anode. The Li deposition morphology is dense and homogeneous when the Cu–10 is used, whether the 2 mAh cm^{-2} or 4 mAh cm^{-2} areal capacity is applied (Figure 3e–h). This is benefited from the smoother surface morphology and the lithiophilicity of Cu–10, which is better for the homogenization and the nucleation of Li-ions.

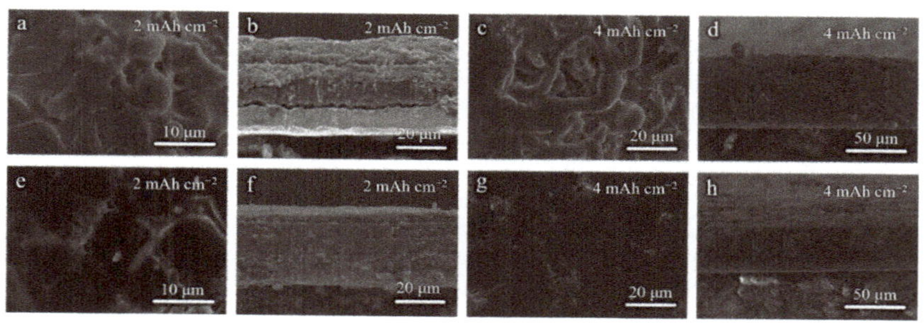

Figure 3. Surface and cross-sectional SEM images of (**a**,**b**,**e**,**f**) 2 mAh cm^{-2} and (**c**,**d**,**g**,**h**) 4 mAh cm^{-2} Li deposition on the (**a**–**d**) commercial Cu foil and (**e**–**h**) Cu–10.

XPS spectra were employed to analyze the components of SEI on the cycled Li-Cu anodes after five cycles. As shown in Figure 4a, the Cu 2p spectrum of the commercial Cu foil electrode shows no obvious peaks, demonstrating that there is no Cu in the SEI and the thickness of the deposited Li is greater than the depth of XPS detection (a few nanometers). However, the Cu 2p spectrum of the Cu–10 electrode shows two peaks at 931.8 and 951.5 eV, corresponding to Cu $2p_{3/2}$ and Cu $2p_{1/2}$, respectively. In addition, the O 1s spectra in Figure 4b show a strong peak at 528.8 eV of Li_2O for the Cu–10 electrode, while showing a negligible peak for the commercial Cu foil electrode. In consideration of the Cu 2p and O 1s spectra, it is believed that after the Li deposition, the Cu_2O on the Cu–10 transforms into Cu and Li_2O, which further participate in the formation of SEI.

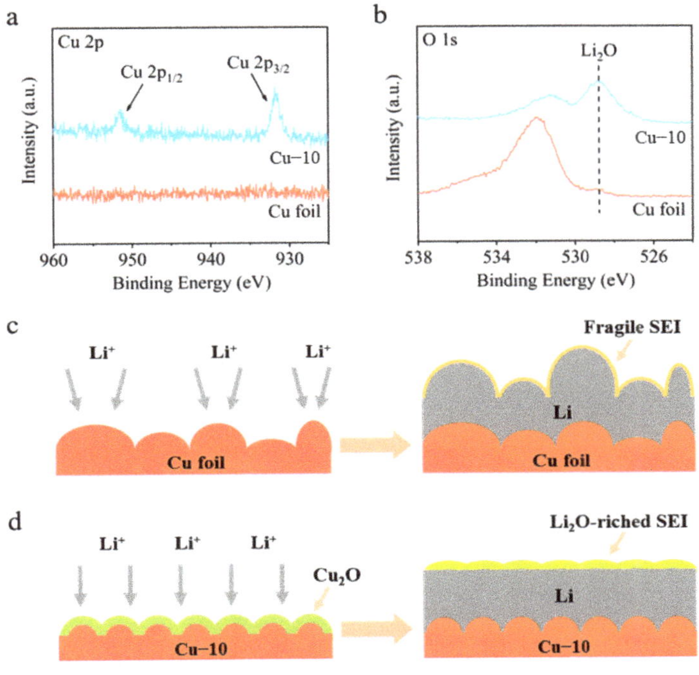

Figure 4. (**a**) Cu 2p and (**b**) O 1s XPS spectra of the surface of the commercial Cu foil and Cu–10 after Li deposition. Schematic illustrations of Li deposition on the (**c**) commercial Cu foil and (**d**) Cu–10.

Based on the above analysis, a comprehensive modification mechanism of the current collector is proposed when the commercial Cu foil is heat-treated (Figure 4c,d). On the one hand, the heat treatment makes the surface morphology smoother before the Li deposition, which is beneficial to the homogenization of charges and Li-ions' flux. Moreover, the Cu_2O generated on the surface increases the lithiophilicity, improving the Li nucleation. On the other hand, the Cu_2O can transform into Cu and Li_2O after Li deposition, which promoted the formation of Li_2O-riched SEI on the Cu–10 electrode rather than the fragile SEI on the commercial Cu foil electrode. Li_2O is reported to be a favorable component of SEI to facilitate Li^+ transportation [37], prevent excessive decomposition of the electrolyte [42], and provide the required mechanical strength [43]. Consequently, uniform Li deposition and stable anode interphase are gained on the Cu–10 electrode.

The effectiveness of the modified current collector in improving the stability of the Li-Cu anode was further verified by the electrochemical tests. Figure 5 shows the CE tests of different Cu foil electrodes at a current density of 1 mA cm^{-2} with various areal capacities. In Figure 5a, the CE values of the commercial Cu foil electrode reach 98.9% at 1 mAh cm^{-2} in the initial cycles but then drop to 98.4%, and finally fail in 130 cycles. The CE values of the Cu–10 electrode stabilize at 98.7% in a few cycles and eventually maintain at 99.0% after 400 cycles. As shown in Figure 5b–d, when the areal capacity is increased to 4 mAh cm^{-2}, the cell with the commercial Cu foil can only stabilize for 40 cycles and the plating/stripping overpotential gradually grow larger, while stable 100 cycles are achieved and almost constant overpotential is maintained in the cell with the Cu–10. The huge difference in the cycling stability is connected to the interphase between the deposited Li and the electrolyte. Poor interphase causes the growth of Li dendrites, the formation of the dead Li and the continuous side reaction, which accelerates the failure of the cell using the commercial Cu foil in the CE test. The smooth surface morphology and the lithiophilicity of the Cu–10, as well as the Li_2O-riched SEI on the Cu–10 electrode, facilitate stable interphase, further guaranteeing the outstanding performance in the CE test.

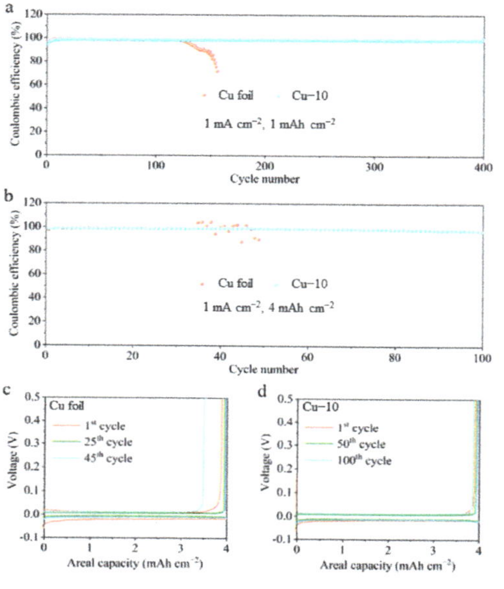

Figure 5. Coulombic efficiency of the cells using the commercial Cu foil and Cu–10 at 1 mA cm^{-2} with the capacity of (**a**) 1 mAh cm^{-2} and (**b**) 4 mAh cm^{-2}. The corresponding voltage profiles of the cells using the (**c**) commercial Cu foil and (**d**) Cu–10 at different cycles with the capacity of 4 mAh cm^{-2}.

The cycling stability of the anode with a limited Li source is evaluated through the symmetric cell, as shown in Figure 6a. The Cu foil electrodes were pre-deposited at 6 mAh cm^{-2} Li before being assembled for symmetric cells. Under a current density of 1 mA cm^{-2} with a capacity of 2 mAh cm^{-2}, the polarization of the symmetric cell with the commercial Cu foil electrodes obviously increases after cycling for 400 h. The overpotential even reaches 0.4 V after 600 h, demonstrating the formation of the unstable interphase. However, the symmetric cell with the Cu–10 electrodes can run for 1100 h and always maintain the overpotential below 25 mV. The improved interphase stability derives from the modified Cu–10 current collectors. To further verify the modification effect, the anode-free full cells with 3.5 mAh cm^{-2} LFP cathodes and Cu foil anodes were assembled for the test. All the Li source comes from the cathode and the N/P ratio is 1 in the anode-free battery; thus, any Li loss caused by the unstable interphase will reflect in the capacity drop of the full cell. In the Cu–10| |LFP full cell, the Li released from the LFP cathode reacts with the Cu_2O on the Cu–10 firstly and then forms metal Li during the first charging process. The Cu_2O transforms into Cu and Li_2O after Li deposition. It is irreversible during the first discharging process. Thus, the initial coulombic efficiency of Cu–10| |LFP cell is lower than that of the Cu| |LFP cell. As shown in Figure 6b,c, the anode-free full cell with the commercial Cu foil delivers an initial discharge capacity of 137 mAh g^{-1}, and it retains at 59 and 42 mAh g^{-1} with capacity retentions of 43% and 31% after 100 and 200 cycles, implying that the active Li is sharply consumed during the cycling process. As for the anode-free full cell with the Cu–10, it delivers a lower initial discharge capacity of 129 mAh g^{-1}, but it maintains 80 and 61 mAh g^{-1} with capacity retentions of 62% and 47% after 100 and 200 cycles. The average values of the coulombic efficiencies of Cu–10 | | LFP cell are 99.1% and 99.3% during the first 100 and 200 cycles, higher than that of 98.7% and 99.1% of the Cu | | LFP cell. The great promotion of the electrochemical performance of anode-free batteries demonstrates that the heat treatment of the Cu foil is beneficial to the establishment of the stable anode interphase although a small amount of active Li will be lost at the initial cycles. In addition, the electrolyte of 1M LiTFSI in DME-DOL + 2%$LiNO_3$ we used here is friendly to lithium metal anode for the moderate side reaction and favorable SEI. Thus, the dead lithium formation is reduced to some extent and better electrochemical performance is obtained. However, the electrolyte is still restricted by the inflammability and narrow electrochemical window. For better application, other electrolytes with modified lithium salts or solvents should be tested and less Li loss is expected during the Li plating/stripping in the future works [44–46].

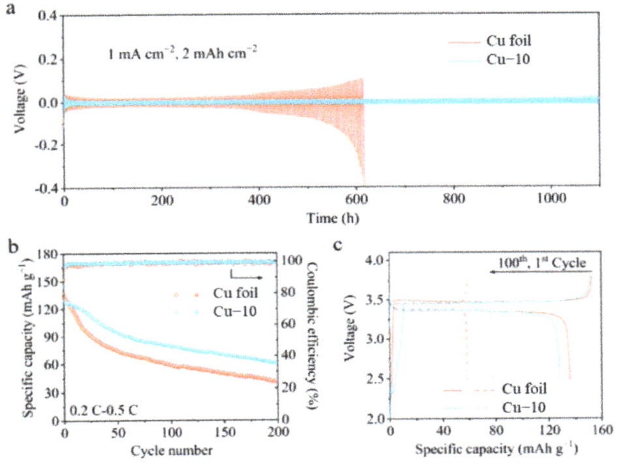

Figure 6. (**a**) Voltage profiles of the symmetric cells using the commercial Cu foil and Cu–10. (**b**) Cycling performance and (**c**) galvanostatic charge-discharge profiles of the anode-free full cells using the commercial Cu foil and Cu–10.

3. Materials and Methods

3.1. Heat Treatment of Cu Foil

The commercial Cu foil in a crucible was transferred into the pre-heated muffle furnace and heated at 320 °C for 5, 10, and 15 min in the atmosphere, respectively (denoted by Cu–5, Cu–10, and Cu–15 in the following). Then, the Cu foil was immediately removed from the muffle furnace so as to avoid excessive oxidation. The heat-treated Cu current collectors were obtained after they were cut into circular pieces.

3.2. Characterization

The morphology of Cu foils and Li deposition morphology on the Cu foils were characterized by a field-emission scanning electron microscope (FSEM, Sirion 200 and Quanta 650 FEG, Eindhoven, Holland). The Raman spectra of the heat-treated Cu foils were performed by a confocal Raman microscope (LabRAM HR800, Paris, France) with a 532 nm excitation from an argon-ion laser. The X-ray diffraction (XRD) patterns of the Cu foils were obtained on the equipment (PANalytical B.V., x'pert3 powder) using Cu Kα radiation. An atomic force microscope (AFM, SPM9700, Shimadzu, Japan) was used to measure the roughness of the Cu foils. X-ray photoelectron spectroscopy (XPS, AXIS SUPRA+, Shimadzu-Kratos, Japan) was performed to explore the composition of SEI.

3.3. Electrochemical Measurements

CR2032 coin cells were assembled for all electrochemical tests at 25 °C. Celgard 2400 polypropylene membranes were used as separators and 1 M lithium bis (trifluoromethanesulfonyl)imide (LiTFSI) solution in 1,3-dioxolane (DOL) and dimethoxymethane (DME) (1:1 by volume) with 2 wt% $LiNO_3$ was used as the electrolyte. The same amount of 50 μL electrolyte was added to each cell. For cathode preparation, the LFP power, conductive carbon and polyvinylidene fluoride (PVDF) binder were mixed at the weight ratio of 93:2.5:4.5 with the solvent of N-methylpyrrolidone. The homogeneous slurry was then cast on carbon-coated Al foils and dried at 70 °C overnight. The mass loading of LFP was about 20–24 mg cm^{-2}. For the Li | Cu half-cells in the Coulombic efficiency (CE) tests, 3 cycles cycling between 0.01–0.5 V were performed at 0.1 mA cm^{-2} for activation, then a certain amount of Li was deposited on different Cu foils and stripped up to 0.5 V at different current densities. For the symmetric cells, two Cu foil electrodes with 6 mAh cm^{-2} pre-deposited Li were assembled for tests. The anode-free LFP | Cu full cells were charged at 0.2 C and discharged at 0.5 C between 2.5 and 3.8 V.

4. Conclusions

In summary, a high-performance AFLMB with a stable anode interphase is realized through the modification of the Cu foil current collector via a facile one-step heat treatment method. The more stable anode interphase is established on the modified Cu–10 due to the following reasons: (1) The smoother surface morphology after the heat treatment homogenizes the charges and the Li-ions flux. (2) The promoted lithiophilicity may improve the Li nucleation and Li deposition; (3) the generated Cu_2O on the surface facilitates the Li_2O-riched SEI, while Li_2O can regulate Li^+ transfer and strengthen the structural modulus of SEI. As a result, the cycling stability of the Li | Cu half-cell in the CE test and the symmetric cell is greatly enhanced. Noticeably, the anode-free LFP | Cu full cell with the modified current collector exhibits superior electrochemical performance in the case of the limited Li source, realizing high capacity retentions of 62% and 47% after 100 and 200 cycles. This work provides insights into a potential route toward the current collector design for AFLMB or the other Li metal batteries.

Author Contributions: Conceptualization, P.H.; methodology, P.H. and Z.L.; software, J.C. and L.D.; validation, J.C.; formal analysis, J.C., L.D. and P.H.; investigation, J.C.; resources, P.H. and Z.L.; writing—original draft preparation, J.C.; writing—review and editing, J.C., P.H. and Z.L.; supervision, P.H., L.D. and Z.L.; project administration, P.H. and Z.L.; funding acquisition, P.H. and Z.L. All authors have read and agreed to the published version of the manuscript.

Funding: This research was funded by School Research Startup fund of Hubei University of Technology (Grant No. XJ2022002101) and Scientific research project fund of Hubei University of Technology (Grant No. GCRC2021005).

Institutional Review Board Statement: Not applicable.

Informed Consent Statement: Not applicable.

Data Availability Statement: Not applicable.

Acknowledgments: The authors acknowledge the Analytical and Testing Center of Huazhong University of Science and Technology (HUST) for the XRD, AFM, Raman, and XPS tests and the State Key Laboratory of Materials Processing and Die & Mould Technology of HUST for the SEM test.

Conflicts of Interest: The authors declare no conflict of interest.

References

1. Zhang, J.G.; Xu, W.; Xiao, J.; Cao, X.; Liu, J. Lithium metal anodes with nonaqueous electrolytes. *Chem. Rev.* **2020**, *120*, 13312–13348. [CrossRef] [PubMed]
2. Zhang, W.; Long, J.; Wang, H.; Lan, J.; Yu, Y.; Yang, X. Novel in situ growth of ZIF-8 in porous epoxy matrix for mechanically robust composite electrolyte of high-performance, long-life lithium metal batteries. *Molecules* **2022**, *27*, 7488. [CrossRef] [PubMed]

3. Ma, Y.; Wu, F.; Chen, N.; Yang, T.; Liang, Y.; Sun, Z.; Luo, G.; Du, J.; Shang, Y.; Feng, M.; et al. A dual functional artificial SEI layer based on a facile surface chemistry for stable lithium metal anode. *Molecules* **2022**, *27*, 5199. [CrossRef]
4. Xiang, J.; Yang, L.; Yuan, L.; Yuan, K.; Zhang, Y.; Huang, Y.; Lin, J.; Pan, F.; Huang, Y. Alkali-metal anodes: From lab to market. *Joule* **2019**, *3*, 2334–2363. [CrossRef]
5. Nanda, S.; Gupta, A.; Manthiram, A. Anode-free full cells: A pathway to high-energy density lithium-metal batteries. *Adv. Energy Mater.* **2020**, *11*, 2000804. [CrossRef]
6. Tian, Y.; An, Y.; Wei, C.; Jiang, H.; Xiong, S.; Feng, J.; Qian, Y. Recently advances and perspectives of anode-free rechargeable batteries. *Nano Energy* **2020**, *78*, 105344. [CrossRef]
7. Yao, W.; Zou, P.; Wang, M.; Zhan, H.; Kang, F.; Yang, C. Design principle, optimization strategies, and future perspectives of anode-free configurations for high-energy rechargeable metal batteries. *Electrochem. Energy Rev.* **2021**, *4*, 601–631. [CrossRef]
8. Chen, J.; Xiang, J.; Chen, X.; Yuan, L.; Li, Z.; Huang, Y. Li_2S-based anode-free full batteries with modified Cu current collector. *Energy Storage Mater.* **2020**, *30*, 179–186. [CrossRef]
9. Park, S.H.; Jun, D.; Lee, G.; Lee, S.; Lee, Y. Toward high-performance anodeless batteries based on controlled lithium metal deposition: A review. *J. Mater. Chem. A* **2021**, *9*, 14656–14681. [CrossRef]
10. Li, P.; Kim, H.; Ming, J.; Jung, H.-G.; Belharouak, I.; Sun, Y.-K. Quasi-compensatory effect in emerging anode-free lithium batteries. *EScience* **2021**, *1*, 3–12. [CrossRef]
11. Tong, Z.; Bazri, B.; Hu, S.-F.; Liu, R.-S. Interfacial chemistry in anode-free batteries: Challenges and strategies. *J. Mater. Chem. A* **2021**, *9*, 7396–7406. [CrossRef]
12. Huang, C.J.; Thirumalraj, B.; Tao, H.; Shitaw, K.; Sutiono, H.; Hagos, T.; Beyene, T.; Kuo, L.; Wang, C.; Wu, S.; et al. Decoupling the origins of irreversible Coulombic efficiency in anode-free lithium metal batteries. *Nat. Commun.* **2021**, *12*, 1452. [CrossRef] [PubMed]
13. Zhang, S.; Fan, X.; Wang, C. A tin-plated copper substrate for efficient cycling of lithium metal in an anode-free rechargeable lithium battery. *Electrochim. Acta* **2017**, *258*, 1201–1207. [CrossRef]
14. Kwon, H.; Lee, J.; Roh, Y.; Baek, J.; Shin, D.; Yoon, J.; Ha, H.; Kim, J.; Kim, H. An electron-deficient carbon current collector for anode-free Li-metal batteries. *Nat. Commun.* **2021**, *12*, 5537. [CrossRef] [PubMed]
15. Chen, W.; Salvatierra, R.; Ren, M.; Chen, J.; Stanford, M.; Tour, J. Laser-induced silicon oxide for anode-free lithium metal batteries. *Adv. Mater.* **2020**, *32*, 2002850. [CrossRef]
16. Jin, C.; Liu, T.; Sheng, O.; Li, M.; Liu, T.; Yuan, Y.; Nai, J.; Ju, Z.; Zhang, W.; Liu, Y.; et al. Rejuvenating dead lithium supply in lithium metal anodes by iodine redox. *Nat. Energy* **2021**, *6*, 378–387. [CrossRef]
17. Hagos, T.M.; Berhe, G.; Hagos, T.; Bezabh, H.; Abrha, L.; Beyene, T.; Huang, C.; Yang, Y.; Su, W.; Dai, H.; et al. Dual electrolyte additives of potassium hexafluorophosphate and tris(trimethylsilyl) phosphite for anode-free lithium metal batteries. *Electrochim. Acta* **2019**, *316*, 52–59. [CrossRef]
18. Yu, Z.; Rudnicki, P.; Zhang, Z.; Huang, Z.; Celik, H.; Oyakhire, S.; Chen, Y.; Kong, X.; Kim, S.; Xiao, X.; et al. Rational solvent molecule tuning for high-performance lithium metal battery electrolytes. *Nat. Energy* **2022**, *7*, 94–106. [CrossRef]
19. Louli, A.J.; Eldesoky, A.; Weber, R.; Genovese, M.; Coon, M.; Degooyer, J.; Deng, Z.; White, R.; Lee, J.; Rodgers, T.; et al. Diagnosing and correcting anode-free cell failure via electrolyte and morphological analysis. *Nat. Energy* **2020**, *5*, 693–702. [CrossRef]
20. Liu, Y.; Meng, X.; Wang, Z.; Qiu, J. Development of quasi-solid-state anode-free high-energy lithium sulfide-based batteries. *Nat. Commun.* **2022**, *13*, 4415. [CrossRef]
21. Qiao, Y.; Yang, H.; Chang, Z.; Deng, H.; Li, X.; Zhou, H. A high-energy-density and long-life initial-anode-free lithium battery enabled by a Li_2O sacrificial agent. *Nat. Energy* **2021**, *6*, 653–662. [CrossRef]
22. Lin, L.; Qin, K.; Zhang, Q.; Gu, L.; Suo, L.; Hu, Y.; Li, H.; Huang, X.; Chen, L. Li-rich $Li_2[Ni_{0.8}Co_{0.1}Mn_{0.1}]O_2$ for anode-free lithium metal batteries. *Angew. Chem. Int. Ed.* **2021**, *60*, 8289–8296. [CrossRef] [PubMed]
23. Lin, L.; Qin, K.; Li, M.; Hu, Y.-s.; Li, H.; Huang, X.; Chen, L.; Suo, L. Spinel-related $Li_2Ni_{0.5}Mn_{1.5}O_4$ cathode for 5-V anode-free lithium metal batteries. *Energy Storage Mater.* **2021**, *45*, 821–827. [CrossRef]
24. Louli, A.J.; Genovese, M.; Weber, R.; Hames, S.; Logan, E.; Dahn, J. Exploring the impact of mechanical pressure on the performance of anode-free lithium metal cells. *J. Electrochem. Soc.* **2019**, *166*, A1291–A1299. [CrossRef]
25. Zhou, C.; Samson, A.; Garakani, M.; Thangadurai, V. Communication—Anode-free lithium metal batteries: A case study of compression effects on coin cell performance. *J. Electrochem. Soc.* **2021**, *168*, 060532. [CrossRef]
26. Genovese, M.; Louli, A.; Weber, R.; Martin, C.; Taskovic, T.; Dahn, J. Hot formation for improved low temperature cycling of anode-free lithium metal batteries. *J. Electrochem. Soc.* **2019**, *166*, A3342–A3347. [CrossRef]
27. Abrha, L.H.; Nikodimos, Y.; Weldeyohannes, H.; Hagos, T.; Wang, D.-Y.; Huang, C.-J.; Jiang, S.-K.; Wu, S.-H.; Su, W.-N.; Tsai, M.-C.; et al. Effects of a thermally electrochemically activated β-PVDF fiber on suppression of Li dendrite growth for anode-free batteries. *ACS Appl. Energy Mater.* **2021**, *4*, 3240–3248. [CrossRef]
28. Louli, A.J.; Coon, M.; Genovese, M.; Degooyer, J.; Eldesoky, A.; Dahn, J. Optimizing cycling conditions for anode-free lithium metal cells. *J. Electrochem. Soc.* **2021**, *168*, 020515. [CrossRef]
29. Chen, X.R.; Zhao, B.; Yan, C.; Zhang, Q. Review on Li deposition in working batteries: From nucleation to early growth. *Adv. Mater.* **2021**, *33*, 2004128. [CrossRef]
30. Zhu, Y.; Pande, V.; Li, L.; Wen, B.; Pan, M.; Wang, D.; Ma, Z.-F.; Viswanathan, V.; Chiang, Y.-M. Design principles for self-forming interfaces enabling stable lithium-metal anodes. *Proc. Natl. Acad. Sci. USA* **2020**, *117*, 27195–27203. [CrossRef]

31. Hao, F.; Verma, A.; Mukherjee, P. Mechanistic insight into dendrite–SEI interactions for lithium metal electrodes. *J. Mater. Chem. A* **2018**, *6*, 19664–19671. [CrossRef]
32. Zhu, W.; Demers, H.; Girard, G.; Clement, D.; Zimin, F.; Guerfi, A.; Trudeau, M.; Vijh, A.; Paolella, A. Monitoring lithium metal plating/stripping in anode free//NMC811 battery by in-situ X-rays diffraction. *J. Power Sources* **2022**, *546*, 231941. [CrossRef]
33. Wang, C.; Wang, A.; Ren, L.; Guan, X.; Wang, D.; Dong, A.; Zhang, C.; Li, G.; Luo, J. Controlling Li ion flux through materials innovation for dendrite-free lithium metal anodes. *Adv. Funct. Mater.* **2019**, *29*, 1905940. [CrossRef]
34. Lee, D.; Sun, S.; Kwon, J.; Park, H.; Jang, M.; Park, E.; Son, B.; Jung, Y.; Song, T.; Paik, U. Copper nitride nanowires printed Li with stable cycling for Li metal batteries in carbonate electrolytes. *Adv. Mater.* **2020**, *32*, 1905573. [CrossRef] [PubMed]
35. Ding, F.; Xu, W.; Graff, G.; Zhang, J.; Sushko, M.; Chen, X.; Shao, Y.; Engelhard, M.; Nie, Z.; Xiao, J.; et al. Dendrite-free lithium deposition via self-healing electrostatic shield mechanism. *J. Am. Chem. Soc.* **2013**, *135*, 4450–4456. [CrossRef] [PubMed]
36. Zheng, X.; Huang, L.; Ye, X.; Zhang, J.; Min, F.; Luo, W.; Huang, Y. Critical effects of electrolyte recipes for Li and Na metal batteries. *Chem* **2021**, *7*, 2312–2346. [CrossRef]
37. Liu, Y.; Zhang, S.; Qin, X.; Kang, F.; Chen, G.; Li, B. In-plane highly dispersed Cu_2O nanoparticles for seeded lithium deposition. *Nano Lett.* **2019**, *19*, 4601–4607. [CrossRef]
38. Yan, C.; Cheng, X.; Yao, Y.; Shen, X.; Li, B.; Li, W.; Zhang, R.; Huang, J.; Li, H.; Zhang, Q. An armored mixed conductor interphase on a dendrite-free lithium-metal anode. *Adv. Mater.* **2018**, *30*, 1804461. [CrossRef]
39. Wan, G.; Guo, F.; Li, H.; Cao, Y.; Ai, X.; Qian, J.; Li, Y.; Yang, H. Suppression of dendritic lithium growth by in situ formation of a chemically stable and mechanically strong solid electrolyte interphase. *ACS Appl. Mater. Interfaces* **2018**, *10*, 593–601. [CrossRef]
40. Rahimi, M.G.; Wang, A.; Ma, G.; Han, N.; Chen, Y. A one-pot synthesis of a monolithic Cu_2O/Cu catalyst for efficient ozone decomposition. *RSC Adv.* **2020**, *10*, 40916–40922. [CrossRef]
41. Zhang, R.; Li, X.; Ni, L.; Xie, A.; Li, P.; Shen, Y.; Lao, L. Octagonal flower-like CuO/C/NF nanocomposite as a self-supporting anode for high-performance lithium-ion batteries. *ChemElectroChem* **2020**, *7*, 4038–4046. [CrossRef]
42. Adams, B.D.; Carino, E.; Connell, J.; Han, K.; Cao, R.; Chen, J.; Zheng, J.; Li, Q.; Mueller, K.; Henderson, W.; et al. Long term stability of Li-S batteries using high concentration lithium nitrate electrolytes. *Nano Energy* **2017**, *40*, 607–617. [CrossRef]
43. Hwang, J.-Y.; Park, S.-J.; Yoon, C.; Sun, Y.-K. Customizing a Li–metal battery that survives practical operating conditions for electric vehicle applications. *Energy Environ. Sci.* **2019**, *12*, 2174–2184. [CrossRef]
44. Yang, G.; Li, Y.; Liu, S.; Zhang, S.; Wang, Z.; Chen, L. LiFSI to improve lithium deposition in carbonate electrolyte. *Energy Storage Mater.* **2019**, *23*, 350–357. [CrossRef]
45. Mauger, A.; Julien, C.; Paolella, A.; Armand, M.; Zaghib, K. A comprehensive review of lithium salts and beyond for rechargeable batteries: Progress and perspectives. *Mater. Sci. Eng. R* **2018**, *134*, 1–21. [CrossRef]
46. Xu, H.; Han, C.; Li, W.; Li, H.; Qiu, X. Quantification of lithium dendrite and solid electrolyte interphase (SEI) in lithium-ion batteries. *J. Power Source* **2022**, *529*, 231219. [CrossRef]

molecules

MDPI

Review

Advances and Challenges in Electrolyte Development for Magnesium–Sulfur Batteries: A Comprehensive Review

Lin Sheng [1], Junrun Feng [2,*], Manxi Gong [3], Lun Zhang [3], Jonathan Harding [4], Zhangxiang Hao [2,*] and Feng Ryan Wang [3,*]

1 School of Mechanical and Electronic Engineering, Suzhou University, Suzhou 234000, China; shenglin@ahszu.edu.cn
2 School of Science, School of Chip Industry, Hubei University of Technology, Wuhan 430068, China
3 Materials and Catalysis Laboratory, Department of Chemical Engineering, University College London, London WC1E 7JE, UK; m.gong@ucl.ac.uk (M.G.); lun.zhang.19@ucl.ac.uk (L.Z.)
4 Department of Electrical Engineering and Electronics, University of Liverpool, Liverpool L69 3GJ, UK; jonathan.harding2@liverpool.ac.uk
* Correspondence: fengjunrun@hbut.edu.cn (J.F.); haozx@hbut.edu.cn (Z.H.); ryan.wang@ucl.ac.uk (F.R.W.)

Abstract: Magnesium–sulfur batteries are an emerging technology. With their elevated theoretical energy density, enhanced safety, and cost-efficiency, they have the ability to transform the energy storage market. This review investigates the obstacles and progress made in the field of electrolytes which are especially designed for magnesium–sulfur batteries. The primary focus of the review lies in identifying electrolytes that can facilitate the reversible electroplating and stripping of Mg^{2+} ions whilst maintaining compatibility with sulfur cathodes and other battery components. The review also addresses the critical issue of managing the shuttle effect on soluble magnesium polysulfide by looking at the innovative engineering methods used at the sulfur cathode's interface and in the microstructure design, both of which can enhance the reaction kinetics and overall battery efficiency. This review emphasizes the significance of reaction mechanism analysis from the recent studies on magnesium–sulfur batteries. Through analysis of the insights proposed in the latest literature, this review identifies the gaps in the current research and suggests future directions which can enhance the electrochemical performance of Mg-S batteries. Our analysis highlights the importance of innovative electrolyte solutions and provides a deeper understanding of the reaction mechanisms in order to overcome the existing barriers and pave the way for the practical application of Mg-S battery technology.

Keywords: magnesium–sulfur; battery; electrolyte; non-nucleophilic; nucleophilic

1. Introduction

The successful application of rechargeable Li-ion batteries (LIBs) has lead to the dramatic development of portable electronic devices, electric vehicles, and large-scale energy storage systems in recent decades as the world attempts to tackle increasingly intensive climate and environmental problems [1–4]. However, LIBs are failing to catch up with the ever-growing safety and energy density demands of emerging applications, such as electric vehicles with long endurances and unmanned aerial vehicles [3,5,6]. The high price of the corresponding materials for LIBs also introduces concerns with the requirement for new rechargeable battery systems to have an economical cost alongside their high-energy density to achieve a sustainable future [7–10]. In recent decades, magnesium batteries have attracted growing interest as a promising candidate for post-lithium-ion battery systems [11–15]. Divalent Mg^{2+} enables two electron transfers per Mg atom, resulting in a high theoretical specific capacity of 2205 mAh g^{-1} [16–18]. Considering the density of magnesium, the Mg anode enables a high volumetric capacity of 3833 mAh mL^{-1}, which is almost double that of lithium (2062 mAh mL^{-1}) [17,19–22]. Moreover, a Mg anode displays advantages such as low cost, high Earth abundance, and easy operation under air [23–25].

In spite of these advantages, the use of Mg as an anode in rechargeable batteries is still hampered by the limited choice of electrolytes and cathode materials in view of facile passivation of the Mg's surface and the sluggish solid-state diffusion of the highly polar divalent cations in the lattices [26–31]. Up until now, pure Mg-ion systems based on the intercalation mechanism have had a limited energy density, usually less than 300 Wh kg^{-1} without counting the weight of the solvent [32–34]. Therefore, it is highly desirable to develop novel and safe systems and materials based on Mg^{2+}/Mg metal conversion chemistry with a higher energy density, where anode dendrite growth can be effectively suppressed [35–37]. Sulfur is one of the most promising candidates for conversion cathodes because of its high theoretical capacity (1672 mAh g^{-1}) and volumetric capacity (3459 mAh mL^{-1}), as well as its reserve abundance [38–41]. Sulfur cathodes have been well developed in Li-S battery systems. The cell exhibits a high theoretical volumetric energy density of 2856 Wh kg^{-1}, which is over four times higher than that of LIBs [42,43]. However, Li-S batteries suffer from poor long-term stability due to the formation of unwanted solid electrolyte interphases (SEIs), shuttle effects, and the growth of lithium dendrites during their operation [44–46]. Compared to the intensive research process for Li-S batteries, Mg-S batteries show a higher volumetric capacity of 3221 mAh mL^{-1} but they are still in their nascent stage [19,47]. Development has been impeded due to the numerous issues present in the system. Firstly, suitable electrolytes are required. Commonly used Mg-ion battery electrolytes such as magnesium perchlorate acetonitrile are nucleophilic in nature and therefore cannot support the reversible Mg-ion redox reaction in a Mg-S battery [23]. Numerous efforts have been made to find compatible electrolytes for the system. For example, Muldoon et al. reported on a Hauser base electrolyte that is produced according to the reaction of $AlCl_3$ with hexamethyldisilazide magnesium chloride (HMDSMgCl) in a tetrahydrofuran (THF) solvent to form a non-nucleophilic $[Mg_2(\mu\text{-}Cl)_3 6THF][HMDSAlCl_3]$ complex, which showed a high voltage stability of up to 3.2 V [48]. Furthermore, electrolyte additives such as salts, ethereal solvents, and ionic liquid additives have been used to improve the performance of the electrolyte used by Muldoon et al. above [49–51]. Non-nucleophilic electrolytes are pivotal in augmenting the stability and efficiency of Mg-S batteries. Researchers have developed both chloride-inclusive and chloride-exempt non-nucleophilic electrolyte systems. For example, chloride-inclusive hexamethyldisilazane magnesium chloride (HMDSMgCl)-based non-nucleophilic electrolytes have exhibited favorable compatibility with sulfur cathodes [52]. Moreover, notwithstanding the typical incompatibility of nucleophilic organic magnesium compounds with sulfur-based conversion cathodes, scholars have effectively modified nucleophilic $(PhMgCl)_2\text{-}AlCl_3$/THF electrolytes to conform to the needs of Mg-S batteries [53,54]. These systems benefit from their similarity to the sulfur reduction mechanism in Li-S batteries, which means it is possible to try to use the same base cathode materials [44,55–57], electrolytes [58–60], separators [61–63], and methodologies [64,65] for the development of Mg-S batteries.

The purpose of this review is to summarize the most up-to-date understanding in the field of electrolytes for Mg-S batteries, highlighting the typical methods and examples that have contributed to the development of these electrolytes. Furthermore, since the research into Mg-S batteries is still in its early stages, the principle of Mg-S batteries will also be discussed, as it has not yet been analyzed in the results of a definitive study. Finally, the remaining challenges and future perspectives will be given in an attempt to inspire researchers in this area.

2. Principle of Mg-S Batteries

The construction of a Mg-S battery (Figure 1a) comprises a magnesium anode, a sulfur cathode, and electrolytes, which together illustrate the structure and working principle of the battery. The magnesium atom loses two electrons during discharge, becomes Mg^{2+} at the anode, and dissolves in the electrolyte. The Mg^{2+} migrates through the cell separator and reaches the sulfur on the electrolyte's cathode side surface [66]. The electrons then proceed to the sulfur cathode through an external connection, where elemental sulfur generates electrons. In the sulfur cathode's continuous reduction process, elemental sulfur is transformed into long-chain magnesium polysulfide (MgS_x, $4 \leq x \leq 8$), which dissociates into short-chain magnesium polysulfide (MgS_x, $2 \leq x < 4$). This short-chain magnesium polysulfide is then transformed into magnesium sulfide (MgS) [55,67–69]. The external circuit applies a current or voltage to the Mg-S battery during the charging process. In the MgS, S^{2-} loses electrons and transforms into elemental sulfur, followed by the dissolution of Mg^{2+} in the electrolyte. The magnesium ions in the electrolyte migrate from the cathode to the magnesium anode, driven by the current, which completes the Mg-S battery's discharge/charge cycle. In the anode of magnesium, Mg is oxidized during the discharging process. The cathode is restored during the charging process to complete the battery's internal migration and forms a complete battery path [26,56,70]. The main equations for the reactions are as follows:

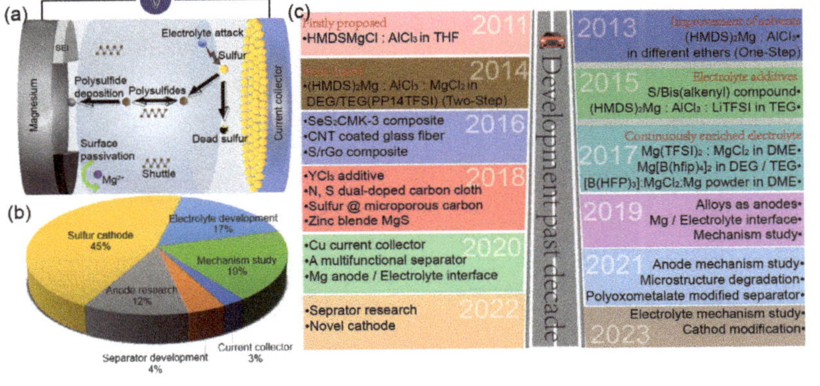

Figure 1. (**a**) Structure and working principles of Mg-S battery; (**b**) proportional representation of research directions, including electrolytes, sulfur-based cathodes, magnesium anodes, separators, current collectors, and mechanistic studies; (**c**) significant advancements in the Mg-S battery field since 2011.

Magnesium anode: $Mg - 2e^- = Mg^{2+}$,
Sulfur cathode: $S + 2e^- + Mg^{2+} = MgS$,
Total response: $Mg + S = MgS$.

So far, the research on Mg-S batteries has focused on the cathode, anode, and electrolyte (Figure 1b). The materials of the anode and cathode in Mg-S batteries include the following types:

Anode: Mg metal is the most commonly used anode material for Mg-S batteries; depending on the type of battery, its form may vary between foils [17,67,71,72], plates [51,73], and discs [74–77]. In addition to Mg metal, Mg–carbon composite materials are also used as anode modifiers [55,78,79]. The carbon provides the anode with good electronic conductivity and a high surface area, meaning a higher capacity and current rate capacity [80].

Cathode: The material of the cathode is a primary research priority for Mg-S battery performance enhancement [81]. The composition of the cathode materials mainly includes S/carbon and S/MOF. The role of the carbon [17,52,82–84] and Metal–Organic Frameworks (s) [72] in the cathode is mainly that of a conductive agent, due to the low electrical conductivity of sulfur. Since 2014, when Zhao-Karger et al. used CMK-3 [55], a mesoporous carbon material, as the conductive agent and as a container to fix the sulfur and mitigate the shuttle effect [85], increasing numbers of researchers are now implementing mesoporous-structure conductive agents in the cathode material of Mg-S batteries [54,86,87].

Furthermore, the study of electrolytes also accounts for a significant proportion of the research, and this is the focus of this review.

3. Recent Developments in Electrolytes

The performance of Mg-S batteries is significantly influenced by the characteristics of the electrolyte used. In recent years, there has been an increasing focus on research pertaining to the electrolytes and mechanisms of Mg-S batteries (Figure 1b,c). This shift in focus is attributed to the realization among researchers that advancements in the electrode materials alone are insufficient to significantly enhance the electrochemical performance of Mg-S batteries. Only by elucidating the working mechanism of these batteries, addressing the shuttle effect, and enhancing the mass transfer efficiency can Mg-S batteries progress beyond the laboratory stage [88–90]. The ideal electrolyte for Mg-S batteries should facilitate the reversible deposition and stripping of magnesium at the anode and enable the efficient utilization of sulfur at the cathode [59,61,91]. Based on the chemical reactions, they can be classified into two categories: non-nucleophilic and nucleophilic electrolytes. A comparison of the key performance statistics of notable nucleophilic and non-nucleophilic electrolytes is shown at the end of Table 1.

Table 1. Electrolyte systems in Mg-S batteries (according to electrolyte type and in chronological order).

Author/Year	Electrolyte Type	Solute	Solvent	Additives	Coulombic Efficiency [%]	Capacity [mAh g^{-1}sulfur]/ Current Rate/ Cycle Number
H.S. Kim et al., 2011 [52]	Non-nucleophilic Cl-containing	HMDSMgCl	THF	$AlCl_3$	95–100	394/no data/2nd
Zhao-Karger, Z et al., 2013 [92]		$(HMDS)_2Mg$/ $(i\text{-}Pr_2N)_2Mg$	THF/Diglyme/ Tetraglyme	$AlCl_3$	97–98	90/10 mA g^{-1}/30th
Zhao-Karger et al., 2014 [55]		$(HMDS)_2Mg$	Diglyme/ Tetraglyme	$AlCl_3$+ $PP_{14}TFSI$	100	150/0.01 C/20th (PVDF, diglyme) 200/0.01 C/20th (CMC, diglyme) 250/0.01 C/20th (PVDF, tetraglyme) 260/0.01 C/20th (CMC, tetraglyme)
Gao et al., 2015 [17]		$(HMDS)_2Mg$	No Data	$AlCl_3$ + LiTFSI	92	1000/0.03 C/30th
Du et al., 2017 [71]		$B(HFP)_3$/OMBB	DME	$MgCl_2$	80.4% (100th)	1000/0.1 C/100th
Zhao et al., 2019 [93]		Magnesium bis(diisopropyl) amide	THF	$AlCl_3$ + LiCl	94	400/0.04 C/100th
Yang et al., 2018 [53]		$Mg(CF_3SO_3)_2$ + anthracene	THF + Tetraglyme	$AlCl_3$ + LiCl/ $LiCF_3SO_3$	100	300/0.05 C/55th 400/0.05 C/55th
Sun et al., 2021 [94]		$Mg(TFSI)_2$	DME	$MgCl_2$ + rPDI	99.4 98	110 (1 mg cm^{-2} loading)/15 C/1000th 100 (10 mg cm^{-2} loading)/1 C/200th
Xu et al., 2019 [95]	Non-nucleophilic Cl-free	$Mg(BPh_4)_2$	PYR14TFSI	YCl_3	98.7	1000/0.04 C/50th
Li et al., 2016 [77]		$[Mg(THF)_6]^{2+}$	PYR14TFSI + THF	No data	100	63/0.02 C/20th
Zhao-Karger, Z et al., 2017 [51]		$Mg[B(hfip)_4]_2$	DME + TEG	No data	100	200/0.1 C/100th
Zhao-Karger, Z et al., 2018 [84]		$Mg[B(hfip)_4]_2$	DME	No data	100	200/0.1 C/100th
Zhang et al., 2017 [50]		THFPB	DME	MgF_2	100	900/0.03 C/30th
Ren et al., 2021 [60]		MBA + AlF_3	THF	LiTFSI + $PP_{14}TFSI$	100	260/0.2 C/70th

Table 1. *Cont.*

Author/Year	Electrolyte Type	Solute	Solvent	Additives	Coulombic Efficiency [%]	Capacity [mAh g^{-1}sulfur]/ Current Rate/ Cycle Number
Zeng et al., 2017 [96]	Nucleophilic	$(PhMgCl)_2$	THF	$AlCl_3$	100	300/0.005 C/40th
Wang et al., 2018 [54]		$(PhMgCl)_2$	THF	$AlCl_3$	100	368/0.1 C/200th

3.1. Non-Nucleophilic Electrolytes

Owing to the electrophilic nature of elemental sulfur, it reacts with the nucleophilic substances in nucleophilic electrolytes, adversely affecting the stability and efficiency of batteries [52,73,74,77]. Therefore, in the design and application of Mg-S batteries and other sulfur-based batteries, researchers predominantly employ electrolytes with non-nucleophilic properties. In this section, there are two types of non-nucleophilic electrolyte systems in Mg-S batteries, which are chloride-containing and chloride-free.

3.1.1. Chloride-Containing

A moderate concentration of Cl^- is regarded as beneficial, not only for stabilizing Mg^{2+} but also for dissolving the passivating species on the Mg anode, thereby enhancing the Mg plating/stripping process. A hexamethyldisilazide magnesium chloride (HMDSMgCl)-based non-nucleophilic electrolyte, which exhibits good compatibility with sulfur cathodes, was synthesized in 2011 by H.S. Kim et al. [52]. The interaction between the HMDSMgCl electrolyte and the Lewis acid $AlCl_3$ was investigated, with the objective of enhancing the electrochemical performance. By varying the acid-to-base ratio and reaction time, it was observed that the electrolyte's electrochemical performance peaked after a 24-hour reaction period, specifically when the HMDSMgCl-to-$AlCl_3$ ratio was 3:1. As illustrated by the green and blue lines in Figure 2a, the current density for Mg deposition exhibited an approximate sevenfold increase following the addition of $AlCl_3$. However, the voltage stability of the HMDSMgCl electrolyte did not show any improvement (Figure 2b). To identify the reaction products of HMDSMgCl with $AlCl_3$, crystals were harvested through the slow diffusion of hexane. The crystal structure $[Mg_2(\mu\text{-}Cl)_3{\cdot}6THF][HMDSAlCl_3]$ (Figure 2c) was determined using single-crystal X-ray diffraction. This structure was found to feature a cation with two octahedrally coordinated Mg centers, each containing three chlorine atoms. THF molecules, through oxygen coordination, occupy the remaining three sites on each Mg center. According to the constant current charging and discharging data, the specific capacity during the first discharge was approximately 1200 mAh g^{-1}. However, the overpotential reached approximately 1.1 V, and the capacity rapidly decayed to around 395 mAh g^{-1} (Figure 2d). This study offers a novel perspective on the development of non-nucleophilic electrolytes for Mg-S batteries.

In order to synthesize the magnesium-bis-(hexamethyldisilazide) $[(HMDS)_2Mg]$ and $AlCl_3$-based non-nucleophilic electrolyte in different ethers, a number of one-step strategies were employed by Z. Z. Karger et al. in 2013 [92]. Through the means of chemical processes involving magnesium bisamide and Lewis acids in aprotic solvents, the non-nucleophilic electrolyte for magnesium batteries was synthesized, which had an excellent electrochemical performance. The in situ process-generated electrolyte possessed excellent characteristics, such as a high anode stability, excellent ionic conductivity, good cycling efficiency, and feasibility of preparation. With these advantageous properties, it holds great promise in the area of rechargeable magnesium batteries.

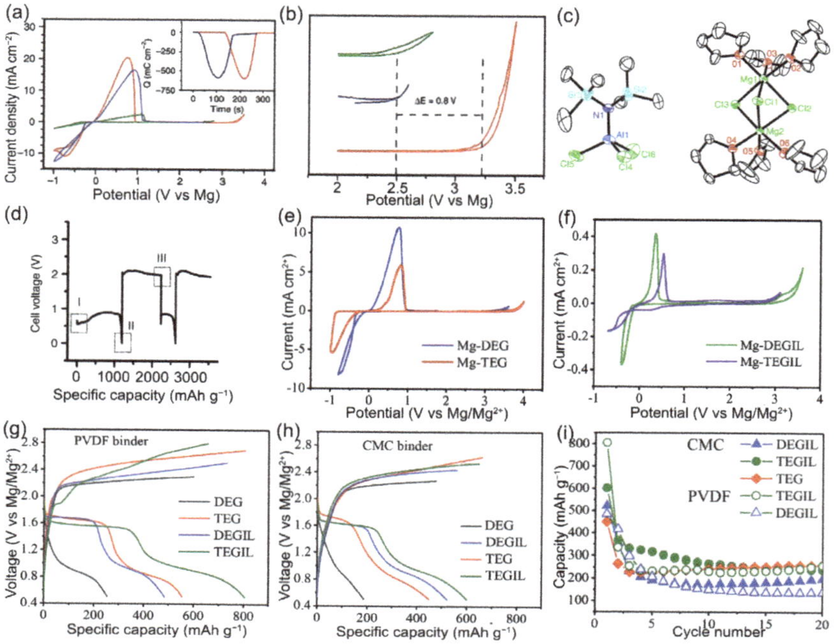

Figure 2. (**a**) Cyclic voltammograms of HMDSMgCl (green), the reaction product generated in situ from a 3:1 mixture of HMDSMgCl to $AlCl_3$ (blue), and the crystal obtained from a 3:1 mixture of HMDSMgCl to $AlCl_3$ (red). Inset shows the charge balance during the deposition and subsequent dissolution of Mg. (**b**) Enlargement of 2–3.5 V region of (**a**) highlighting the oxidative stability of the electrolytes. (**c**) ORTEP plot (25% thermal probability ellipsoids) of $[Mg_2Cl_3{\cdot}6THF][HMDSAlCl_3]$. Hydrogen atoms, THF crystallization, and second component of disorder are omitted for clarity. (**d**) Discharge and charge of a Mg-S coin cell at 50 and 25 μA, respectively. XPS spectra were taken from coin cells at various stages of cycling [52]. Cyclic voltammograms of the electrolyte in (**e**) diglyme solution (blue) and tetraglyme solution (red) and (**f**) diglyme/PP_{14}TFSI solution (green) and tetraglyme/PP_{14}TFSI solution (purple), using Pt as electrode at a scan rate of 25 mV s^{-1}. Initial discharge–charge curves of S/CMK400PEG composite using (**g**) PVDF binder and (**h**) CMC binder in the electrolyte in diglyme (gray), tetraglyme (red), diglyme/PP_{14}TFSI (blue), tetraglyme/PP_{14}TFSI (green). (**i**) Cycling performance of S/CMK400PEG cathode in the electrolyte in diglyme/PP_{14}TFSI (denoted as DEGIL in blue), in tetraglyme/PP_{14}TFSI (denoted as TEGIL in green), and in tetraglyme (denoted as TEG in red) using CMC and PVDF as binders, respectively [55]. (**a**–**d**) Copyright © 2011, Hee Soo Kim et al. (**e**–**i**) © Copyright 2014 WILEY-VCH Verlag GmbH & Co. KGaA, Weinheim.

To advance the performance of Mg-S batteries, Z. Z. Karger et al. explored a novel preparation method for non-nucleophilic electrolyte solutions using a two-step reaction in one pot [55]. This study initially employed $(HMDS)_2Mg$-based diglyme and tetraglyme electrolyte solutions for constructing Mg-S batteries, integrating *N*-methyl-*N*-butylpiperidinium bis(trifluoromethanesulfonyl)imide (PP_{14}TFSI) as an additive into these electrolytes. During the testing phase with a sulfur cathode (S/CMK), the electrolytes based on diglyme and tetraglyme demonstrated distinct capacities of 250 and 550 mAh g^{-1}, respectively. However, the Mg-S batteries exhibited low-capacity retention, with a discharge potential of 1.65 V. Furthermore, the study also assessed the role of the ionic liquid (IL) PP_{14}TFSI as a cosolvent in the electrolyte. Utilizing a PVDF binder and a tetraglyme/PP_{14}TFSI solution, the batteries initially delivered approximately 800 mAh g^{-1} in the first cycle, but this value was significantly reduced to around 350 mAh g^{-1} in the subsequent cycle. Conversely, the batteries containing TEGIL (tetraglyme and $PP_{14}TFSI$) with binders of either PVDF or CMC maintained a stable reversible capacity of about 260 mAh g^{-1} after over 20 cycles (Figure 2e–i). This study elucidated that the electrochemical conversion of magnesium and sulfur demonstrates a fundamentally unique battery chemistry compared to Li-S systems. It was observed that the significant hysteresis between the discharge and charge voltages during cell cycling contributes to the capacity degradation of the batteries. This type of electrolyte has garnered considerable interest among researchers, with scholars like Vinayan et al. [79], Yu et al. [82], and many other researchers also utilizing it to investigate Mg-S batteries.

Previous research on lithium-ion and lithium–sulfur batteries has demonstrated that the physicochemical properties of electrolytes can be effectively enhanced with the addition of electrolyte additives. Furthermore, these additives can have a positive impact on the electrochemical performance of battery systems. In 2015, Gao et al. introduced a novel strategy to augment the reversibility of Mg-S chemistry [17]. A non-nucleophilic Mg electrolyte, supplemented with LiTFSI additives, facilitates the integration of a reversible polysulfide redox process into the cathode with Mg deposition/stripping at the anode. Galvanostatic charge–discharge tests revealed that the specific capacity of the primary magnesium electrolytes was approximately 650 mAh g^{-1}. Notably, the electrolyte containing LiTFSI exhibited a reversible capacity of around 1000 mAh g^{-1} at 71 mA g^{-1}, with a stable voltage plateau at 1.75 V, maintaining stability for 30 cycles (Figure 3a–e). This enhanced reversibility is attributed to two main factors: (1) Li^+ ions participating in the cathode reaction, either forming readily rechargeable Li polysulfide (Li-PS) or integrating into the Mg-PS to create hybrid Mg/Li polysulfide (MgLi-PS) during discharge. (2) The hard Lewis acid characteristic of Li^+ coordinating strongly with the surface S^{2-} of the lower-order Mg-PS, thereby increasing its solubility, reducing the reoxidation energy barrier, and rendering it electrochemically active. A reversible capacity of 1000 mAh g^{-1} is one of the highest shown in the Mg-S battery research so far, as shown in Table 1.

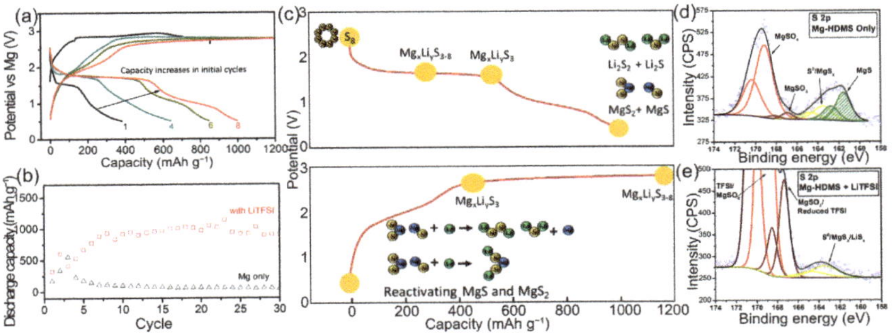

Figure 3. (**a**) Charge/discharge curves of sulfur cathode in 0.1 M Mg-HMDS + 1.0 M LiTFSI electrolyte in a three-electrode cell at a current of 71 mAh g^{-1} at room temperature. Arrow illustrates the capacity-increasing trend of the ACC/S composite cathode as a result of slow electrolyte penetration. (**b**) Cycling stability of the Mg-S battery in electrolyte with and without LiTFSI. (**c**) Working mechanism of the Mg-S battery with LiTFSI additive. Comparison of surface XPS measurements of Mg anode cycled in Mg-HDMS in the absence (**d**) and presence (**e**) of LiTFSI [17]. Copyright © 2015, American Chemical Society.

In 2017, Du et al. proposed an organic magnesium borate-based (OMBB) electrolyte, predominantly comprising a tetrakis(hexafluoroisopropyl) borate anion $[B(HFP)_4]^-$ (Figure 4a) and a solvating cation $[Mg_4Cl_6(DME)_6]^{2+}$ [71]. This electrolyte was synthesized using a simple in situ process involving tris(hexafluoroisopropyl)boronic acid $[B(HFP)_3]$, $MgCl_2$, and Mg powder in 1,2-dimethoxyethane (DME). The overpotential was approximately 0.07 V at a current density of 0.1 mA cm^{-2} and increased marginally, with a rise in the current density from 0.1 mA cm^{-2} to 1 mA cm^{-2} (Figure 4b). Various sulfur–carbon composite materials (S-AMC, S-CNT, and S-CMK), prepared using the melt diffusion method, were employed to evaluate the OMBB electrolyte's compatibility. Among these sulfur–carbon composite cathodes, tested at a current density of 160 mA g^{-1}, the highest specific discharge capacity achieved was 1247 mAh g^{-1} (Figure 4c). Additionally, the S-CNT cathode exhibited a capacity retention rate of 80.4% after 100 cycles. The specific capacity of the sulfur–carbon composite cathodes increased initially, which may be attributed to the 0.5 M OMBB electrolyte's self-conditioning effect, as it gradually permeated into the sulfur–carbon composite cathode. Both the discharge and charge curves feature two distinct voltage plateaus (Figure 4d). Remarkably, the Mg-S-CNT battery maintained a discharge capacity of about 500 mAh g^{-1} even at a current rate of 500 mA g^{-1}. Electrochemical impedance spectroscopy (EIS) measurements (Figure 4f) were conducted to investigate the reasons for the initial increase in specific capacity. These measurements revealed a sharp decrease in the charge transfer resistance (R_{ct}) during the initial cycles, elucidating the enhanced discharge capacity observed.

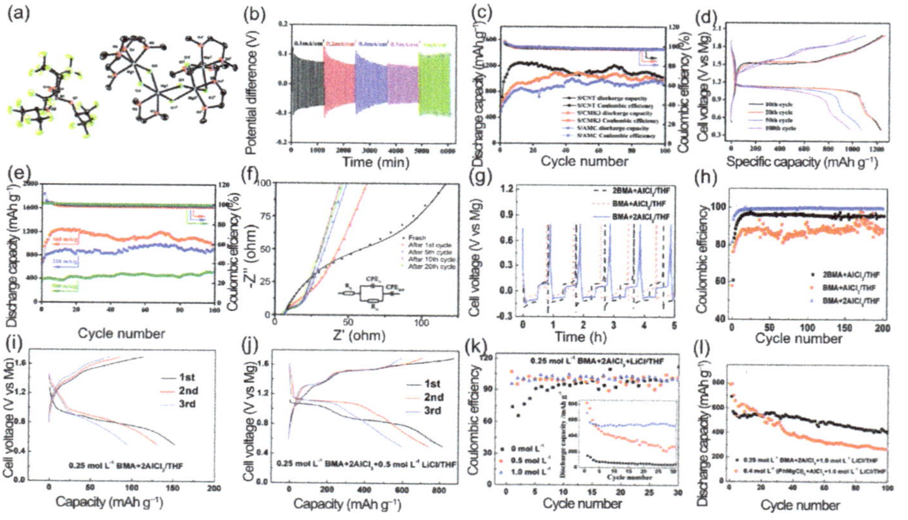

Figure 4. (**a**) ORTEP plot (50% thermal probability ellipsoids) of the molecular structure of crystalline $[Mg_4Cl_6(DME)_6][B(HFP)_4]_2$. Hydrogen atoms are omitted for clarity. (**b**) Polarization properties of Mg/Mg symmetrical cells with the 0.5 M OMBB electrolyte at current densities of 0.1, 0.2, 0.4, 0.5, and 1 mA cm^{-2}; the cycling time was 1 h per cycle (30 min charging and 30 min discharging). (**c**) Discharge capacities and Coulombic Efficiency as a function of the cycle number for different S-C composite cathodes at a current rate of 160 mA g^{-1} in the 0.5 M OMBB electrolyte. (**d–f**) Electrochemical characterization of the Mg-S-CNT battery in the 0.5 M OMBB electrolyte: (**d**) galvanostatic discharge–charge profiles for different cycles at a current rate of 160 mA g^{-1}; (**e**) discharge capacities and Coulombic Efficiency at different charge–discharge current rates; (**f**) EIS measurements after different cycles (charge–discharge at a current rate of 160 mA g^{-1}). The colored points represent the results of the tests, and the colored lines are the corresponding fitting curves [71]. The cycling curves (**g**) and cycling efficiency (**h**) of Mg plating/stripping on the SS substrate from 0.25 mol L^{-1} MBA–$AlCl_3$/THF electrolytes with 2:1, 1:1, and 1:2 MBA-to-$AlCl_3$ molar ratios. Discharge–charge profiles at 0.04C of the S@MC | Mg coin cell with 0.25 mol L^{-1} BMA + $_2AlCl_3$/THF electrolyte (**i**) and 0.25 mol L^{-1} BMA + $2AlCl_3$ + 0.5 mol L^{-1} LiCl/THF electrolyte (**j**). Cycling performance (inset) and Coulombic Efficiency at 0.04C of S@MC | Mg coin cells with 0.25 mol L^{-1} BMA + $2AlCl_3$/THF electrolytes containing LiCl at different concentrations (**k**). The cycling performance at 0.04 C of S@MC | Mg coin cells with 0.4 mol L^{-1} $(PhMgCl)_2$ + $AlCl_3$/THF and 0.25 mol L^{-1} BMA + $2AlCl_3$/THF electrolytes containing 1.0 mol L^{-1} LiCl (**l**) [93]. Copyright © The Royal Society of Chemistry 2017.

In 2019, Zhao et al. synthesized a remarkable magnesium electrolyte through the reaction of a magnesium salt, magnesium bis(diisopropyl)amide (MBA), and $AlCl_3$ in THF [93]. For this Mg electrolyte, achieving a low overpotential and high Coulombic Efficiency during long-term cycling for Mg electrochemical plating/stripping is paramount. From the initial cycle, the overpotentials for Mg plating/stripping in the three electrolytes were consistently below −0.2 V and 0.1 V, respectively (Figure 4g). The corresponding Coulombic Efficiency (Figure 4h) is based on the ratio of the charge amount for magnesium plating to that of magnesium stripping. Notably, when the concentration of the MBA + $2AlCl_3$/THF electrolyte is 0.25 mol L^{-1}, it exhibits the highest and most stable Coulombic Efficiency. The Coulombic Efficiency is, however, lower than other notable electrolytes, as demonstrated in Table 1.

To verify the compatibility of the electrolyte with the sulfur cathode, researchers assembled a coin cell using a 0.25 mol L^{-1} BMA + $2AlCl_3$/THF electrolyte, a S@MC cathode, and a Mg anode. This cell provided an initial discharge capacity of 152.0 mAh g^{-1} and a charge capacity of 112.8 mAh g^{-1}, with a Coulombic Efficiency of 74.2%, (Figure 4i). The existing research indicates that incorporating lithium ions into magnesium electrolytes typically activates the electrochemistry of magnesium–sulfur battery systems. The first three galvanostatic discharge–charge curves of the S@MC | Mg coin cell with a 0.25 mol L^{-1} BMA + $2AlCl_3$ + 0.5 mol L^{-1} LiCl/THF electrolyte at 0.04C show an initial discharge capacity of about 815.6 mAh g^{-1} (Figure 4j). The capacities for the second and third cycles are approximately 747.5 mAh g^{-1} and 602.2 mAh g^{-1}, respectively. The presence of Li^+ has notably enhanced the reversibility of the sulfur cathode. Additionally, the inclusion of LiCl significantly improves the cycling stability (Figure 4k). The Coulombic Efficiency of S@MC | Mg coin cells with 0.25 mol L^{-1} BMA + $2AlCl_3$/THF electrolytes containing varying concentrations of LiCl are displayed in Figure 4k. The Coulombic Efficiency's stability improves with an increasing LiCl concentration, especially at 1.0 mol L^{-1}, where the charge capacity gradually aligns with, but does not exceed, the discharge capacity. These results suggest that the cathode's reversibility is contingent on the presence of Li^+. The enhanced interfacial compatibility and improved electrochemical performance may result from increased solution conductivity (1144 and 1185 μS cm^{-1} for 0.5 and 1.0 mol L^{-1} LiCl, respectively). The S@MC | Mg coin cell with a 0.25 mol L^{-1} BMA + $2AlCl_3$ + 1.0 mol L^{-1} LiCl/THF electrolyte demonstrates better cycling stability than one with a 0.4 $(PhMgCl)_2$ + $AlCl_3$ + 1.0 mol L^{-1} LiCl/THF electrolyte (Figure 4l). After 100 cycles, the capacity retention is approximately 57.7% and 34.4% of the initial capacity, respectively. However, this experiment could be considered controversial, as the Mg-S battery's performance gains also correlate with the increasing concentration of lithium ions. It is important to consider what level of contribution the lithium ions are making to the performance and to question whether the battery can still be called a Mg-S battery and still comes with the associated benefits of a Mg-S battery compared to a Li^+ battery.

In 2019, Yang et al. introduced a novel electrolyte based on magnesium trifluoromethanesulfonate ($Mg(CF_3SO_3)_2$)–$AlCl_3$–$MgCl_2$–anthracene–LiCl dissolved in THF and tetraglyme [53]. $Mg(SO_3CF_3)_2$, which served as the source of the Mg^{2+} ions, is non-nucleophilic, easier to handle, and more cost-effective compared to $Mg(TFSI)_2$ (where TFSI = bis(trifluoromethanesulfonyl)imide). However, challenges in Mg deposition/dissolution were observed, marked by a high overpotential, attributed to an inherent oxide layer on the Mg anode. In response, inspired by transmetalation reactions, where magnesium compounds react with Lewis acids, a typical Lewis acid, $AlCl_3$ containing Cl^-, was added. This addition reacted with $Mg(CF_3SO_3)_2$ to generate effective active species within the solutions.

A Mg-S@microporous carbon cell, incorporating a 0.125 M $Mg(CF_3SO_3)_2$ + 0.25 M $AlCl_3$ + 0.25 M $MgCl_2$ + 0.025 M anthracene/THF and tetraglyme (1:1, v/v) electrolyte, was cycled at 0.05 C (Figure 5a,b). The Coulombic Efficiency of the cell initially reached almost 100% after a few cycles. However, due to substantial unreacted sulfur and inefficient Mg ion dissociation within the cathode, the discharge capacity sharply declined to just 50 mAh $g_{sulfur}{}^{-1}$ after 50 cycles.

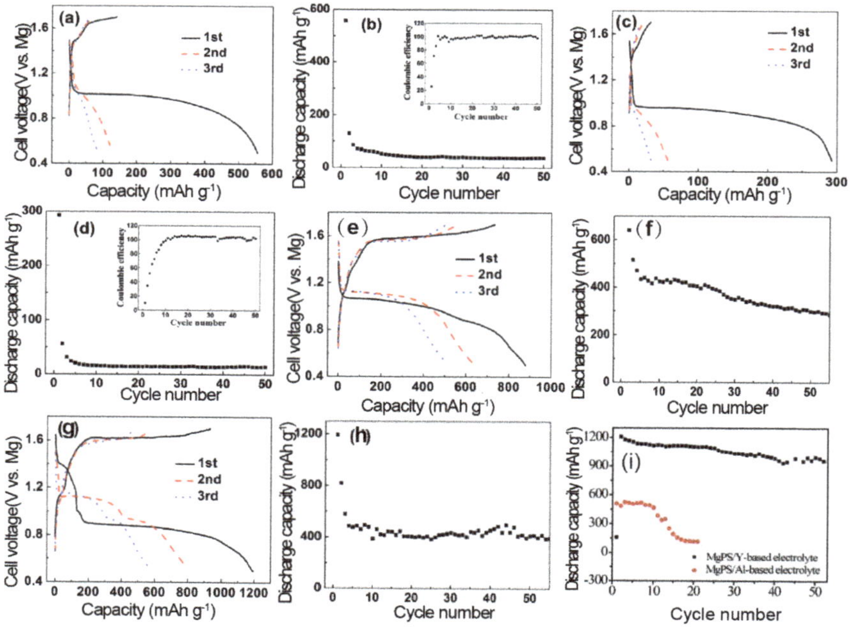

Figure 5. Discharge–charge curves and cycling performance of S@MC | Mg coin cell with 0.125 M $Mg(CF_3SO_3)_2$ + 0.25 M $AlCl_3$ + 0.25 M $MgCl_2$ + 0.025 M anthracene/THF + TG (1:1 volume ratio) electrolyte (**a**,**b**) and 0.25 M $Mg(CF_3SO_3)_2$ + 0.5 M $AlCl_3$ + 0.25 M $MgCl_2$ + 0.025 M anthracene/THF + TG (1:1 volume ratio) electrolyte (**c**,**d**) between 0.5 and 1.7 V at 0.05 C. Discharge–charge profiles and cycling performance of S@MC | Mg coin cell with 0.125 M $Mg(CF_3SO_3)_2$ + 0.25 M $AlCl_3$ + 0.25 M $MgCl_2$ + 0.025 M anthracene/THF + TG (1:1 volume ratio) electrolyte adding 0.5 M LiCl (**e**,**f**) or $LiCF_3SO_3$ (**g**,**h**) between 0.5 and 1.7 V at 0.05 C [53]. (**i**) Cycling stability of the MgPS/Y-based electrolyte and MgPS/Al-based electrolyte cells under a current density of 80 mA g^{-1} [95]. Copyright © 2019, American Chemical Society.

The previous research indicates that the issue of polysulfide shuttling in electrolytes can be mitigated by increasing the electrolyte concentration. In such high-concentration electrolytes, fewer sulfur molecules, either as elemental sulfur or polysulfides, dissolve into the electrolyte during cycling. Consequently, this approach helps in curtailing the loss of active material. Furthermore, the discharge–charge performance of a S@MC | Mg coin cell containing a concentrated electrolyte (0.25 M $Mg(CF_3SO_3)_2$ + 0.5 M $AlCl_3$ + 0.25 M $MgCl_2$ + 0.025 M anthracene/THF + TG (1:1 volume ratio)) was evaluated. This configuration not only led to a significant reduction in the specific capacity of the coin cell but also resulted in a poorer cycle performance compared to using a dilute electrolyte (Figure 5c). Notably, the Coulombic Efficiency exceeded 100% during stable cycling (Figure 5d).

The incorporation of Li^+ ions into the electrolyte, with the aim of facilitating the dissolution of Mg^{2+} ions and thereby reducing the kinetic barriers while increasing the solubility of low-order polysulfides, was undertaken to enhance the electrochemical performance. The researchers introduced LiCl (Figure 5e,f) and $LiCF_3SO_3$ (Figure 5g,h), respectively, into a 0.125 M $Mg(CF_3SO_3)_2$ + 0.25 M $AlCl_3$ + 0.25 M $MgCl_2$ + 0.025 M anthracene/THF and tetraglyme (1:1, v/v) solution. The results (Figure 5e,f) demonstrate that the addition of LiCl prolonged the discharge plateau at 1.05 V, leading to an enhanced discharge capacity and reversibility, confirming that Li^+ ions effectively promote Mg dissolution. Notably, adding 0.5 M $LiCF_3SO_3$ as an additive further improved the cell performance. The cell achieved a discharge capacity of approximately 400 mAh ${g_{sulfur}}^{-1}$ at 0.05 C over 50 cycles, suggesting that the addition of CF_3SO_3Li also mitigates the detrimental effects on Mg plating. However, this method also faces the same controversy mentioned before: can the battery still be classified as a Mg-S battery, which maintains a high level of safety and is less prone to dendrites than one with the addition of Li^+? In recent years, the majority of researchers have encountered this problem and have stopped adding Li^+ into Mg-S batteries.

It is significant to note that in 2014, Zhao-Karger et al. reported in their study the effect of augmenting the electrolyte viscosity by incorporating high-viscosity $PP_{14}TFSI$ (173 mPa s), which can partially impede the movement of polysulfides toward the anode [55]. This method demonstrates a novel strategy for mitigating the shuttle effect in Mg-S batteries. By enhancing the electrolyte's viscosity, the mobility of the polysulfides is decreased, thus curtailing their transference and subsequently reducing the adverse impacts on the battery performance.

It is noteworthy that in 2021, Sun et al. described reduced perylene diimide-ethylenediamine (rPDI) as an efficacious electrolyte additive. They added 0.2 mM rPDI into a $Mg(TFSI)_2$–$MgCl_2$-based electrolyte, which adsorbed onto the Mg and repelled the $TFSI^-$ anions away from the Mg surface, preventing $TFSI^-$ decomposition and Mg passivation [94]. The full cell showed a highly stable cycle life at 15 C (>1000 cycles), revealing the mitigation of the shuttle effect. Although the capacity is low at 110 mAh g^{-1} after 1000 cycles, it is a significant improvement over other electrolyte systems, as can be seen in Table 1, where it has the largest number of cycles reported.

In recent years, there have been significant advancements in the field of novel Lewis acids. In 2018, Xu et al. developed an yttrium (Y)-based electrolyte by replacing $AlCl_3$ with YCl_3 [95]. Their findings revealed several key advantages: firstly, the standard electrode potential of Y ions (−2.372 V vs. SHE) is higher than that of Al ions (−1.66 V vs. SHE), and secondly, YCl_3 effectively facilitates the removal of water from the electrolyte. In their synthesis, $MgCl_2$ (1 molar equivalent) and YCl_3 (2 molar equivalent) were reacted with diglyme in an ionic liquid solvent, PYR14TFSI, at a temperature of 120 °C.

To investigate the distinct impacts of the YCl_3 and $AlCl_3$ additives, Xu et al. compared the electrochemical performances of two Mg-S cells: one with a MgPS cathode in an aluminum-based electrolyte and the other with a MgPS cathode in an yttrium-based electrolyte. The cell featuring the yttrium-type electrolyte demonstrated stable cycling for 50 cycles with a discharge capacity of approximately 900 mAh g^{-1} (Figure 5i). In contrast, the cell containing the aluminum-based electrolyte only managed 20 cycles before experiencing a sharp decline in capacity. Furthermore, the electrochemical impedance spectroscopy (EIS) data indicated that the yttrium-based electrolyte exhibited a lower impedance [95].

The inclusion of chloride ions has been shown to provide a multitude of benefits, such as stabilization of the Mg^{2+} ions and amplification of the dissolution of passivating substances on the Mg anode. This subsequently facilitated improved efficiency in the plating and stripping of Mg. The impact is particularly pronounced when employing HMDSMgCl-based electrolytes, where the introduction of $AlCl_3$ results in a significant rise in current density for Mg deposition, although it does not enhance the voltage stability. Additionally, the advancement of organic magnesium borate-based (OMBB) electrolytes and the exploration of innovative techniques for preparing non-nucleophilic electrolyte solutions have further elevated the performance of Mg-S batteries.

3.1.2. Chloride-Free

Muldoon et al. highlighted the adverse effects associated with the presence of chlorine in electrolytes, pointing out that the chlorides in the electroactive species $[Mg_2(\mu\text{-}Cl)_3 \cdot 6THF]$ are a primary cause of corrosion [97]. Additionally, the bulky structure of the cation, characterized by two octahedrally coordinated Mg atoms linked by three chlorides, impedes Mg ion mobility. This analysis underscores the urgent need to develop and synthesize new types of chlorine-free salts for use in Mg-S batteries. In 2016, Li et al. pioneered a method for synthesizing a simple chloride-free $[Mg(THF)_6]^{2+}$ cation salt with $AlCl^{4-}$ as a counteranion, utilizing a straightforward heating process in an ionic liquid solvent [77]. The crystal structure of the $[Mg(THF)_6][AlCl_4]_2$ salt is illustrated in Figure 6a. Li et al. underscored the benefits of using ionic liquids, including their high boiling point, low melting point, exceptional chemical and thermal stability, nonflammability, and low vapor pressure [77,98]. They proposed a stoichiometric reaction of $MgCl_2$ (1 molar equivalent) with $AlCl_3$ (2 molar equivalents), using the ionic liquid n-methyl-(n-butyl) pyrrolidinium bis(trifluoromethanesulfonyl)imide (PYR14TFSI) as a reaction medium. It was postulated that the chloride ions in $MgCl_2$ could be completely displaced by $AlCl_3$ under elevated temperatures.

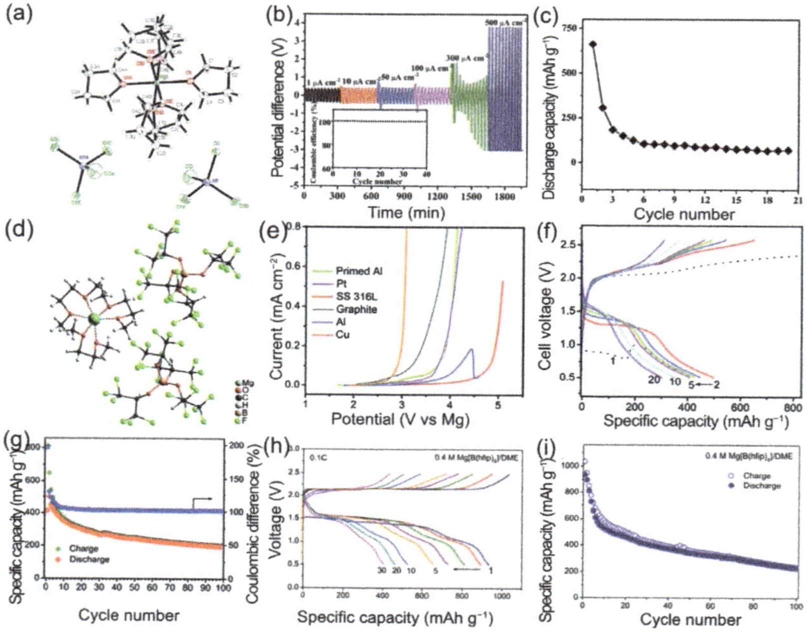

Figure 6. (**a**) Proposed formation mechanism and X-ray crystal structure of Mg salt. (**b**) Cycling behavior of a symmetrical cell with electrolyte $[Mg(THF)_6][AlCl_4]_2$ in PYR14TFSI/THF (1:1 v/v) at different current densities of 1 $\mu A\,cm^{-2}$ to 500 $\mu A\,cm^{-2}$; the cycle time was 30 min per cycle (15 min charging and 15 min discharging). (**c**) Discharge and charge profile for 20 cycles of discharging and charging, cathode: 50% S loading, NG/SP/commercial S/PVDF = 4:5:10:1. Rate: 0.01 C discharging, 0.02 C charging. Anode: Mg disk [77]. (**d**) Ball-and-stick representation of the $Mg[B(hfip)_4]_2{\cdot}3DME$ crystal. (**e**) Linear sweep voltammograms of various electrodes at a scan rate of 5 mV s^{-1}. Battery performance of the Mg-S-CMK-3 cell: (**f**) discharge–charge profiles, (**g**) cycling behavior [51]. (**h**) Charge/discharge profiles; (**i**) cycling performance of the ACCS–Mg cell with 0.4 M electrolytes [84]. (**a**–**c**) © 2014 WILEY-VCH Verlag GmbH & Co. KGaA, Weinheim. (**d**–**g**) Copyright © The Royal Society of Chemistry 2017. (**h**,**i**) Copyright © 2019, American Chemical Society.

Initially, a magnesium stripping/plating experiment was conducted, with the current density varying from 1 $\mu A\ cm^{-2}$ to 500 $\mu A\ cm^{-2}$ and the cycling time for each charge and discharge set at 15 min. At lower current densities, the Coulombic Efficiency of the cell is nearly 100% (Figure 6b). However, as the current density escalates to 300 $\mu A\ cm^{-2}$, an increase in polarization occurs, leading to irregular potential fluctuations. For further analysis, electrochemical impedance spectroscopy (EIS) was employed to examine the formation and stability of the interface between the electrolyte solution and the magnesium electrode over the course of the cycling. Initially, the interface between the magnesium electrode and the electrolyte solution is unstable, but it stabilizes after a certain number of cycles, evidenced by the gradual stabilization of the solid electrolyte interface (SEI) layer on the magnesium electrode. In the cycling performance of the Mg-S cells (Figure 6c), it was observed that the cell maintains stability for over 20 cycles, although its capacity decreases sharply in the first five cycles from approximately 700 mAh g^{-1} to 130 mAh g^{-1}, before stabilizing at a relatively steady capacity of around 70 mAh g^{-1}.

In 2017, Zhao-Karger et al. developed a fluorinated magnesium alkoxyborate-based electrolyte for Mg-S batteries [51]. They synthesized conductive salts via the reaction of $Mg[BH_4]_2$ with fluorinated alcohols (RF-OH) in ethereal solvents, such as DME. Upon removing the solvent from Hexafluoro-2-propanol (hfip), the conductive salt $Mg[B(hfip)_4]_2{\cdot}3DME$ was obtained. Its crystal structure, determined using X-ray crystallography, is $Mg[B(hfip)_4]_2{\cdot}3DME$ (Figure 6d). The crystal unit consists of typical ion pairs, with the octahedral coordination geometry around the Mg^{2+} ions being slightly deformed due to the solvation by the three DME molecules. In the $[B(hfip)_4]^-$ counteranion, the boron (B) atom is tetrahedrally bonded to four hexafluoroisopropyloxy groups. The O–B–O angles (107.8° and 107.2°) approximate the ideal tetrahedral angle. The anodic stability of the MgBOR(hfip)/DME electrolyte on conventional electrode substrates, including stainless steel (SS), Al, primed Al, and Cu, was examined using linear sweep voltammetry (LSV). The voltammogram indicates that the oxidation stability of the electrolyte on Pt is about 3.5 V, possibly limited by the DME oxidation (Figure 6e). Swagelok-type cells, containing a MgBOR(hfip)/DEG–TEG electrolyte, S/CMK-3 cathode, and Mg foil anode, were cycled at room temperature at 0.1 C (167 mA cm^{-2}). The galvanostatic discharge/charge curves from the 1st to the 5th, 10th, and 20th cycles are displayed in Figure 6f. After the initial cycles, the discharge voltage stabilized at around 1.5V. From the second cycle onward, the discharge capacity decreased but remained above approximately 200 mAh g^{-1} sulfur up to the 100th cycle (Figure 6g). The overcharging behavior, evident in the first five cycles, led to a reduced efficiency. To optimize the electrochemical performance, the concentration of $Mg[B(hfip)_4]_2$ in the DME solvent varied from 0.1 to 0.4 M. Cyclic voltammetry revealed that higher concentrations of $Mg[B(hfip)_4]_2$ increased the current densities [84]. Based on these findings, the authors concluded that fluorinated alkoxyborate-based electrolytes hold promise for Mg-S batteries.

The same research team also explored Swagelok-type cells consisting of a 0.4 M MgB-hfip/DME electrolyte, an ACCS cathode, and a Mg foil anode. These cells were cycled at a current rate of 0.1 C (167 mA cm^{-2}) at 25 °C [84]. The initial galvanostatic discharge/charge curves displayed a relatively flat discharge voltage plateau at approximately 1.5 V, followed by a sloped region until a cutoff voltage of 0.5 V, suggesting a stepwise reaction pathway. The cells achieved a discharge capacity of about 930 mAh g^{-1}, while the charge capacity marginally surpassed the discharge capacity (Figure 6h,i). The Coulombic Efficiency for the first cycle, calculated by dividing the charge capacity by the discharge capacity, was approximately 110%. This gradual decline in capacity is attributed to the dissolution of the magnesium polysulfide (MgSx) and the ongoing loss of active material.

In 2017, Zhang et al. introduced a boron-centered anion-based magnesium electrolyte (BCM electrolyte) characterized by its ease of synthesis, high ionic conductivity, broad potential window (3.5 V vs. Mg), compatibility with electrophilic sulfur, and non-corrosiveness toward the cell components [50]. They analyzed the chemical and electrochemical properties of various anionic forms in non-nucleophilic electrolytes, detailing the specific challenges encountered in their practical applications, as highlighted in the colored boxes of Figure 7a. In their approach, the properties of the BCM electrolytes, including the electrochemical window, salt concentration, and compatibility with the Mg anode, could be finely tuned by selecting specific anionic groups. Notably, the use of tris(2*H*-hexafluoroisopropyl) borate (THFPB)/MgF_2 salts in the DME solvent led to the formation of the $[Mg(DME)_n][FTHB]_2$ complex (Figure 7b), which demonstrated a wide operating window of 3.5 V. With a magnesium anode and a sulfur cathode containing 85 wt.% sulfur and a sulfur loading of 1.5 mg cm^{-2}, the cell delivered a discharge capacity of 1081 mAh g^{-1} and a stable voltage plateau at 1.1 V and exhibited no overcharging in subsequent cycles (Figure 7c). The excellent capacity retention over 30 cycles, with an initial discharge capacity of 86.4%, underscores the success of this novel electrolyte design concept.

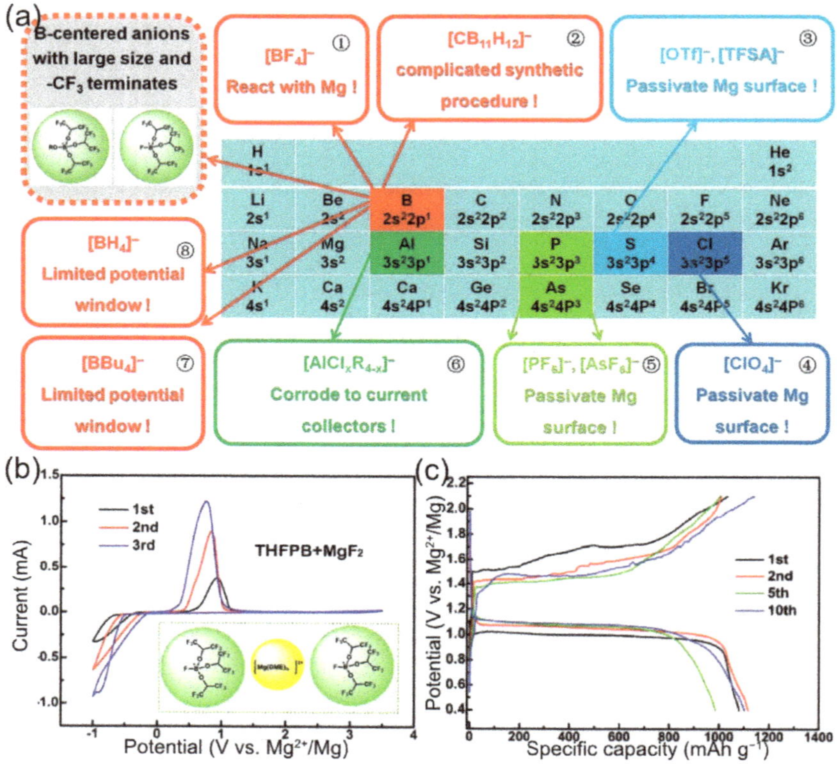

Figure 7. (**a**) Decorated periodic table for directing efficient Mg-ion electrolytes. Electrolytes derived from anions of ① to ⑧ suffer from specific problems. Electrolytes from ①, ③, ④, and ⑤ can react with Mg metals, forming impervious surface layers. Anions of ⑦ and ⑧ display insufficient anodic stability, resulting in limited potential windows of the determined Mg-ion electrolytes. Electrolytes containing anions of ⑥ corrode typical current collectors and show incompatibility with sulfur and oxide cathodes. The complicated synthetic procedure and incompatibility with sulfur cathodes of ②-constructed Mg-ion electrolyte makes it less promising for practical Mg-ion electrolytes. (**b**) CV curves of SS electrode in BCM electrolyte containing THFPB and MgF_2 at 5 mV s^{-1}. (**c**) The electrochemical performances of the Se/C and S/C electrodes in BCM electrolytes: galvanostatic charge/discharge profile [50]. Copyright © 2017 WILEY-VCH Verlag GmbH & Co. KGaA, Weinheim.

In 2021, Ren et al. demonstrated for the first time the use of a PP_{14}TFSI ionic liquid as a co-solvent with THF in a chlorine-free MBA-based electrolyte [60]. This approach significantly enhanced the ionic conductivity and increased the oxidative stability potential on stainless steel to 2.2 V vs. Mg/Mg^{2+}. The experimental results indicated that this electrolyte exhibited a low overpotential below 200 mV and maintained approximately 90% Coulombic Efficiency in the reversible electrochemical plating/stripping of magnesium. Following the addition of PP_{14}TFSI, the current density for magnesium plating/stripping was escalated by a factor of 238. Additionally, the MBA-2AlF_3 electrolyte showed good compatibility with the Mo_6S_8 cathode. Furthermore, Ren et al. also reported the performance of a Se@pPAN | Mg full cell, which delivered an initial capacity of 447.8 mAh g^{-1} at 0.2 C, with a minimal capacity decay of about 0.66% per cycle over more than 70 cycles.

The chloride-free electrolytes highlighted in this section, such as the fluorinated alkoxyborate-based electrolytes and boron-centered anion-based (BCM) electrolytes, demonstrate wide operating windows, high ionic conductivities, and good compatibility with Mg anodes and sulfur cathodes. Chloride-free electrolytes mitigate the disadvantages shown in chloride-containing electrolytes of corrosion and limited Mg ion mobility. The synthesis of simple cation salts, such as $[Mg(THF)_6]^{2+}$ with non-chloride counterions, and the development of fluorinated magnesium alkoxyborate-based electrolytes signify great strides toward the goal of achieving high-performance Mg-S batteries. These advancements are not merely academic; they offer practical pathways to enhance the electrochemical performance, stability, and safety of Mg-S batteries.

3.2. Nucleophilic Electrolytes

The reaction between PhMgCl and $AlCl_3$ facilitates the synthesis of Grignard reagent-based nucleophilic all-phenyl complex (APC) electrolytes. Initially, these APC electrolytes were not tailored to use in Mg-S batteries due to the high nucleophilicity of organomagnesium compounds, which generally renders them incompatible with sulfur-based conversion cathodes. However, in 2017, Linqi Zeng et al. adapted an APC electrolyte for Mg-S battery applications [96]. Utilizing copper as the current collector for a sulfur cathode and integrating it with the nucleophilic $(PhMgCl)_2$-$AlCl_3$/THF electrolyte, they achieved an initial discharge capacity of 659 mAh g^{-1}. Additionally, at a current density of 10 mA g^{-1}, the reversible capacity of the cell stabilized at 113 mAh g^{-1} after 20 cycles.

The cyclic voltammetry (CV) curves of the Mg-S coin cell at a scan rate of 0.05 mV s^{-1} (Figure 8a) feature Cu as the cathode current collector. During the first cathodic scan, two significant reduction peaks at 1.0 V and 1.35 V were observed, accompanied by shoulder peaks at 0.9 V. These peaks are indicative of the formation of higher-order magnesium polysulfides (MgSx, where $4 \leq X \leq 8$). These high-order polysulfides undergo reduction into their lower-order counterparts, subsequently leading to the formation of MgS_2 and MgS. In the anodic scan, the re-oxidation of MgS and MgS_2 back into polysulfide MgSx ($x > 2$) is characterized by a distinct oxidation peak at 1.43 V. They observed an increase in current above 1.6 V, which suggests further oxidation into higher-order polysulfides or elemental sulfur. A marked decrease in the peak currents during the second cathodic scan points to the dissolution of some of the sulfur or polysulfides into the electrolyte, contributing to a decline in cell capacity. The shift in the peak voltage aligns with these electrochemical processes.

The discharge–charge cycling performance and corresponding Coulombic Efficiency of a Mg-S coin cell with a Cu cathode collector operating at a current density of 10 mA g^{-1} are displayed in Figure 8b. The cell's capacity exhibited a steady decline until the 20th cycle, stabilizing at approximately 113 mAh g^{-1} thereafter. Notably, after a few initial cycles, there was a gradual increase in the charging capacity compared to the discharge capacity. The Coulombic Efficiency, exceeding 100% (Figure 8b), is attributed to the polysulfide shuttle effect, a recognized parasitic side reaction. Furthermore, elevating the current density to 20 mA g^{-1} proved effective in maintaining the discharge capacity above 99 mAh g^{-1} after 20 cycles, indicating a mitigation of the dissolution and shuttle effects. This improvement suggests a robust interaction between Cu and S, along with the enhanced electronic conductivity of copper sulfides, positively affecting the sulfur utilization and the cyclic stability of the sulfur cathode in the nucleophilic electrolyte. Additionally, incorporating LiCl into the $(PhMgCl)_2$-$AlCl_3$/THF electrolyte further enhanced its cycling stability and rate performance. The capacity initially decreased from 512 mAh g^{-1} to 388 mAh g^{-1} and then was consistently held at 384 mAh g^{-1} (Figure 8c).

In 2018, expanding on their previous work, W. Wang et al. from the same research group introduced an innovative sulfur@microporous carbon (S@MC) electrode [54]. This electrode, employing copper as the current collector, was designed as a novel cathode for advanced Mg-S batteries. It effectively utilizes APC-based nucleophilic electrolytes, marking a significant advancement in the development of high-performance Mg-S battery technology.

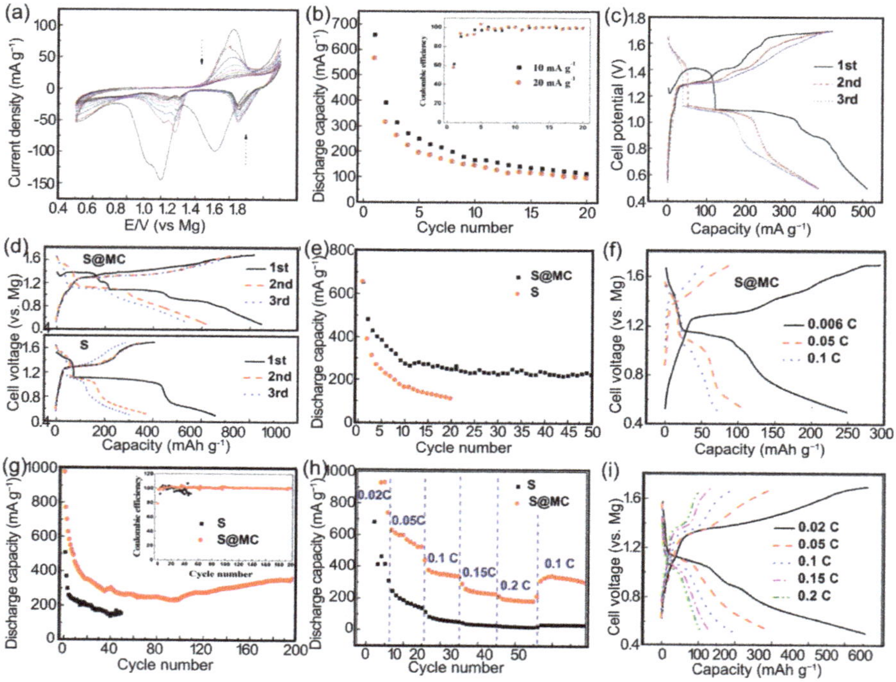

Figure 8. (**a**) CVs of the Mg-S coin cell with Cu as the cathode current collector at 0.05 mV s^{-1}. (**b**) The cycling performance of Mg-S coin cells with Cu as the cathode current collector at 10 mA g^{-1} and 20 mA g^{-1}. (**c**) Initial three discharge–charge curves of Mg-S cell with Cu as the cathode current collector in 0.4 mol L^{-1} $(PhMgCl)_2$-$AlCl_3$ + 1.0 mol L^{-1} LiCl/THF electrolyte at 10 mA g^{-1} [96]. Initial three discharge–charge curves (**d**) and cycling performance (**e**) of elemental sulfur and S@MC at a rate of 0.006 C. (**f**) The 20th discharge–charge curves of S@MC at different rates; the electrolyte is 0.4 mol L^{-1} $(PhMgCl)_2$–$AlCl_3$/THF. (**g**) Cycling performance of elemental sulfur and S@MC composite at 0.1 C; inset is the Coulombic Efficiency upon cycling. (**h**) The rate performance of elemental sulfur and S@MC composite at different rates. (**i**) Typical discharge–charge curves of S@MC at different rates; the electrolyte is 0.4 mol L^{-1} $(PhMgCl)_2$–$AlCl_3$ + 1.0 mol L^{-1} LiCl/THF [54]. (**a–c**) © Linqi et al., 2017. Published by ECS. (**d–i**) Copyright © 2018, American Chemical Society.

Microporous carbon (MC), serving as the host material, enhances the chemical kinetics of the electrode and adsorbs sulfur and polysulfides effectively. At 50 °C, the formation of copper sulfide occurs when sulfur is coated onto a Cu collector, creating a robust chemical interaction between the Cu and sulfur. This interaction plays a crucial role in preventing the sulfur from being eroded by the electrolyte and reducing the rate of polysulfide dissolution. Utilizing the unique properties of microporous carbon and Cu current collectors, the initial discharge capacity of the S@MC electrode is around 979.0 mAh g^{-1}. After undergoing 200 cycles at a 0.1 C rate, the capacity reaches a stable value of 368.8 mAh g^{-1}, indicating improved sulfur utilization and enhanced cycle stability. This retention rate is on par with several non-nucleophilic electrolytes [52,79]. Even at a higher rate of 0.2 C, the composite maintains a capacity of about 200 mAh g^{-1}, achieving a Coulombic Efficiency of up to 200%. The strategy of reinforcing the chemical bonds between smaller $S_{2\text{-}4}$ molecules and larger S_8 molecules within the MC framework has been demonstrated to be an effective approach to boosting the cyclic stability, rate performance, and sulfur utilization in Mg-S batteries with nucleophilic electrolytes (Figure 8d–i).

Research on electrolyte systems suitable for Mg-S batteries is still ongoing; nonetheless, some achievements have already been accomplished. Figure 9 summarizes all the clearly identified electroactive species in the electrolytes investigated so far. According to the structure of these electroactive species, we have divided them into cations: O^-, cations: Cl^-, anions: non-metal, anions: metal, size: small, size: medium, and size: large.

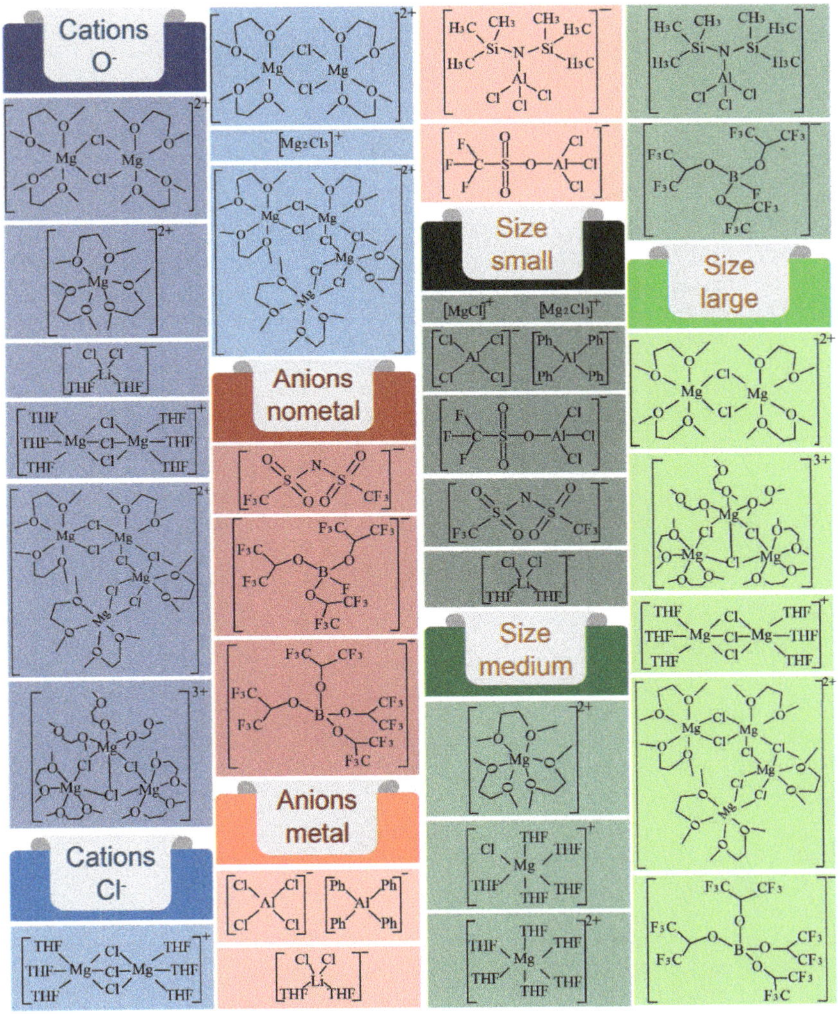

Figure 9. All clearly identified electroactive species in the electrolytes investigated so far.

The electrolyte plays a pivotal role in the performance of Mg-S batteries, directly influencing their efficiency, capacity, cycling stability, and safety. The choice between non-nucleophilic and nucleophilic electrolytes remains unclear. The addition of chloride ions then adds a further layer of complexity. Finding the ideal electrolyte that balances conductivity, reactivity, and stability with the Mg anode and sulfur cathode is still a challenging goal.

Non-nucleophilic electrolytes have shown promise due to their ability to stabilize Mg^{2+} ions and facilitate efficient plating/stripping processes. The shuttle effect and the need for a higher Coulombic Efficiency remain as the biggest issues. Chloride-free electrolytes address the corrosion issues associated with chloride ions but require careful synthesis to maintain high ionic conductivity and compatibility with sulfur cathodes. Nucleophilic electrolytes offer unique pathways for Mg-S battery configurations, but the reactivity of organic magnesium compounds with sulfur-based cathodes needs to be addressed. The use of copper current collectors and microporous carbon in the literature has shown that there is great promise for the future of these electrolytes.

This review has analyzed a number of notable electrolytes which have been used in the field of Mg-S battery research. Table 1 provides a comparison of these electrolytes, highlighting key statistics such as Coulombic Efficiency, capacity, and current rate. It clearly reveals a major difficulty that magnesium–sulfur batteries are currently facing with low cycling performances. This is mainly due to the sharp drop in the battery capacity caused by the shuttle effect.

4. The Mechanism of Sulfur Reduction

In 2012, Muldoon et al. emphasized the necessity of using non-nucleophilic magnesium organohaloaluminate electrolytes to achieve an effective pairing between magnesium and sulfur [48]. In 2014, Zhirong et al. used modified electrolytes in tetraglyme or a binary solvent of glyme and $PP_{14}TFSI$. The use of this additive aims to adjust the viscosity of the electrolyte and may reduce the solubility of magnesium polysulfides [55]. The result shows the ability of the glyme polyether chains to permit multidentate cation coordination through the oxygen atoms, allowing for the adoption of flexible coordination numbers and geometries, which considerably improves the performance, Coulombic Efficiency, and discharge voltage. Then, they propose a redox mechanism for Mg-S batteries:

Step I: Elemental sulfur is reduced into MgS_8 at the solid electrolyte interface (SEI), followed by the dissolution of MgS_8 into a liquid cathode, transitioning into lower-order polysulfides.

$$\text{Step I } S_8 + 4e^- + Mg^{2+} \rightarrow 2MgS_4 \tag{1}$$

Step II: Low-order polysulfides such as MgS_4 are reduced into MgS_2, corresponding to the second discharge platform.

$$\text{Step II } MgS_4 + 2e^- + Mg^{2+} \rightarrow 2MgS_2 \tag{2}$$

Step III: MgS_2 is reduced into MgS, a process characterized by high kinetic barriers and polarization.

$$\text{Step III } MgS_2 + 2e^- \rightarrow 2MgS \tag{3}$$

In the work of Robba et al., 2017, they used a non-nucleophilic electrolyte solution prepared from $MgCl_2$ and $Mg(TFSI)_2$ salts dissolved in a binary mixture of ether solvents [67]. The results show the battery exhibited two clear discharge plateaus during the first discharge process, corresponding to the conversion of MgS_x (high-voltage plateau) and the further conversion of polysulfides into MgS (low-voltage plateau). In the same year, Gao et al. determined the specific discharge plateau in a $MgTFSI_2$–DME electrolyte:

Step I: The transformation of elemental sulfur into long-chain polysulfides (2.4–1.5 V potential slope)

$$\text{Step I } S_8 + 4e^- + Mg^{2+} \rightarrow MgS_8 \text{ 2.5–1.5 V} \tag{4}$$

Step II: Shortening of the polysulfide chain (1.5 V potential plateau)

$$\text{Step II } MgS_8 + 6e^- + 3Mg^{2+} \rightarrow 4MgS_2 \text{ 1.5 V} \tag{5}$$

Step III: The solid-state transition from short-chain polysulfides to magnesium sulfide (1.5–0.5 V potential slope)

$$\text{Step III } MgS_2 + 2e^- \rightarrow 2MgS \text{ 1.5–0.5 V} \tag{6}$$

In 2019, Yan et al. further investigated the sulfur reduction reaction route according to the in situ method using an electrolyte of $Mg(HMDS)_2$–$AlCl_3$ [68], also including three stages (Figure 10b):

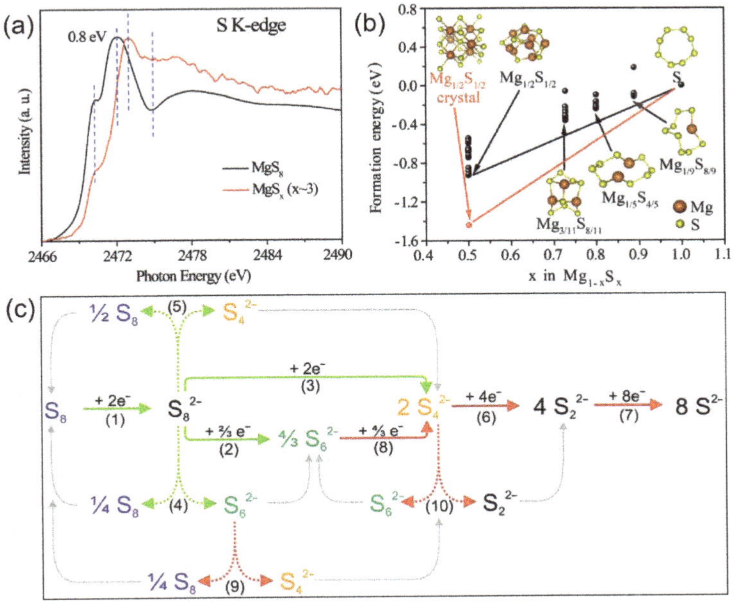

Figure 10. (**a**) The S K-edge XAS spectra of the MgS_8 and MgS_x (x∼3) reference samples. (**b**) Calculated structures of magnesium polysulfides and their formation energies [68]. (**c**) Proposed reduction pathway for sulfur species in glyme-based electrolytes [99]. (**a**,**b**) Copyright © 2019, American Chemical Society. (**c**) Copyright © 2021 Joachim et al. Published by American Chemical Society.

Stage I: Rapid reaction stage for the formation of high-order MgS_x (MgS_8, MgS_4).

$$\text{Stage I } S_8 + Mg^{2+} + e^- \rightarrow MgS_8 + MgS_4 \tag{7}$$

Stage II: Reduction stage from high-order MgS_4 into Mg_3S_8

$$\text{Stage II } MgS_8 + MgS_4 + e^- \rightarrow Mg_3S_8 \tag{8}$$

Stage III: Sluggish further reduction of Mg_3S_8 into MgS; this reaction predominantly occurs in the solid phase, contributing to the rapid capacity decay of the Mg-S battery.

$$\text{Stage III } Mg_3S_8 + e^- \rightarrow MgS \tag{9}$$

In situ XAS analysis (Figure 10a) reveals that Mg_3S_8 and MgS are electrochemically inert and difficult to convert back into high-order polysulfides. This leads to a rapid decline in the battery capacity and a shortened cycle life.

Forrest et al. investigated the impact of dissolved sulfur on the passivation of Mg anodes in Mg-S batteries, revealing that interactions between the dissolved polysulfides and the Mg anode led to the formation of a passivating MgS layer on the anode surface [25]. This layer readily reforms during reduction but can be removed under oxidative conditions. The research highlights that the concentration of dissolved S_8 influences the rate of MgS layer formation by altering the equilibrium of polysulfide disproportionation.

In 2022, Joachim et al. characterized the magnesium polysulfide dissolution behavior in a different electrolyte of $Mg[B(hfip)_4]_2$ in tetraethylene glycol dimethyl ether (G4, tetraglyme) [99]. By applying operando UV/Vis spectroscopy, S_8, S_6^{2-}, and S_4^{2-} were identified as the species present in the electrolyte, while S_8^{2-} and $S_3^{\bullet-}$ were not detected. A reduction pathway is proposed, with the previously gained insights summarized in Figure 10c. In solvents with a high dielectric permittivity and donor number (e.g., DMSO, DMF, or ACN), the low-charge-density polysulfides S_8^{2-}, S_6^{2-}, and $S_3^{\bullet-}$ are dominant, while in solvents with a low dielectric permittivity and donor number (G1, G2, G4, THF), the high-charge-density polysulfide S_4^{2-} is well stabilized.

The intermediate magnesium polysulfides are the main reason for a low Coulombic Efficiency and low cycle performance, which is termed the shuttle effect. The mechanisms by which the shuttle effect diminishes the electrochemical performance of Mg-S batteries can be summarized in three main points:

Reduced Coulombic Efficiency: Coulombic Efficiency refers to the effective utilization of charge during the battery's charging and discharging processes. Magnesium polysulfides form and dissolve in the electrolyte, with these dissolved polysulfides shuttling between the anode and cathode. This shuttling leads to a portion of the charge in the battery being inefficiently used for energy storage and release, thus reducing the Coulombic Efficiency.

Self-discharge: Self-discharge is a phenomenon where a battery naturally loses its charge when not in use [99]. This process occurs in three stages: (I) the dissolution and reduction into S_6^{2-} and S_4^{2-} of S_8 in the electrolyte, (II) the stabilization of the sulfur concentration and increase in polysulfides (S_6^{2-}/S_4^{2-}), (III) equilibrium of the sulfur and polysulfide ($S_8/S_6^{2-}/S_4^{2-}$) concentrations. This self-discharge behavior leads to the loss of active materials in the battery, thereby affecting the battery's capacity and cycle stability.

Anode Passivation: Due to the low solubility of the sulfides, not all the formed polysulfides can be dissolved, resulting in the formation of insoluble MgS_x and MgS at the interface, which poses a significant barrier to further magnetization [68]. This leads to premature termination of discharge, especially at high sulfur/carbon ratios, resulting in lower sulfur utilization.

The extensive research into the sulfur reduction mechanism within Mg-S batteries underscores the pivotal role of the electrolyte composition and solvent dynamics in dictating the battery performance. There is a need for innovative solutions to address challenges such as the shuttle effect and anode passivation, highlighted by the work to enhance Coulombic Efficiency and mitigate self-discharge. This review highlights the key literature in this field, with the aim of advancing the understanding of Mg-S battery chemistry. It has also set a clear direction for the future, with a need to develop electrolyte and battery designs that address these specific challenges, thereby unlocking the potential of Mg-S batteries for high-efficiency, sustainable energy storage.

5. Concluding Remarks and Outlook

In this comprehensive review of the development of Mg-S battery electrolytes, we take an in-depth look at the progress, challenges, and future directions in this field. Mg-S batteries have the potential to become a viable alternative to conventional Li-ion systems, offering advantages in terms of energy density, safety, and sustainability. However, their current development is hampered by critical issues such as electrolyte compatibility and anode passivation, as well as being limited by the shuttle effect. Innovations in non-nucleophilic electrolytes, advanced cathode materials, and anode protection strategies are considered key to overcoming these obstacles. This review highlights the need for interdisciplinary research and technological breakthroughs in materials science and electrochemistry to address the complex challenges of Mg-S batteries.

The recent research on Mg-S batteries indicates that, while they have an impressive theoretical energy density of up to 2856 Wh kg^{-1} and rank highly in terms of safety due to the minimal formation of dendrites during charge/discharge, the electrochemical performance of the actual Mg-S batteries is far from ideal. This includes practical metrics such as energy density, cycling stability, and charge/discharge rates, which currently fall significantly short of those in Li-ion batteries. The primary reason for this under-performance, similar to that in Li-S batteries, is the formation and uncontrollability of polysulfides during the charge/discharge cycles. These polysulfides can freely move across the separator between the anode and cathode, a phenomenon known as the "shuttle effect". This shuttle effect leads to inefficient utilization of some of the battery's charge in energy storage and release. Moreover, the insoluble polysulfides can form a passivation layer on the electrode surfaces, impeding electrode reactions. These cumulative adverse effects make it challenging for Mg-S batteries to achieve their theoretical electrochemical performance.

To mitigate the shuttle effect, researchers have employed various strategies, including, but not limited to, 1. Non-Nucleophilic Electrolytes: These do not react with sulfur and thus help reduce the shuttle effect. For instance, non-nucleophilic electrolytes synthesized from hexamethyldisilazane magnesium chloride (HMDSMgCl) and $AlCl_3$, as well as organic magnesium borate-based electrolytes containing tetrafluoroborate ammonium ($[B(HFP)_4]^-$), are used. 2. Chloride Additives: Adding an appropriate amount of chloride, such as lithium chloride (LiCl), to non-nucleophilic electrolytes can enhance their stability and electrochemical performance, thereby reducing the shuttle effect. 3. Electrolyte Additives: For example, adding certain ionic liquids, such as $PP_{14}TFSI$, to the electrolyte can improve its conductivity and oxidation stability, which, in turn, helps reduce the shuttle effect. 4. High-Concentration Electrolytes: High-concentration electrolytes can decrease the solubility of the sulfur molecules or polysulfides in the electrolyte, thereby reducing the loss of active material and the shuttle effect. These methods address the shuttle effect according to different mechanisms and have indeed improved the overall performance of Mg-S batteries to a certain extent.

Despite various efforts, breakthrough advancements in the overall performance of Mg-S batteries remain elusive. A significant factor contributing to this stagnation is the lack of a clear understanding of the internal reaction pathways in Mg-S batteries, specifically the sulfur reduction pathways. Although numerous studies have been published investigating the mechanisms of Mg-S batteries, the complexity of the sulfur reduction reactions and limitations in the research methodologies make it challenging for researchers to delineate a complete reaction pathway for each type of electrolyte. There has been some progress, with certain studies proposing relatively comprehensive reaction pathways. However, the authors of these studies also acknowledge discrepancies between their proposed pathways and the characterization results of other research. These inconsistencies might be attributed to the transient nature of some intermediate products or the difficulty of detecting them using traditional methods. This highlights the ongoing challenges in fully understanding and optimizing the electrochemistry of Mg-S batteries.

Non-nucleophilic electrolytes, magnesium borate-based electrolytes ($Mg(BPh_4)_2$, and $Mg[B(hfip)_4]_2$) exhibit a higher capacity, Coulombic Efficiency, and cycle performance compared to the other types of electrolytes, as was shown in Table 1. Benefitting from their high solvation capacity and electrochemical stability, DME and TEG have become the most popular solvents for use with these electrolytes. Looking at the future prospects, there are a number of interesting topics to explore. Additive reduced perylene diimide–ethylenediamine (rPDI) forms a protecting layer on the Mg anode and leads to a high number of stable cycle lives, opening up a promising research prospect. In addition, research into the reaction mechanism, especially in-depth investigations into the mechanisms of the sulfur reduction reaction pathways, is a very important task, and findings in this area will significantly contribute to the mitigation of the shuttle effect, which is key to the future of Mg-S battery use.

In summary, rechargeable Mg-S batteries represent a battery system with immense potential for development, characterized by their high theoretical energy density and proven high safety performance. It is reasonable to anticipate that, in the foreseeable future, a significant breakthrough in the practical electrochemical performance of Mg-S batteries will be achieved.

Author Contributions: Conceptualization, J.F., Z.H. and F.R.W.; Funding acquisition, L.S., Z.H. and F.R.W.; Formal analysis, J.F. and M.G.; Investigation, L.S., M.G. and L.Z.; Resources, J.F. and L.Z.; Project administration, L.S., Z.H. and F.R.W.; Writing—original draft, L.S.; Writing—review & editing, L.S., J.F., J.H., Z.H. and F.R.W.; Supervision, Z.H. and F.R.W. All authors have read and agreed to the published version of the manuscript.

Funding: This research was funded by the startup fund at Hubei University of Technology, high-level talent grant from Hubei province, China Scholarship Council (Grant No. 201908060125, CSC), the Doctor of Suzhou University Scientific Research Foundation Project (Grant No. 2023BSK056), the Anhui Province Higher Education Innovation Team: Key Technologies and Equipment Innovation Team for Clean Energy (Grant No. 2023AH010055).

Institutional Review Board Statement: Not applicable.

Informed Consent Statement: Not applicable.

Data Availability Statement: All data generated or analyzed during this study are included in the article.

Conflicts of Interest: The authors declare no conflicts of interest.

References

1. Li, M.; Lu, J.; Chen, Z.; Amine, K. 30 Years of Lithium-Ion Batteries. *Adv. Mater.* **2018**, *30*, e1800561. [CrossRef] [PubMed]
2. Rivera-Barrera, J.; Muñoz-Galeano, N.; Sarmiento-Maldonado, H. SoC Estimation for Lithium-ion Batteries: Review and Future Challenges. *Electronics* **2017**, *6*, 102. [CrossRef]
3. Xu, J.; Cai, X.; Cai, S.; Shao, Y.; Hu, C.; Lu, S.; Ding, S. High-Energy Lithium-Ion Batteries: Recent Progress and a Promising Future in Applications. *Energy Environ. Mater.* **2023**, *6*, e12450. [CrossRef]
4. Placke, T.; Kloepsch, R.; Dühnen, S.; Winter, M. Lithium ion, lithium metal, and alternative rechargeable battery technologies: The odyssey for high energy density. *J. Solid State Electrochem.* **2017**, *21*, 1939–1964. [CrossRef]
5. Shahjalal, M.; Roy, P.K.; Shams, T.; Fly, A.; Chowdhury, J.I.; Ahmed, M.R.; Liu, K. A review on second-life of Li-ion batteries: Prospects, challenges, and issues. *Energy* **2022**, *241*, 122881. [CrossRef]
6. Yang, Z.; Zhu, P.; Ullah, Z.; Zheng, S.; Yu, S.; Zhu, S.; Liu, L.; Li, Q. Synchronous Light Harvesting and Energy Storing Organic Cathode Material 1,4-Dihydroxyanthraquinone for Lithium-Ion Batteries. *Chem. Eng. J.* **2023**, *468*, 143787. [CrossRef]
7. Wu, J.; Zheng, M.; Liu, T.; Wang, Y.; Liu, Y.; Nai, J.; Zhang, L.; Zhang, S.; Tao, X. Direct recovery: A sustainable recycling technology for spent lithium-ion battery. *Energy Storage Mater.* **2023**, *54*, 120–134. [CrossRef]
8. Li, J.; Fleetwood, J.; Hawley, W.B.; Kays, W. From Materials to Cell: State-of-the-Art and Prospective Technologies for Lithium-Ion Battery Electrode Processing. *Chem. Rev.* **2022**, *122*, 903–956. [CrossRef]
9. Li, G.; Zheng, X. Thermal energy storage system integration forms for a sustainable future. *Renew. Sustain. Energy Rev.* **2016**, *62*, 736–757. [CrossRef]
10. Schneider, S.F.; Novak, P.; Kober, T. Rechargeable Batteries for Simultaneous Demand Peak Shaving and Price Arbitrage Business. *IEEE Trans. Sustain. Energy* **2021**, *12*, 148–157. [CrossRef]
11. Choi, J.W.; Aurbach, D. Promise and reality of post-lithium-ion batteries with high energy densities. *Nat. Rev. Mater.* **2016**, *1*, 16013. [CrossRef]
12. Bella, F.; De Luca, S.; Fagiolari, L.; Versaci, D.; Amici, J.; Francia, C.; Bodoardo, S. An Overview on Anodes for Magnesium Batteries: Challenges towards a Promising Storage Solution for Renewables. *Nanomaterials* **2021**, *11*, 810. [CrossRef] [PubMed]
13. Chen, X.; Liu, X.; Le, Q.; Zhang, M.; Liu, M.; Atrens, A. A comprehensive review of the development of magnesium anodes for primary batteries. *J. Mater. Chem. A* **2021**, *9*, 12367–12399. [CrossRef]
14. Lei, X.; Liang, X.; Yang, R.; Zhang, F.; Wang, C.; Lee, C.S.; Tang, Y. Rational Design Strategy of Novel Energy Storage Systems: Toward High-Performance Rechargeable Magnesium Batteries. *Small* **2022**, *18*, e2200418. [CrossRef] [PubMed]
15. Kong, L.; Yan, C.; Huang, J.-Q.; Zhao, M.-Q.; Titirici, M.-M.; Xiang, R.; Zhang, Q. A Review of Advanced Energy Materials for Magnesium-Sulfur Batteries. *Energy Environ. Mater.* **2018**, *1*, 100–112. [CrossRef]
16. Benmayza, A.; Ramanathan, M.; Arthur, T.S.; Matsui, M.; Mizuno, F.; Guo, J.; Glans, P.-A.; Prakash, J. Effect of Electrolytic Properties of a Magnesium Organohaloaluminate Electrolyte on Magnesium Deposition. *J. Phys. Chem. C* **2013**, *117*, 26881–26888. [CrossRef]
17. Gao, T.; Noked, M.; Pearse, A.J.; Gillette, E.; Fan, X.; Zhu, Y.; Luo, C.; Suo, L.; Schroeder, M.A.; Xu, K.; et al. Enhancing the reversibility of Mg/S battery chemistry through Li^+ mediation. *J. Am. Chem. Soc.* **2015**, *137*, 12388–12393. [CrossRef] [PubMed]
18. Razaq, R.; Li, P.; Dong, Y.; Li, Y.; Mao, Y.; Bo, S.H. Practical energy densities, cost, and technical challenges for magnesium-sulfur batteries. *EcoMat* **2020**, *2*, e12056. [CrossRef]
19. Zhang, R.; Cui, C.; Xiao, R.; Li, R.; Mu, T.; Huo, H.; Ma, Y.; Yin, G.; Zuo, P. Interface regulation of Mg anode and redox couple conversion in cathode by copper for high-performance Mg-S battery. *Chem. Eng. J.* **2023**, *451*, 138663. [CrossRef]

20. Hong, W.; Ge, P.; Jiang, Y.; Yang, L.; Tian, Y.; Zou, G.; Cao, X.; Hou, H.; Ji, X. Yolk-Shell-Structured Bismuth@N-Doped Carbon Anode for Lithium-Ion Battery with High Volumetric Capacity. *ACS Appl. Mater. Interfaces* **2019**, *11*, 10829–10840. [CrossRef]
21. Togonon, J.J.H.; Esparcia, E.A., Jr.; Del Rosario, J.A.D.; Ocon, J.D. Development of Magnesium Anode-Based Transient Primary Batteries. *ChemistryOpen* **2021**, *10*, 471–476. [CrossRef]
22. Panigrahi, P.; Mishra, S.B.; Hussain, T.; Nanda, B.R.K.; Ahuja, R. Density Functional Theory Studies of Si_2BN Nanosheets as Anode Materials for Magnesium-Ion Batteries. *ACS Appl. Nano Mater.* **2020**, *3*, 9055–9063. [CrossRef]
23. Saha, P.; Datta, M.K.; Velikokhatnyi, O.I.; Manivannan, A.; Alman, D.; Kumta, P.N. Rechargeable magnesium battery: Current status and key challenges for the future. *Prog. Mater. Sci.* **2014**, *66*, 1–86. [CrossRef]
24. Fan, H.; Zheng, Z.; Zhao, L.; Li, W.; Wang, J.; Dai, M.; Zhao, Y.; Xiao, J.; Wang, G.; Ding, X.; et al. Extending Cycle Life of Mg/S Battery by Activation of Mg Anode/Electrolyte Interface through an LiCl-Assisted $MgCl_2$ Solubilization Mechanism. *Adv. Funct. Mater.* **2020**, *30*, 1909370. [CrossRef]
25. Laskowski, F.A.L.; Stradley, S.H.; Qian, M.D.; See, K.A. Mg Anode Passivation Caused by the Reaction of Dissolved Sulfur in Mg-S Batteries. *ACS Appl. Mater. Interfaces* **2021**, *13*, 29461–29470. [CrossRef] [PubMed]
26. Wang, P.; Buchmeiser, M.R. Rechargeable Magnesium–Sulfur Battery Technology: State of the Art and Key Challenges. *Adv. Funct. Mater.* **2019**, *29*, 1905248. [CrossRef]
27. Zhang, Z.; Dong, S.; Cui, Z.; Du, A.; Li, G.; Cui, G. Rechargeable Magnesium Batteries using Conversion-Type Cathodes: A Perspective and Minireview. *Small Methods* **2018**, *2*, 1800020. [CrossRef]
28. Guo, Z.; Zhao, S.; Li, T.; Su, D.; Guo, S.; Wang, G. Recent Advances in Rechargeable Magnesium-Based Batteries for High-Efficiency Energy Storage. *Adv. Energy Mater.* **2020**, *10*, 1903591. [CrossRef]
29. Cheng, X.; Zhang, Z.; Kong, Q.; Zhang, Q.; Wang, T.; Dong, S.; Gu, L.; Wang, X.; Ma, J.; Han, P.; et al. Highly Reversible Cuprous Mediated Cathode Chemistry for Magnesium Batteries. *Angew. Chem. Int. Ed. Engl.* **2020**, *59*, 11477–11482. [CrossRef]
30. Zhang, Z.; Chen, B.; Xu, H.; Cui, Z.; Dong, S.; Du, A.; Ma, J.; Wang, Q.; Zhou, X.; Cui, G. Self-Established Rapid Magnesiation/De-Magnesiation Pathways in Binary Selenium-Copper Mixtures with Significantly Enhanced Mg-Ion Storage Reversibility. *Adv. Funct. Mater.* **2018**, *28*, 1701718. [CrossRef]
31. Zhao-Karger, Z.; Fichtner, M. Magnesium–sulfur battery: Its beginning and recent progress. *MRS Commun.* **2017**, *7*, 770–784.
32. Yang, Y.; Yang, H.; Wang, X.; Bai, Y.; Wu, C. Multivalent metal–sulfur batteries for green and cost-effective energy storage: Current status and challenges. *J. Energy Chem.* **2022**, *64*, 144–165. [CrossRef]
33. Shi, F.; Yu, J.; Chen, C.; Lau, S.P.; Lv, W.; Xu, Z.-L. Advances in understanding and regulation of sulfur conversion processes in metal–sulfur batteries. *J. Mater. Chem. A* **2022**, *10*, 19412–19443. [CrossRef]
34. Zhang, Z.; Li, Y.; Zhao, G.; Zhu, L.; Sun, Y.; Besenbacher, F.; Yu, M. Rechargeable Mg-Ion Full Battery System with High Capacity and High Rate. *ACS Appl. Mater. Interfaces* **2021**, *13*, 40451–40459. [CrossRef]
35. Zhao-Karger, Z.; Fichtner, M. Beyond Intercalation Chemistry for Rechargeable Mg Batteries: A Short Review and Perspective. *Front. Chem.* **2018**, *6*, 656. [CrossRef] [PubMed]
36. Romio, M.; Surace, Y.; Mautner, A.; Hamid, R.; Jahn, M.; Cupid, D.M.; Abrahams, I. A Comparative Mechanistic Study on the Intercalation Reactions of Mg^{2+} and Li^+ Ions into $(Mg_{0.5}Ni_{0.5})_3(PO_4)_2$. *Batteries* **2023**, *9*, 342. [CrossRef]
37. Yang, J.; Li, J.; Gong, W.; Geng, F. Genuine divalent magnesium-ion storage and fast diffusion kinetics in metal oxides at room temperature. *Proc. Natl. Acad. Sci. USA* **2021**, *118*, e2111549118. [CrossRef]
38. Yu, X.-G.; Xie, J.-Y.; Yang, J.; Huang, H.-J.; Wang, K.; Wen, Z.-S. Lithium storage in conductive sulfur-containing polymers. *J. Electroanal. Chem.* **2004**, *573*, 121–128.
39. Zhu, J.; Mu, S. Defect Engineering in Carbon-Based Electrocatalysts: Insight into Intrinsic Carbon Defects. *Adv. Funct. Mater.* **2020**, *30*, 2001097. [CrossRef]
40. Sheha, E.M.; Farrag, M.; Refai, H.S.; El-Desoky, M.M.; Abdel-Hady, E. Positron Annihilation Spectroscopy as a Diagnostic Tool for Probing the First-Cycle Defect Evolution in Magnesium–Sulfur Battery Electrodes. *Phys. Status Solidi A* **2023**, *220*, 2200661. [CrossRef]
41. Zhang, S.; Ren, W.; NuLi, Y.; Wang, B.; Yang, J.; Wang, J. Sulfurized-Pyrolyzed Polyacrylonitrile Cathode for Magnesium-Sulfur Batteries Containing Mg^{2+}/Li^+ Hybrid Electrolytes. *Chem. Eng. J.* **2022**, *427*, 130902. [CrossRef]
42. Zu, C.-X.; Li, H. Thermodynamic analysis on energy densities of batteries. *Energy Env. Mater.* **2011**, *4*, 2614–2624. [CrossRef]
43. Bieker, G.; Küpers, V.; Kolek, M.; Winter, M. Intrinsic differences and realistic perspectives of lithium-sulfur and magnesium-sulfur batteries. *Commun. Mater.* **2021**, *2*, 37. [CrossRef]
44. Wang, M.; Bai, Z.; Yang, T.; Nie, C.; Xu, X.; Wang, Y.; Yang, J.; Dou, S.; Wang, N. Advances in High Sulfur Loading Cathodes for Practical Lithium-Sulfur Batteries. *Adv. Energy Mater.* **2022**, *12*, 2201585. [CrossRef]
45. Zhao, F.; Xue, J.; Shao, W.; Yu, H.; Huang, W.; Xiao, J. Toward high-sulfur-content, high-performance lithium-sulfur batteries: Review of materials and technologies. *J. Energy Chem.* **2023**, *80*, 625–657. [CrossRef]
46. Zhang, Z.; Wang, B.; Ju, S.; Wu, Z.; Yang, Y.; Pan, H.; Yu, X. Progress and prospects for solving the "shuttle effect" in magnesium-sulfur batteries. *Energy Storage Mater.* **2023**, *62*, 102933. [CrossRef]
47. Bieker, G.; Diddens, D.; Kolek, M.; Borodin, O.; Winter, M.; Bieker, P.; Jalkanen, K. Cation-Dependent Electrochemistry of Polysulfides in Lithium and Magnesium Electrolyte Solutions. *J. Phys. Chem. C* **2018**, *122*, 21770–21783. [CrossRef]
48. Muldoon, J.; Bucur, C.B.; Oliver, A.G.; Sugimoto, T.; Matsui, M.; Kim, H.S.; Allred, G.D.; Zajicek, J.; Kotani, Y. Electrolyte roadblocks to a magnesium rechargeable battery. *Energy Environ. Mater.* **2012**, *5*, 5941–5950. [CrossRef]

49. Nguyen, D.T.; Horia, R.; Eng, A.Y.S.; Song, S.W.; Seh, Z.W. Material design strategies to improve the performance of rechargeable magnesium-sulfur batteries. *Mater. Horiz.* **2021**, *8*, 830–853. [CrossRef] [PubMed]
50. Zhang, Z.; Cui, Z.; Qiao, L.; Guan, J.; Xu, H.; Wang, X.; Hu, P.; Du, H.; Li, S.; Zhou, X.; et al. Novel Design Concepts of Efficient Mg-Ion Electrolytes toward High-Performance Magnesium-Selenium and Magnesium-Sulfur Batteries. *Adv. Energy Mater.* **2017**, *7*, 1602055. [CrossRef]
51. Zhao-Karger, Z.; Gil Bardaji, M.E.; Fuhr, O.; Fichtner, M. A new class of non-corrosive, highly efficient electrolytes for rechargeable magnesium batteries. *J. Mater. Chem. A* **2017**, *5*, 10815–10820. [CrossRef]
52. Kim, H.S.; Arthur, T.S.; Allred, G.D.; Zajicek, J.; Newman, J.G.; Rodnyansky, A.E.; Oliver, A.G.; Boggess, W.C.; Muldoon, J. Structure and compatibility of a magnesium electrolyte with a sulphur cathode. *Nat. Commun.* **2011**, *2*, 427. [CrossRef] [PubMed]
53. Yang, Y.; Wang, W.; Nuli, Y.; Yang, J.; Wang, J. High Active Magnesium Trifluoromethanesulfonate-Based Electrolytes for Magnesium–Sulfur Batteries. *ACS Appl. Mater. Interfaces* **2019**, *11*, 9062–9072. [CrossRef] [PubMed]
54. Wang, W.; Yuan, H.; Nuli, Y.; Zhou, J.; Yang, J.; Wang, J. Sulfur@microporous Carbon Cathode with a High Sulfur Content for Magnesium–Sulfur Batteries with Nucleophilic Electrolytes. *J. Phys. Chem. C* **2018**, *122*, 26764–26776. [CrossRef]
55. Zhao-Karger, Z.; Zhao, X.; Wang, D.; Diemant, T.; Behm, R.J.; Fichtner, M. Performance Improvement of Magnesium Sulfur Batteries with Modified Non-Nucleophilic Electrolytes. *Adv. Energy Mater.* **2014**, *5*, 1401155. [CrossRef]
56. Yao, Y.Y.; Zhan, Y.; Sun, X.Y.; Li, Z.; Xu, H.; Laine, R.M.; Zou, J.X. Advances in Cathodes for High-Performance Magnesium-Sulfur Batteries: A Critical Review. *Batteries* **2023**, *9*, 203. [CrossRef]
57. Wang, C.; Ji, Q.; Chu, R.; Ullah, Z.; Zheng, M.; Dong, X.; Sun, Y.; Li, Q.; Liu, L. High-Performance PDB Organic Cathodes Reinforced by 3D Flower-like Carbon for Lithium-/Sodium-Ion Batteries. *ACS Appl. Energy Mater.* **2021**, *4*, 12641–12648. [CrossRef]
58. Zheng, S.; Yu, S.; Ullah, Z.; Liu, L.; Chen, L.; Sun, H.; Chen, M.; Liu, L.; Li, Q. π-d conjugation regulates the cathode/electrolyte interface in all-solid-state lithium-ion batteries. *J. Mater. Chem. A* **2024**, *12*, 3967–3976. [CrossRef]
59. Li, Y.; Cheng, M.; Liu, Q.; Wang, R.; Ma, W.; Li, X.; Hu, J.; Wei, T.; Liu, C.; Ling, Y.; et al. Toward High-Performance Mg/S Batteries with M4-Assisted $Mg(AlCl_{(4))(2)}$ /PYR14TFSI/DME Electrolyte and MoS_2@CMK/S Cathode. *Small* **2023**, e2307396. [CrossRef]
60. Ren, W.; Wu, D.; NuLi, Y.; Zhang, X.; Yang, J.; Wang, J. A Chlorine-Free Electrolyte Based on Non-nucleophilic Magnesium Bis(diisopropyl)amide and Ionic Liquid for Rechargeable Magnesium Batteries. *ACS Appl. Mater. Interfaces* **2021**, *13*, 32957–32967. [CrossRef]
61. Yang, Y.; Fu, W.; Zhang, D.; Ren, W.; Zhang, S.; Yan, Y.; Zhang, Y.; Lee, S.J.; Lee, J.S.; Ma, Z.F.; et al. Toward High-Performance Mg-S Batteries via a Copper Phosphide Modified Separator. *ACS Nano* **2022**, *17*, 1255–1267. [CrossRef]
62. Ji, Y.; Liu-Théato, X.; Xiu, Y.; Indris, S.; Njel, C.; Maibach, J.; Ehrenberg, H.; Fichtner, M.; Zhao-Karger, Z. Polyoxometalate Modified Separator for Performance Enhancement of Magnesium–Sulfur Batteries. *Adv. Funct. Mater.* **2021**, *31*, 2100868. [CrossRef]
63. Zhou, Z.F.; Chen, B.B.; Fang, T.T.; Li, Y.; Wang, Q.J.; Zhang, J.J.; Zhao, Y.F. A Multifunctional Separator Enables Safe and Durable Lithium/Magnesium-Sulfur Batteries under Elevated Temperature. *Adv. Energy Mater.* **2020**, *10*, 1902023. [CrossRef]
64. Guan, Z.; Ullah, Z.; Zheng, S.; Yang, R.; Zhu, P.; Cheng, Q.; Song, P.; Li, Q.; Liu, L. Single Iron Atom Anchored on ZIF-8 Derived Carbon Framework to Directionally Regulate Lithium Deposition with Minimum Nucleation Overpotential. *Adv. Mater. Interfaces* **2022**, *9*, 2201278. [CrossRef]
65. Zhao, L.; Ning, Y.; Dong, Q.; Ullah, Z.; Zhu, P.; Zheng, S.; Xia, G.; Zhu, S.; Li, Q.; Liu, L. Longer cycle life and higher discharge voltage of a small molecular indanthrone resulting from the extended conjugated framework. *J. Power Sources* **2023**, *556*, 232518. [CrossRef]
66. Chen, S.; Wang, Y.; Sun, Y.; Zhang, D.; Zhang, S.; Zhao, Y.; Wang, J.; Yang, J.; NuLi, Y. Research status and prospect of separators for magnesium-sulfur batteries. *J. Energy Chem.* **2023**, *87*, 225–246. [CrossRef]
67. Robba, A.; Vizintin, A.; Bitenc, J.; Mali, G.; Arčon, I.; Kavčič, M.; Žitnik, M.; Bučar, K.; Aquilanti, G.; Martineau-Corcos, C.; et al. Mechanistic Study of Magnesium–Sulfur Batteries. *Chem. Mater.* **2017**, *29*, 9555–9564. [CrossRef]
68. Xu, Y.; Ye, Y.; Zhao, S.; Feng, J.; Li, J.; Chen, H.; Yang, A.; Shi, F.; Jia, L.; Wu, Y.; et al. In Situ X-ray Absorption Spectroscopic Investigation of the Capacity Degradation Mechanism in Mg/S Batteries. *Nano Lett.* **2019**, *19*, 2928–2934. [CrossRef]
69. Luo, T.; Wang, Y.; Elander, B.; Goldstein, M.; Mu, Y.; Wilkes, J.; Fahrenbruch, M.; Lee, J.; Li, T.; Bao, J.L.; et al. Polysulfides in Magnesium-Sulfur Batteries. *Adv. Mater.* **2023**, *36*, e2306239. [CrossRef]
70. Rashad, M.; Asif, M.; Ali, Z. Quest for magnesium-sulfur batteries: Current challenges in electrolytes and cathode materials developments. *Coord. Chem. Rev.* **2020**, *415*, 213312. [CrossRef]
71. Du, A.; Zhang, Z.; Qu, H.; Cui, Z.; Qiao, L.; Wang, L.; Chai, J.; Lu, T.; Dong, S.; Dong, T.; et al. An efficient organic magnesium borate-based electrolyte with non-nucleophilic characteristics for magnesium–sulfur battery. *Energy Env. Mater.* **2017**, *10*, 2616–2625. [CrossRef]
72. Zhou, X.; Tian, J.; Hu, J.; Li, C. High Rate Magnesium-Sulfur Battery with Improved Cyclability Based on Metal-Organic Framework Derivative Carbon Host. *Adv. Mater. Lett.* **2018**, *30*, 1704166. [CrossRef]
73. Itaoka, K.; Kim, I.-T.; Yamabuki, K.; Yoshimoto, N.; Tsutsumi, H. Room temperature rechargeable magnesium batteries with sulfur-containing composite cathodes prepared from elemental sulfur and bis(alkenyl) compound having a cyclic or linear ether unit. *J. Power Sources* **2015**, *297*, 323–328. [CrossRef]
74. Gao, T.; Ji, X.; Hou, S.; Fan, X.; Li, X.; Yang, C.; Han, F.; Wang, F.; Jiang, J.; Xu, K.; et al. Thermodynamics and Kinetics of Sulfur Cathode during Discharge in $MgTFSI_2$–DME Electrolyte. *Adv. Mater. Lett.* **2018**, *30*, 1704313. [CrossRef]

75. Muthuraj, D.; Ghosh, A.; Kumar, A.; Mitra, S. Nitrogen and Sulfur Doped Carbon Cloth as Current Collector and Polysulfide Immobilizer for Magnesium-Sulfur Batteries. *Chemelectrochem* **2019**, *6*, 684–689. [CrossRef]
76. Du, H.; Zhang, Z.; He, J.; Cui, Z.; Chai, J.; Ma, J.; Yang, Z.; Huang, C.; Cui, G. A Delicately Designed Sulfide Graphdiyne Compatible Cathode for High-Performance Lithium/Magnesium-Sulfur Batteries. *Small* **2017**, *13*, 1702277. [CrossRef] [PubMed]
77. Li, W.; Cheng, S.; Wang, J.; Qiu, Y.; Zheng, Z.; Lin, H.; Nanda, S.; Ma, Q.; Xu, Y.; Ye, F.; et al. Synthesis, Crystal Structure, and Electrochemical Properties of a Simple Magnesium Electrolyte for Magnesium/Sulfur Batteries. *Angew. Chem.* **2016**, *128*, 6516–6520. [CrossRef]
78. Sievert, B.; Häcker, J.; Bienen, F.; Wagner, N.; Friedrich, K.A. Magnesium Sulfur Battery with a New Magnesium Powder Anode. *ECS Trans.* **2017**, *77*, 413–424. [CrossRef]
79. Vinayan, B.P.; Zhao-Karger, Z.; Diemant, T.; Chakravadhanula, V.S.K.; Schwarzburger, N.I.; Cambaz, M.A.; Behm, R.J.; Kubel, C.; Fichtner, M. Performance study of magnesium-sulfur battery using a graphene based sulfur composite cathode electrode and a non-nucleophilic Mg electrolyte. *Nanoscale* **2016**, *8*, 3296–3306. [CrossRef]
80. Lu, Y.; Wang, C.; Liu, Q.; Li, X.; Zhao, X.; Guo, Z. Progress and Perspective on Rechargeable Magnesium-Sulfur Batteries. *Small Methods* **2021**, *5*, e2001303. [CrossRef]
81. Kotobuki, M.; Yan, B.; Lu, L. Recent progress on cathode materials for rechargeable magnesium batteries. *Energy Storage Mater.* **2023**, *54*, 227–253. [CrossRef]
82. Yu, X.; Manthiram, A. Performance Enhancement and Mechanistic Studies of Magnesium–Sulfur Cells with an Advanced Cathode Structure. *ACS Energy Lett.* **2016**, *1*, 431–437. [CrossRef]
83. Chadha, U.; Bhardwaj, P.; Padmanaban, S.; Kabra, D.; Pareek, G.; Naik, S.; Singh, M.; Banavoth, M.; Sonar, P.; Singh, S.; et al. Review—Carbon Electrodes in Magnesium Sulphur Batteries: Performance Comparison of Electrodes and Future Directions. *J. Electrochem. Soc.* **2021**, *168*, 120555. [CrossRef]
84. Zhao-Karger, Z.; Liu, R.; Dai, W.; Li, Z.; Diemant, T.; Vinayan, B.P.; Bonatto Minella, C.; Yu, X.; Manthiram, A.; Behm, R.J.; et al. Toward Highly Reversible Magnesium–Sulfur Batteries with Efficient and Practical $Mg[B(hfip)_4]_2$ Electrolyte. *ACS Energy Lett.* **2018**, *3*, 2005–2013. [CrossRef]
85. Ji, X.; Lee, K.T.; Nazar, L.F. A highly ordered nanostructured carbon–sulphur cathode for lithium–sulphur batteries. *Nat. Mater.* **2009**, *8*, 500–506. [CrossRef]
86. Wu, D.; Ren, W.; Yang, Y.; Wang, J.; NuLi, Y. A Se-Doped $S@CMK_3$ Composite as a High-Performance Cathode for Magnesium–Sulfur Batteries with Mg^{2+}/Li^+ Hybrid Electrolytes. *J. Phys. Chem. C* **2021**, *125*, 25959–25967. [CrossRef]
87. Zheng, C.; Liu, M.; Chen, W.; Zeng, L.; Wei, M. An in situ formed Se/CMK-3 composite for rechargeable lithium-ion batteries with long-term cycling performance. *J. Mater. Chem. A* **2016**, *4*, 13646–13651. [CrossRef]
88. Qian, M.D.; Laskowski, F.A.L.; Ware, S.D.; See, K.A. Effect of Polysulfide Speciation on Mg Anode Passivation in Mg-S Batteries. *ACS Appl. Mater. Interfaces* **2023**, *15*, 9193–9202. [CrossRef] [PubMed]
89. Hu, S.; Hu, Y.; Liu, X.; Zhang, J. Simultaneously enhancing redox kinetics and inhibiting the polysulfide shuttle effect using MOF-derived CoSe hollow sphere structures for advanced Li-S batteries. *Nanoscale* **2021**, *13*, 10849–10861. [CrossRef] [PubMed]
90. Wang, L.; Jankowski, P.; Njel, C.; Bauer, W.; Li, Z.; Meng, Z.; Dasari, B.; Vegge, T.; Lastra, J.M.G.; Zhao-Karger, Z.; et al. Dual Role of $Mo_{(6)}$ $S_{(8)}$ in Polysulfide Conversion and Shuttle for Mg-S Batteries. *Adv Sci.* **2022**, *9*, e2104605. [CrossRef] [PubMed]
91. Gao, T.; Hou, S.; Wang, F.; Ma, Z.; Li, X.; Xu, K.; Wang, C. Reversible $S_{(0)}/MgS_{(x)}$ Redox Chemistry in a $MgTFSI_{(2)}/MgCl_{(2)}$/DME Electrolyte for Rechargeable Mg/S Batteries. *Angew. Chem. Int. Ed. Engl.* **2017**, *56*, 13526–13530. [CrossRef] [PubMed]
92. Zhao-Karger, Z.; Zhao, X.; Fuhr, O.; Fichtner, M. Bisamide based non-nucleophilic electrolytes for rechargeable magnesium batteries. *RSC Adv.* **2013**, *3*, 16330. [CrossRef]
93. Zhao, X.; Yang, Y.; Nuli, Y.; Li, D.; Wang, Y.; Xiang, X. A new class of electrolytes based on magnesium bis(diisopropyl)amide for magnesium–sulfur batteries. *Chem. Commun.* **2019**, *55*, 6086–6089. [CrossRef] [PubMed]
94. Sun, Y.; Zou, Q.; Wang, W.; Lu, Y.-C. Non-passivating Anion Adsorption Enables Reversible Magnesium Redox in Simple Non-nucleophilic Electrolytes. *ACS Energy Lett.* **2021**, *6*, 3607–3613. [CrossRef]
95. Xu, Y.; Zhou, G.M.; Zhao, S.Y.; Li, W.F.; Shi, F.F.; Li, J.; Feng, J.; Zhao, Y.X.; Wu, Y.; Guo, J.H.; et al. Improving a Mg/S Battery with YCl3 Additive and Magnesium Polysulfide. *Adv. Sci.* **2019**, *6*, 1800981. [CrossRef]
96. Zeng, L.; Wang, N.; Yang, J.; Wang, J.; Nuli, Y. Application of a Sulfur Cathode in Nucleophilic Electrolytes for Magnesium/Sulfur Batteries. *J. Electrochem. Soc.* **2017**, *164*, A2504–A2512. [CrossRef]
97. Muldoon, J.; Bucur, C.B.; Oliver, A.G.; Zajicek, J.; Allred, G.D.; Boggess, W.C. Corrosion of magnesium electrolytes: Chlorides—The culprit. *Energy Environ. Sci.* **2013**, *6*, 482–487. [CrossRef]
98. Ruß, C.; König, B. Low melting mixtures in organic synthesis—An alternative to ionic liquids? *Green. Chem.* **2012**, *14*, 2969. [CrossRef]
99. Häcker, J.; Nguyen, D.H.; Rommel, T.; Zhao-Karger, Z.; Wagner, N.; Friedrich, K.A. Operando UV/vis Spectroscopy Providing Insights into the Sulfur and Polysulfide Dissolution in Magnesium–Sulfur Batteries. *ACS Energy Lett.* **2021**, *7*, 1–9. [CrossRef]

molecules

MDPI

Review

Recent Advances on F-Doped Layered Transition Metal Oxides for Sodium Ion Batteries

Hao Wang [1,2], Lifeng Zhou [1], Zhenyu Cheng [1], Liying Liu [1], Yisong Wang [1,2] and Tao Du [1,2,*]

1 State Environmental Protection Key Laboratory of Eco-Industry, School of Metallurgy, Northeastern University, Shenyang 110819, China; 2201664@stu.neu.edu.cn (H.W.)
2 Engineering Research Center of Frontier Technologies for Low-Carbon Steelmaking (Ministry of Education), School of Metallurgy, Northeastern University, Shenyang 110819, China
* Correspondence: dut@smm.neu.edu.cn

Abstract: With the development of social economy, using lithium-ion batteries in energy storage in industries such as large-scale electrochemical energy storage systems will cause lithium resources to no longer meet demand. As such, sodium ion batteries have become one of the effective alternatives to LIBs. Many attempts have been carried out by researchers to achieve this, among which F-doping is widely used to enhance the electrochemical performance of SIBs. In this paper, we reviewed several types of transition metal oxide cathode materials, and found their electrochemical properties were significantly improved by F-doping. Moreover, the modification mechanism of F-doping has also been summed up. Therefore, the application and commercialization of SIBs in the future is summarized in the ending of the review.

Keywords: sodium-ion batteries; transition metal oxides; F-doping; mechanism; electrochemical performance

1. Introduction

Since the industrial revolution [1], the greenhouse effect and extreme weather events more frequently take place, caused by the increasing CO_2 emissions from the fossil fuels [2]. As such, we urgently need to find clean renewable energy sources to avoid climate deterioration [3]. However, renewable energy sources such as wind, solar, wave, and geothermal energy are intermittent, unstable, uncontrollable, and have other disadvantages [4]. Aiming to use these energy sources efficiently, large-scale electrochemical energy storage systems (EESSs) are needed to store this energy and convert it into electricity for continuable use [5]. Therefore, the development of advanced energy storage techniques to store intermittent renewable energy has gradually become more important [6]. Owing to the high safety, the high efficiency of conversion, the low cost, and environmental friendliness, rechargeable batteries are regarded as one of the most competent representatives of energy storage and conversion technologies. In addition, the peaks and valleys of the daily electricity consumption usually do not coincide with the peaks and valleys of renewable energy generation, thus storing electricity via batteries can perfectly solve this problem [7].

Since LIBs were discovered in the 1970s and officially commercialized for powering electronic products in 1991 [8], they have achieved great success with great advantages in practical applications, but they are limited by the extremely uneven distribution and the high price of lithium ore [9]. Their electrolytes contain non-aqueous flammable organic solvents that pose a serious fire hazard in the event of thermal runaway [10]. In the 1980s, researchers discovered that Na^+ could be reversibly (de)inserted from $NaCoO_2$ layered oxides. Since then, SIBs have gradually attracted the attention of researchers. Because sodium and lithium are located in the same main group in the periodic table of elements and have similar physical and chemical properties, SIBs and LIBs have similar charging and discharging operating principles [11]. Due to the uniform distribution of sodium resources in the world from sea water, SIBs have the advantages of low cost, abundant element reserves, environmental friendliness, and high safety [12], and have great potential to replace LIBs as a new generation of energy storage devices. So, SIBs have been considered as one of the most promising energy storage systems to substitute for LIBs in the field of large-scale energy storage [13]. However, at the current stage of SIB research, there are many problems that are still restricting their future commercialization application to fully meet the requirement for future EESSs. (1) Lower energy density: the current commercial LIBs energy storage density is much higher than the actual energy storage density of SIBs [14]; (2) Structural damage: due to the larger atomic radii of Na^+ than that of Li^+, Na^+ repeatedly embedding in and out of the cathode material will cause greater damage to the structure of the cathode material, meaning that the conventional cathode for LIBs cannot be directly applied to SIBs [15]; (3) Complex phase transition: the complex outer ring electronic structure and the wide range of valence states of transition metals lead to crystal structure distortion and complex electrochemical behaviors during the charging and discharging process [16]; (4) Instability in air: SIBs normally show poor air stability, which may originate from insertion into the interlayer sites of water or carbon dioxide and the formation of alkaline species such as NaOH or Na_2CO_3, leading to surface structure degradation or even failure [17]. Therefore, for the development of the energy industry, new electrode materials with less shortcomings for SIBs should be developed to meet the needs of commercial applications. To achieve high performance for SIBs, the best endeavors have been devoted to exploiting a broad range of new materials, pursuing an outstanding cycling stability, ultra-stable structure, and improving Na^+ transport kinetics.

It is obvious that the cathode material is the key to the electrochemical performance of SIBs based on the energy storage mechanism [18]. A variety of compounds have been used as cathode materials for SIBs, such as layered transition metal oxides [19], polyanionic compounds [20], Prussian blue analogues [21], and organic materials [22]. Among the leading cathode candidates, layered transition metal oxide cathodes have aroused much interest due to their higher special capacity, faster Na^+ diffusion rate, and easier industrial synthesis. In recent years, researchers have made great contributions and significant breakthroughs in the optimization of Na-based layered oxide cathodes and the improved performance of various materials are summarized in Table 1 and Figure 1, respectively. Furthermore, ion doping, structural optimization design, and interface optimization as important modification methods have been summarized in other studies [23]. In particular, the ion doping for layered cathode materials should be studied in depth and classified by different kinds of anion/cation doping, different ratios of one ion doping, and different site doping. These methods can bring different performance enhancement effects. Especially, F is considered as a promising dopant anion for improving electrochemical properties due to its extremely high electronegativity [24].

Table 1. Summary of the performances of the recent reported layered oxide cathode materials for SIBs and LIBs [3,4,6,11,17,25–44].

Materials	Synthesis Method	Cycling Performance after F-Doping	Ref.
(SIBs)			
$Na_{2/3}Ni_{1/3}Mn_{2/3}O_{1.95}F_{0.05}$	Solid-state high temperature reaction	88%→95% (2.0–4.0 V 400 cycles at 340 mA g^{-1})	[32]
$Na_{0.67}Ni_{0.15}Fe_{0.2}Mn_{0.65}O_{1.95}F_{0.05}$	Coprecipitation method	38%→88% (1.6–4.2 V 50 cycles at 100 mA g^{-1})	[34]
$Na_{0.44}MnO_{1.93}F_{0.07}$	Oxalate precursor method	36%→86% (2.0–4.0 V 150 cycles at 200 mA g^{-1})	[28]

Table 1. *Cont.*

Materials	Synthesis Method	Cycling Performance after F-Doping	Ref.
$Na_{1.2}Mn_{0.8}O_{1.5}F_{0.5}$	Solid-state method	76%→90% (1.5–4.0 V 300 cycles at 1000 mA g^{-1})	[38]
$Na_{0.67}Ni_{0.33}Ti_{0.3}Mn_{0.37}O_{1.9}F_{0.1}$	Solid-state reaction	62%→82% (2.2–4.5 V 300 cycles at 170 mA g^{-1})	[33]
$Na_{0.6}Ni_{0.3}Mn_{0.7}O_{1.95}F_{0.05}$	Solid-state reaction	65%→78% (1.5–3.8 V 900 cycles at 1000 mA g^{-1})	[43]
$Na_{0.6}Ni_{0.05}Mn_{0.95}O_{1.95}F_{0.05}$	Co-precipitation route and solid-state reaction	71%→75% (2.5–4.0 V 960 cycles at 180 mA g^{-1})	[25]
$Na(Ni_{1/3}Fe_{1/3}Mn_{1/3})_{0.99}Mo_{0.01}O_{1.99}F_{0.01}$	Solid-state reaction	83%→92% (2.0–4.0 V 100 cycles at 130 mA g^{-1})	[29]
$NaNi_{1/3}Fe_{1/3}Mn_{1/3}O_{1.99}F_{0.01}$	Solid-state reaction	63%→96% (2.0–4.0 V 70 cycles at 150 mA g^{-1})	[27]
$Na_{0.66}Ca_{0.01}Ni_{0.33}Mn_{0.67}O_{1.98}F_{0.02}$	Solid-state reaction	72%→94% (2.0–4.2 V 500 cycles at 200 mA g^{-1})	[35]
$Na_{2/3}Ni_{1/3}Mn_{2/3}O_2$-LiF	Sol–gel method	36%→64% (2.0–4.2 V 100 cycles at 100 mA g^{-1})	[40]
$Na_{0.67}Ni_{0.28}Zn_{0.05}Ti_{0.05}Mn_{0.62}O_{1.95}F_{0.05}$	Sol–gel method	83%→90% (2.5–4.2 V 100 cycles at 170 mA g^{-1})	[44]
$Na_{0.67}Fe_{0.1}Cu_{0.1}Mn_{0.8}O_{1.9}F_{0.1}$	Solid-state reaction	52%→88% (1.5–4.5 V 200 cycles at 200 mA g^{-1})	[41]
$Na_{0.67}Li_{0.1}Fe_{0.4}Mn_{0.5}O_2$-F	Co-precipitation route and solid-state reaction	19%→67% (1.5–4.0 V 200 cycles at 200 mA g^{-1})	[30]
$Na_{0.7}MnO_{2.05}$-0.1TF	Solid-state reaction	82%→96% (2.0–4.2 V 200 cycles at 1000 mA g^{-1})	[45]
(LIBs)			
$LiNi_{0.8}Co_{0.1}Mn_{0.1}O_2$-F	Solid-state reaction	62%→96% (2.8–4.2 V 100 cycles at 100 mA g^{-1})	[42]
$LiNi_{0.83}Co_{0.1}1Mn_{0.06}O_2$-MoF	Co-precipitation method	60%→80% (2.8–4.3 V 160 cycles at 200 mA g^{-1})	[36]
$LiNi_{0.8}Co_{0.15}Al_{0.05}O_{1.96}F_{0.04}$	Co-precipitation method	72%→98% (2.8–4.3 V 100 cycles at 400 mA g^{-1})	[39]
$LiNi_{0.8}Co_{0.1}Mn_{0.1}O_2$-$LiPF_6$	Solid-state reaction	77%→94% (3.0–4.3 V 100 cycles at 85 mA g^{-1})	[6]
$LiNi_{0.8}Co_{0.1}Mn_{0.1}O_2$	Solid-state reaction	45%→88% (2.8–4.3 V 200 cycles at 200 mA g^{-1})	[5]
$Li_{1.2}(Ni_{0.13}Co_{0.13}Mn_{0.54})_{0.96}Cr_{0.04}O_{1.985}F_{0.03}$	Sol-gel method with citric acid as chelating agent	79%→95% (2.0–4.8 V 50 cycles at 50 mA g^{-1})	[26]
$Li(Ni_{0.8}Co_{0.15}Al_{0.05})_{0.99}Zr_{0.01}O_{1.99}F_{0.01}$	Solid-state reaction	78%→94% (2.0–4.8 V 200 cycles at 200 mA g^{-1})	[31]
$LiNi_{0.88}Co_{0.09}Al_{0.03}O_2$-F	Solid-state reaction	68%→85% (2.8–4.3 V 150 cycles at 360 mA g^{-1})	[37]
$Li_{1.15}Na_{0.05}Ni_{0.13}Co_{0.13}Mn_{0.54}O_{1.99}F_{0.01}$	Carbonate co-precipitation method	89%→96% (2.5–5.0 V 100 cycles at 200 mA g^{-1})	[7]
$Li_{1.2}Ni_{0.13}Co_{0.13}Mn_{0.54}O_2$-$FATO_2$	Hydro/solvothermal method	79%→92% (2.0–4.8 V 500 cycles at 200 mA g^{-1})	[4]
$Li_{1.15}Na_{0.05}Ni_{0.13}Co_{0.13}Mn_{0.54}O_{1.95}F_{0.05}$	Facile co-precipitation method	73%→85% (2.0–4.7 V 300 cycles at 200 mA g^{-1})	[11]

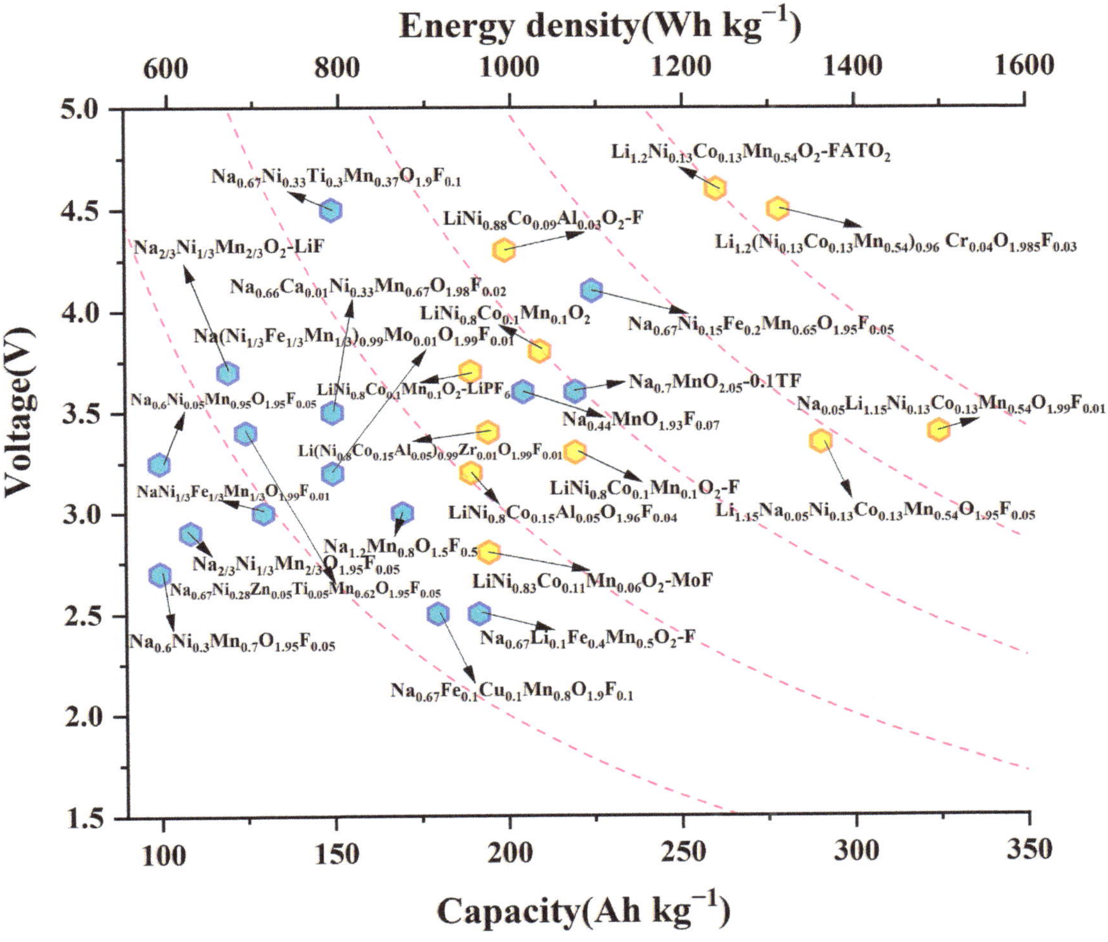

Figure 1. The performance comparison of the reported oxide cathode materials, including plateau potential (V vs. Na^+/Na and V vs. Li^+/Li for the blue and yellow hexagons, respectively) and energy density (Wh kg^{-1}) [3,4,6,11,17,25–45].

Some previous reviews [46–48] have pointed out that F ion-doping contributes to the enhancement of electrochemical performance of materials. But their classification is not detailed, the explanation is not specific, and the content is not perfect enough. In addition, the role of F ion-doping and the mechanisms of performance improvement are not specifically indicated. In the case of a certain elemental composition, none of these reviews has pointed out in detail the effect on the electrochemical performance after doping with F ions. As opposed to categorizing materials from a structural point of view (P2, P3, O2), we believe analyzing materials composed of the same elements is more conducive to understanding the role of F ions. In this paper, we summarize typical F-doped layered oxide cathode materials for SIBs, and then review the modification mechanism of F-doping and the corresponding energy storage mechanism (Figure 2), including (1) increasing Na^+ diffusion rate by widening the layer spacing; (2) enhancing lifespan by reducing the irreversible multiphase transformation and preventing the distortion collapse of crystal structure; (3) accelerating Na^+ transport kinetics by lowering the energy barrier. The systematic summary will provide some support for the future development and application of SIBs. In addition, we make a reasonable outlook on the research of F-doped layered oxide cathode materials and the future application potential for SIBs.

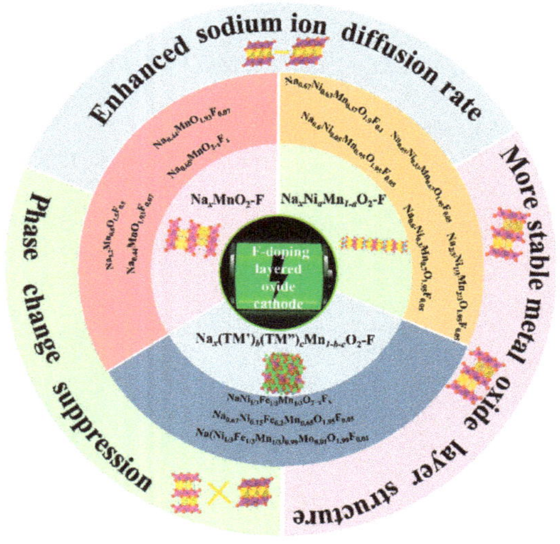

Figure 2. The classification of F-doped layered metal oxides: Na_xMnO_2-F, $Na_xNi_aMn_{1-a}O_2$-F, $Na_x(TM')_b(TM'')_cMn_{1-b-c}O_2$-F, and the corresponding modification effects.

2. F-Doped Layered Oxide Cathode Materials

The roadmap of layered metal oxide materials briefly is shown in Figure 3, including the discovery and development, the energy storage mechanism, and the structure information. With the progress of the in-depth study on it below, layered metal oxides as cathode materials for SIBs have been regarded as one of essential candidates recently.

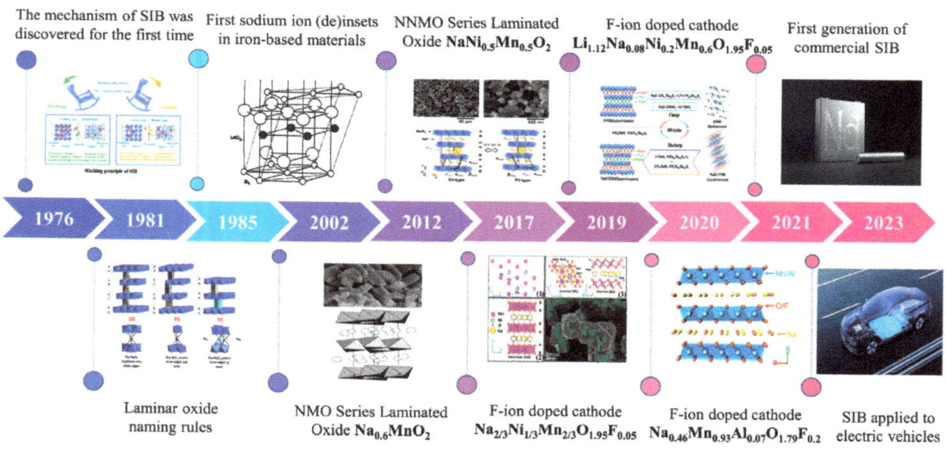

Figure 3. The roadmap of layered metal oxide cathode materials in SIBs [1,30,49–54]. Copyright 1991, Royal Society of Chemistry. Copyright 1994 Published by Elsevier Ltd. Copyright 2012, American Chemical Society. Copyright 2017, American Chemical Society. Copyright 2019, Elsevier Ltd. Copyright 2020, WILEY-VCH Verlag GmbH & Co. KGaA, Weinheim, Germany.

2.1. Na_xMnO_2-F

In the 1970s, a systematic study of the structure and physical properties of A_xMO_2 compounds was conducted (A = Li, Na, K; M = Cr, Mn, Fe, Co) [49]. In the 1980s, the variety of MO_2 layer stackings in A_xMO_2 phases was devised [55]. Specially, $NaMnO_2$(NMO) had been reported earlier and are well-known for their high energy densities and safety [56].

Mn-based layered oxides have many advantages such as being inexpensive, having high special capacity, and long life span, but they always face structural collapse caused by the phase from P2 to O2 at high voltages [57]. Meanwhile, the production of Mn^{3+} is extremely prone to induce Jahn–Teller (J-T) effects [58], further leading to severe distortion and deformation of the lattice structure, reducing the reproducibility of Na^+ (de)insertion, and affecting the electrochemical properties of the material. F ions can help to solve the above problems, and F ions can effectively improve the resistance of the cathode material and electrolyte in the high voltage range, as well as help to improve the structural integrity of the TM layer and reduce the dissolution of Mn^{3+}; many research results have proven the role of F ions. Layered sodium-rich $Na_{1.2}Mn_{0.8}O_{2-y}F_y$ (y = 0–0.5) has been developed based on NMO [38], which was obtained in a basic solid-state mixing method. Under the comparison of the electrochemical performance of samples with different F contents, it was determined that $Na_{1.2}Mn_{0.8}O_{1.5}F_{0.5}$ has the best electrochemical performance with a specific capacity of 174 mAh g^{-1} at a current density of 10 mA g^{-1}. It can also maintain a retention of 68% at a current density of 1000 mA g^{-1} after 300 cycles. However, the increase of Mn^{3+} leads to the crystal structure transformation from the O3 phase to the P2 phase, that makes the poor cycle life. In addition, as shown as in Figure 4a,b, Ti- and F-co-doped cathode oxide P2-$Na_{0.7}MnO_{2.05}$ (Ti 2%, F 8%) (NMO-0.1TF) has been prepared by adjusting different ratios of NH_4TiF_6/NMO [45], and it shows a very high specific capacity (227 mAh g^{-1} at 20 mA g^{-1} in the voltage range of 2.0–4.2 V and a good lifespan (96.2% after 200 cycles at 1000 mA g^{-1}) based on a good structural stability. In the rate test, it has a high specific capacity of 76 mAh g^{-1} at 3000 mA g^{-1}, and when the current density returned to 20 mA g^{-1}, it could return to a very high specific capacity of 227 mAh g^{-1}.

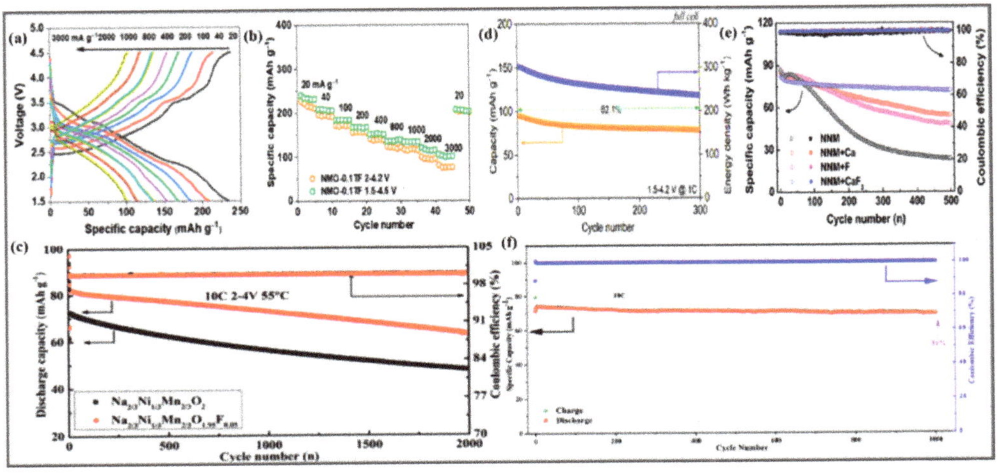

Figure 4. (**a**) Galvanostatic charging and discharging curves and (**b**) rate performance of NMO-0.1TF cathode at different rates in the voltage range of 1.5–4.5 V [45]. Copyright 2023, Chinese Chemical Society. (**c**) High temperature cycling performance (55 °C) of $Na_{2/3}Ni_{1/3}Mn_{2/3}O_2$ and $Na_{2/3}Ni_{1/3}Mn_{2/3}O_{1.95}F_{0.05}$ cathode at 10 C [32]. Copyright 2020, WILEY-VCH Verlag GmbH & Co. KGaA, Weinheim, Germany. (**d**) Cycling stability of NNMTOF as cathode sodium-ion full battery [33].

Copyright 2021, Science Press and Dalian Institute of Chemical Physics, Chinese Academy of Sciences. Published by ELSEVIER B.V. and Science Press. (**e**) Comparison of cycling performance of $Na_{0.67}Ni_{0.33}Mn_{0.67}O_2$ cathodes doped with/without F at 1C in the voltage range of 2.0–4.3 V [35]. Copyright 2021, Royal Society of Chemistry. (**f**) Cycling stability of sodium-ion full batteries based on NNZMTOF and commercial hard carbon at 1C [42]. Copyright 2023, Royal Society of Chemistry.

2.2. $Na_xNi_aMn_{1-a}O_2$-F

As we all know, F-doping can bring some performance improvement of NMO, but some drawbacks of NMO limit performance enhancement [59]. Mn^{3+} easily dissociates from the skeleton of metal oxide and dissolves into the electrolyte during Na^+ (de)insertion, leading to structural damage [60]. In addition, the high voltage might induce irreversible phase change between P2 and O2, resulting in poor cycle stability [61]. Low average voltage affects energy density. Moreover, these problems result in a series of phase transformations and structural evolutions due to the valence changes of transition metal ions during the charging/discharging process [62]. To address these issues, transition metal ions are doped in NMO. The introduction of Ni can effectively improve the average voltage and make use of $Ni^{2+/3+/4+}$ redox couples [63] during the electrochemical cycling. Previous reports have shown that NNMO has excellent electrochemical properties. $Na_{2/3}Ni_{1/3}Mn_{2/3}O_2$ [64] shows a specific capacity of 145 mAh g^{-1} at 0.1C and a capacity retention of 89% after 50 cycles. Compared with NMO, the structure of NNMO contains the higher number of Mn^{4+} ions; however, a large amount of Mn^{4+} without electrochemical activity will make the specific capacity decrease [65]. The introduction of Ni^{2+} will also make the specific capacity decrease and cycling stability deteriorate [66], while F ion-doping can perfectly solve the above contradictory phenomena, which can help to prepare high performance cathode materials. Thus, we review the current research progress of F-doped $NaNi_xMn_{1-x}O_2$ (NNMO) layered oxide cathode material in this section.

$Na_{0.6}Ni_{0.05}Mn_{0.95}O_{2-x}F_x$ (x = 0.00, 0.02, 0.05, 0.08) [25] has been synthesized by a simple co-precipitation method with the molar ratio of Mn:Ni = 95:5. $Na_{0.6}Mn_{0.95}Ni_{0.05}O_{1.95}F_{0.05}$ displays a reversible specific capacity of 80.76 mAh g^{-1} at 2C with a capacity retention of 75.0% after 960 cycles and a capacity decay of 0.026% per cycle. This is performed to further optimize the ratio of Mn:Ni for better electrochemical performance and more remarkable capacity retention, especially to investigate the specific role of F^- in terms of structural stability. A series of $Na_{2/3}Ni_{1/3}Mn_{2/3}O_{2-x}F_x$ (x = 0, 0.03, 0.05, 0.07) [32] have been synthesized by a simple high temperature solid phase reaction that exhibits excellent cycling stability. The $Na_{2/3}Ni_{1/3}Mn_{2/3}O_{1.95}F_{0.05}$ still reaches a specific capacity of 61 mAh g^{-1} at 10C, as shown in Figure 4c. Therefore, the F-doped NNMO reached a reasonable ratio of each element, and the P2-$Na_{2/3}Ni_{1/3}Mn_{2/3}O_2$ has become one of the most promising cathode candidates for SIBs showing a high electrochemical performance due to its high theoretical capacity in a wider voltage window. Similar to the research process of NMO, the bi-ion co-doped NNMO also have been gradually investigated. P2-$Na_{0.67}Ni_{0.33}Mn_{0.67}O_{2-y}F_y$ and P2-$Na_{0.67}Ni_{0.33}Mn_{0.67-x}Ti_xO_{1.9}F_{0.1}$ cathode materials [33] have been synthesized by a simple conventional solid-phase reaction. Among them, the P2-$Na_{0.67}Ni_{0.33}Mn_{0.37}Ti_{0.3}O_{1.9}F_{0.1}$ has a specific capacity of 128.1 mAh g^{-1}at 2C in the high voltage of 2.0–4.4 V, and maintains a 77.2% retention ratio after more than 300 cycles, as shown in Figure 4d. In order to effectively reduce the electron repulsion of O^{2-}-O^{2-} at low capacity and suppress the phenomenon of rapid decay due to the P2-O2 phase transition when charging above 4.2 V, the Zn/Ti/F co-doped $Na_{0.67}Ni_{0.33}Mn_{0.67}O_2$ (NNZMTOF) cathode was prepared by the sol-gel method [44], with excellent capacity retention (86% after 1000 cycles), as shown in Figure 4f. Moreover, Ca/F co-doped $Na_{0.67-x}Ca_xNi_{0.33}Mn_{0.67}O_{2-2x}F_{2x}$ (x = 0, 0.01, 0.03 and 0.05) [35] has been prepared by a high temperature solid phase reaction, as shown in Figure 4e. It is worth noting that the capacity retention after 500 cycles at 1C increased from 27.1% (the undoped material) to 87.2% (the co-doped material). In order to more comprehensively show the synergistic effect of the F substitution with other cations, Min, K et al. [67] conducted a large-scale co-doped experiment and selected more than 12 transition metal cations and F^- to optimize NNMO, determine the optimal doping position, and examine the five co-doped ion pairs by calculating the total energy of the structure. Te-, Sb-, Hf-, Y-, and Ti-F are more favorable to enhance the structural stability of NNMO. As a result, the cations and F^- co-doping materials are more effective at dramatically enhancing the whole performance than the single doped cations as demonstrated via many data comparisons.

2.3. $Na_x(TM')_b(TM'')_cMn_{1-b-c}O_2$-F (TM', TM'' = Transition Metal Elements)

Since it was significantly proven that the F-doping had greater improvement for the performance of SIBs, the iron-doped layered oxide cathodes also gradually became popular [68]. This is because the weak interaction between layers and the hollow two-dimensional space provides great convenience for the migration of Na^+ [11]. Additionally, it suppresses the migration of other transition metal ions into the Na layer, hindering the crystal structure transformation. However, the Fe^{3+} will migrate, leading to structural instability when the Na^+ (de)inserts [67]. So, how to solve this problem has become one of the important research directions to enhance the structural stability of iron-doped cathodes [68].

The O3-type $NaNi_{1/3}Fe_{1/3}Mn_{1/3}O_{2-x}F_x$ (x = 0, 0.005, 0.01, 0.02, NNFeMO-F) cathode materials [27] were prepared by using a simple solid-phase reaction. Its capacity was 110 mAh g^{-1} at 150 mA g^{-1} and the capacity retention rate was 85% after 70 cycles, as shown in Figure 5a. Partial replacement of O by F can significantly improve the cycling performance of materials. As shown in Figure 5b, NNFeMO-F shows a good modification effect. The O3-type co-doping Mo and F increase the reversible capacity and cycling stability [29]. In addition, the P2-type $Na_{0.67}Ni_{0.15}Fe_{0.2}Mn_{0.65}O_{1.95}F_{0.05}$ (NNFeMO-$F_{0.05}$) [34] was firstly synthesized by the co-precipitation method, and in situ Mg substitution was performed by electrochemical method, where Mg was directly introduced into the Na site of the sodium-deficient P2 phase cathode material without occupying the TM layer. The obtained P2-type NNFeMMO-$F_{0.05}$ has an excellent electrochemical performance with a remarkable specific capacity of 229 mAh g^{-1}, as shown in Figure 5c. An appropriate F content greatly improves the cycling performance of the cathode, and the combined effect of F and Mg demonstrates excellent electrochemical performance. Even so, the effect of the F-doping still needs a more in-depth and detailed study in the future.

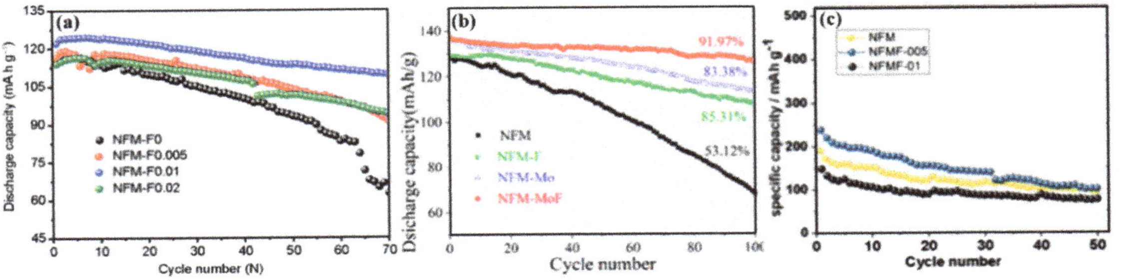

Figure 5. (**a**) Cycling performance of the $NaNi_{1/3}Fe_{1/3}Mn_{1/3}O_{2-x}F_x$ (x = 0, 0.005, 0.01, 0.02) at 150mAg^{-1} [27]. Copyright 2017, Science China Press and Springer-Verlag GmbH Germany. (**b**) Cycling performance at 1 C for 100 cycles of all cathodes in 2.0–4.0 V [29]. Copyright 2022, Elsevier Ltd. All rights reserved. (**c**) cycle performance of P2-type $Na_{0.67}Ni_{0.15}Fe_{0.2}Mn_{0.65}O_{1.95}F_{0.05}$ (x = 0, 0.05, 0.1) [34]. Copyright 2021, Elsevier B.V. All rights reserved.

3. F-Doping Modification Mechanism

3.1. Increasing the Na^+ Diffusion

The F-doped layered oxide cathode can effectively improve Na^+ transport kinetics, and the diffusion rate of Na^+ can be accelerated by more than 100 times if the doping ratio and location are reasonable in Figure 6a [43], as analyzed and calculated by Electrochemical Impedance Spectroscopy (EIS) curve and Constant Current Titration Technique (GITT), showing better electrochemical modification advantages. Boosting electron mobility is mainly due to the following reasons:(1) Increasing the layer spacing as in Figure 6b. With the increasing amount of F in the crystal structure, the angle of the diffraction peak (002) first slightly decreases and then increases [69]. F^- will have a slight effect on the local structure and cell size of the crystal, but it cannot change the overall structure of the crystal due to the fact that a small number of F^- replace the position of the O^{2-}, and the ionic radius of the F^- (1.33 Å) is slightly smaller than that of the O^{2-} (1.40 Å) [34]. Due to the radius change from Mn^{4+} to Mn^{3+}, the crystal changes can lead to performance changes. In detail, the doping of F^- makes the metal oxide layer spacing increase significantly, and the larger layer spacing will facilitate the Na^+ (de)insertion conveniently, as shown in Figure 6c [70], which is conducive to the Na^+ diffusion in metal oxide layers. (2) Reducing the energy barrier of the Na^+ migration is a key factor for enhancing electrochemical performance. Therefore, studying the intensity of the energy barrier of Na^+ migration and the effect of the F-doping are the important ways to analyze the mechanism of material modification [71].Transition state theory and density functional theory (DFT) in Figure 6d,e show that lower migration energy means faster diffusion rate. In the study of Ti- and F-co-doped NMO [45], the Na^+ migration energy (0.80 eV) in the NMO-0.1TF crystal structure was obtained by DFT, while in the NMO crystal structure, the migration energy was as high as 1.26 eV.

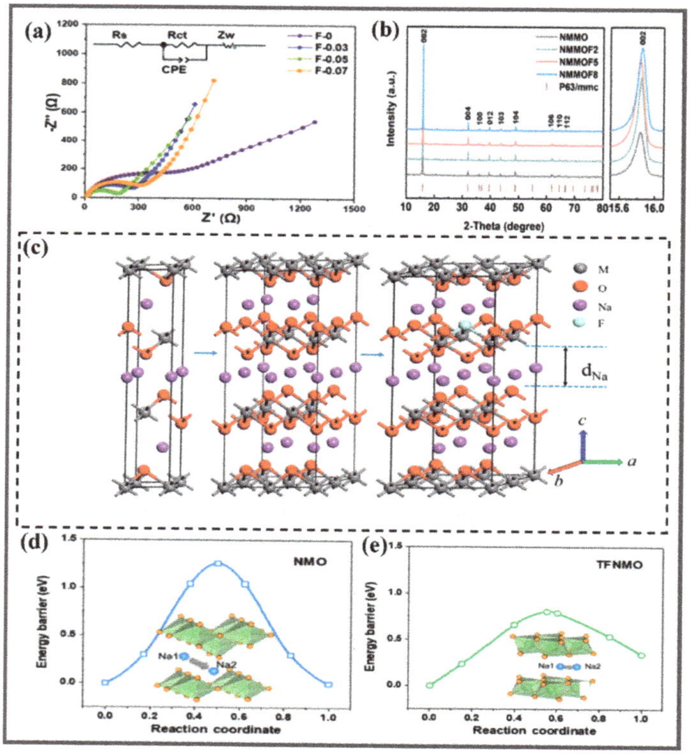

Figure 6. (**a**) EIS and equivalent circuits of NMO with different F-doping contents [43]. Copyright 2021, American Chemical Society. (**b**) X-ray diffraction of the NNMO material with different F-doping contents, and the detailed (002) diffraction peaks [69]. Copyright 2022, Published by Elsevier Ltd. (**c**) Schematic illustration of the supercell NMO and its F-doping [70]. (**d**,**e**) DFT calculation of $Na_{0.7}MnO_{2.05}$ (*P63/mmc*) and NMO (*Pmmn*) [45]. Copyright 2023, Chinese Chemical Society.

3.2. Enhancing the Structural Stability

Due to the strong electronegativity and the single valence state of the F^-, the cathode materials form the ultra-stable structure by the following aspects. (1) Suppressing the J-T effects. As shown in Figure 7a, the results of X-ray photoelectron spectroscopy (XPS) measurements show that the Mn^{3+}/Mn^{4+} ratio increases significantly with the increase of F^- content due to charge balance, and it is further verified that F-doping can reduce the number of Mn^{4+}, and the increased Mn^{3+} will occupy the Ni^{2+} sites and destroy the Ni^{2+}/Mn^{4+} redox ion pairs, thus suppressing the J-T effects to improve structural stability [43]. (2) Increasing the disorder of transition metal sites and preventing Na^+/vacancy ordering. The disorder induced by F-doping can break the crystal structure distortion due to J-T effects and facilitate the structural stability [32]. Based on the DFT results, in the structure of $Na_{0.6}Mg_{0.3}Mn_{0.7}O_{1.95}F_{0.05}$, the chemical bonding between Mn and F strongly increases, and less Mn^{3+} reduces the distorted collapse of the crystal structure and structural transitions [69]. (3) Preventing the collapse of the P2 phase structure. The strong Na-F bond can enhance the excessive Na^+ extraction and intercalation to improve the structural stability [25]. (4) Inducing partial reduction of Mn^{4+} to Mn^{3+} via charge neutrality. In Figure 7b,c, it is confirmed by electron energy loss spectroscopy (EELS) that Ni^{2+}/Ni^{3+} as a pair of oxidation and reduction are mainly responsible for charge compensation during the charge/discharge process, while the redox couples of Mn^{3+}/Mn^{4+} are also involved for increasing the specific capacity [32].

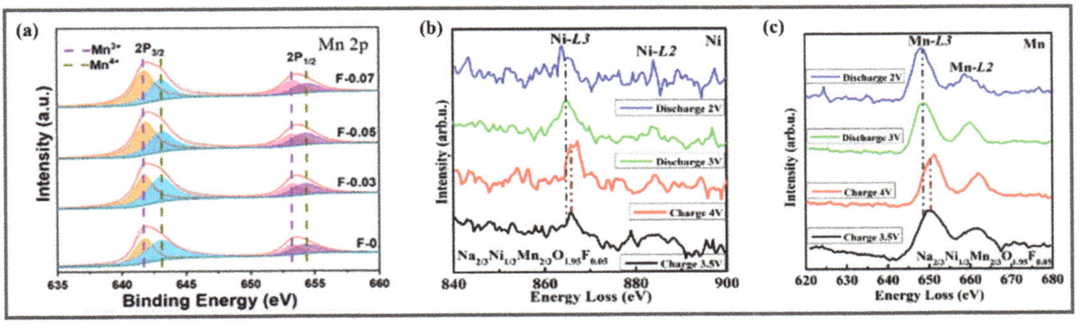

Figure 7. (**a**) Mn 2p high-resolution spectra of XPS patterns of the $Na_{0.6}Mn_{0.7}Ni_{0.3}O_{2-x}F_x$ samples [43]. Copyright 2021, American Chemical Society. EELS spectra of $Na_{2/3}Ni_{1/3}Mn_{2/3}O_2$ [32]: (**b**) Ni L-edge signals, (**c**) Mn L-edge signals. Copyright 2020, WILEY-VCH Verlag GmbH & Co. KGaA, Weinheim, Germany.

3.3. Suppress Phase Transition and Change Surface Morphology

As shown in Figure 8a, in situ X-ray diffraction (XRD) was conducted to monitor the structural change, and the cathode presented highly reversible transitions. During the Na^+ extraction and intercalation process, the reversible P2-P4 phase transition of cathode material after doping replaced the irreversible P2-O2 phase transition in a high voltage window, exhibiting excellent structural reversibility and integrity [33]. Before Na^+ extraction, the overall structure remains intact. After Na^+ were removed, the thermodynamic stability of the O3-type structure usually decreased in virtue of the easy collapse of the hollow part of the structure caused by the removal of Na^+, but Ti-F co-doping has the best effect on inhibiting phase transition and plays the most important role in structural stability and integrity [32]. Meanwhile, due to the strong electronegativity of F, the arrangement and distribution of TM elements (Ni, Fe, Mn) becomes disordered, which is favorable to suppress the structural distortion. As shown in Figure 8b, for $Na_{0.67}Ni_{0.15}Fe_{0.2}Mn_{0.65}O_{1.95}F_{0.05}$ (NFMF-005), the fluctuation range of the lattice volume is smaller than that of $Na_{0.67}Ni_{0.15}Fe_{0.2}Mn_{0.65}O_2$(NFM), indicating that the F substitution effectively suppresses the phase transition and structural distortion during the electrochemical cycling, while the plateau voltage is hysteretic, which can be explained by the fact that the F substitution modification delays the phase transition of the P2 structure, and a more stable structure has better cycling stability [34]. At the same time, doping makes it easier to form lamellar sheets, loosening the microspheres, and forming a hierarchical structure, which is conducive to accelerate the fast transport kinetics of Na^+ [43]. In addition, when we add the amount of F^-, the particle size increases slightly, which is consistent with earlier reports of lamellar oxides [72,73]. An appropriate amount of F-doping can improve the taper density of the layered oxide, while the difference in particle size affects the contact area between the electrode and the electrolyte, which is very important for a high reversible capacity and remarkable capacity retention. However, there are also studies that promote phase transition, as shown in Figure 8c. With F-doping, the number of Mn^{3+} increases close to that of Mn^{4+} in the study of P2-$Na_{1.2}Mn_{0.8}O_{1.5}F_{0.5}$ cathode material, which promotes the transformation of structure from O3 to P2 and thus facilitates Na^+ migration kinetics [38]. Not coincidentally, as shown in Figure 8d, in the study of the modified enhancement of $Na_{0.44}MnO_{1.93}F_{0.07}$ [28], the tunnel structure of the low sodium content is modified into a laminar tunnel composite structure after the F^- are doped into it, which has both the stability of the tunnel structure and the high reversible capacity of the laminar structure. Therefore, the layered tunnel intergrowth structure exhibits exceptional rate performance as well as outstanding cycling stability. In addition, the reaction temperature plays an important role in the formation of the material structure in this research article.

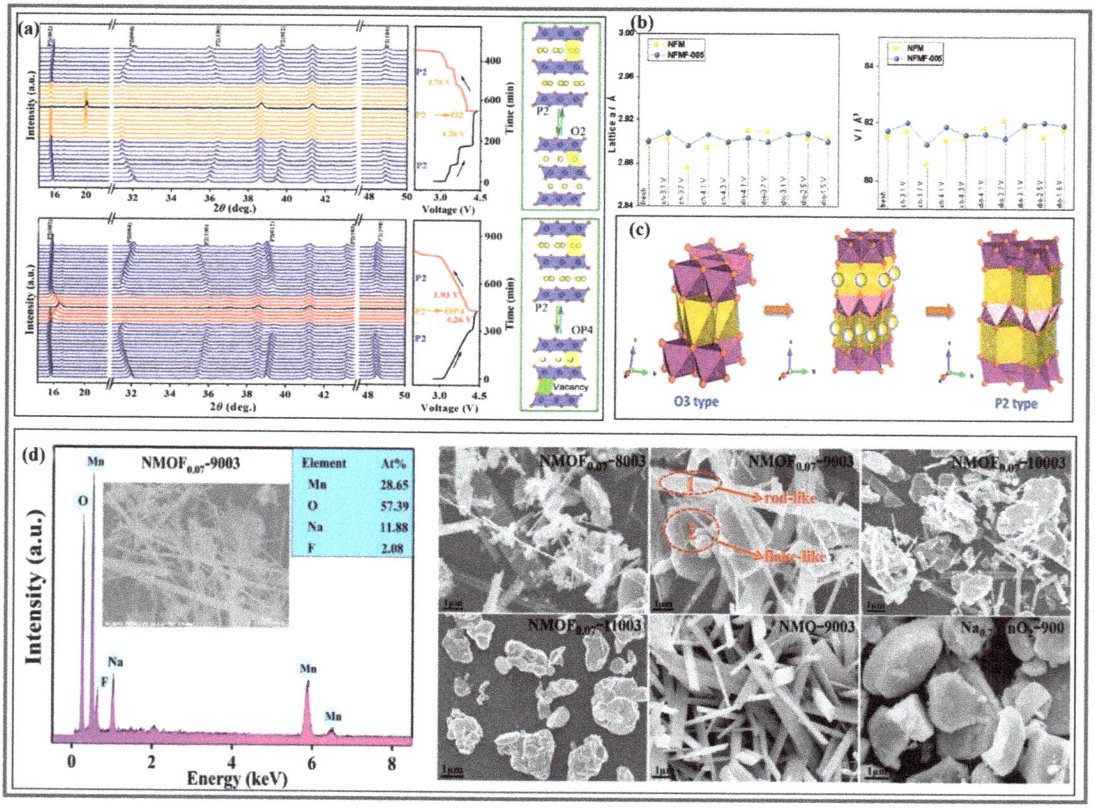

Figure 8. (**a**) In situ XRD patterns of $Na_{0.67}Ni_{0.33}Mn_{0.67}O_2$(NNMO) and $Na_{0.67}Ni_{0.33}Mn_{0.37}Ti_{0.3}O_{1.9}F_{0.1}$(NNMTi$_{0.3}$OF) during the first charge/discharge process [33]. Copyright 2021, Science Press

and Dalian Institute of Chemical Physics, Chinese Academy of Sciences. Published by ELSEVIER B.V. and Science Press. (**b**) Structure parameters of lattice parameter and unit cell volume of $Na_{0.67}Ni_{0.15}Fe_{0.2}Mn_{0.65}O_2$ (NFM) and $Na_{0.67}Ni_{0.15}Fe_{0.2}Mn_{0.65}O_{1.95}F_{0.05}$ (NFMF-005) under different charge/discharge states [34]. Copyright 2021, Elsevier B.V. All rights reserved. (**c**) Schematic of the cathode structure evolution based on F-doping [38]. Copyright 2023, American Chemical Society. (**d**) EDS elemental mapping images of $NMOF_{0.07}$-900 and FE-SEM images of prepared materials for $NMOF_{0.07}$-800, $NMOF_{0.07}$-900, $NMOF_{0.07}$-1000, $NMOF_{0.07}$-1100, NMO-900 and $Na_{0.7}MnO_2$-900 [28]. Copyright 2019, Elsevier B.V. All rights reserved.

4. Summary and Prospects

In summary, recent progress of Mn-based transition metal layered oxide cathodes for SIBs from different aspects has been comprehensively reviewed in this paper, especially focusing on the F-doping modification. Meanwhile, this paper summarizes the significant role of F-doping in improving the diffusion rate of Na+.

Nevertheless, in the process of commercializing SIBs, there are still some difficulties that have not been effectively solved. First of all, upon Na^+ extraction, the electrostatic attraction between the Na layers and the oxide layers disappears during the charge/discharge process, and the repulsion between the TM and the TM layers will gradually enhance. Meanwhile, the deintercalation and intercalation of the large radios of Na^+ will leave structural voids. These phenomena seriously undermine structural stability. Secondly, for balancing the positive charge loss and maintaining charge neutrality due to Na^+ extraction, the transition metal will undergo a redox reaction, and the structure of the cathode material will be twisted and deformed with the different transition metal valence states. Additionally, to achieve the lowest energy steady-state, the internal structure would inevitably change, resulting in the formation of Na^+/vacancy ordering, the transition metal ions migration/dissolution, the series of phase transformations and structural evolution, and so forth. Thirdly, recent studies have also confirmed the migration of some transition metals to the sodium layer during the charging and discharging process, which will bring about many drawbacks, such as poor thermal stability, boundary side reactions, and rapid capacity degradation. Therefore, inhibiting the migration is important to reduce electrode dissolution and particle cracking.Last but not the least, at high voltages (4.2 V and above), it is often accompanied by structural irreversible phase transition, which will bring about the collapse of the overall structure and release of O_2 gas [74]. Although significant achievements have been made in advancing SIBs in recent years, it still has a long way to go to overcome the subsistent obstacles in order to overcome the inherent problems of the material itself and put full-battery SIBs into practical application. We give an outlook on future technologies to improve electrochemical performance and safety in the following areas.

1. Dopant selection and reasonable composition design. At present, the study of F-doping mainly stays in the analysis of experimental results, and it does not clearly address the quantitative optimization of F-doping and does not address the optimal solution of the doping amount from a molecular structural perspective. Meanwhile, there are some directional studies about the selection of dopant and element, but the mechanism explanation from the atomic and molecular perspective needs to be further studied, and the mechanism of interaction between different elemental ions from the quantum perspective needs to be advanced more deeply.

2. Accurate characterization of doping sites. Guided by density functional theory calculations, the F substitution may change crystal structure and the formation energy as well as thermodynamic energy, but how to reasonably design the sites of different elements, especially F^-, is facing great challenge. For future research on layered oxide cathodes [51], it is crucial to explore controlled and precise doping sites and provide intuitive and in-depth proof of the mechanism.

3. Multi-phase combination synergy of P2, O3, O2, O4, and other different classical types. Although the P2-type structure is conducive to the embedding and detachment of Na^+, the P2-type layered oxide cathode usually suffers from short cycling lifespan.

Meanwhile, the O3 structure is conducive to the storage of large amounts of Na^+, but this structure is less reversible due to O3-P3 phase transitions that alter the diffusion mechanism of Na^+ ions [75]. Our recent studies have revealed that different doping amounts of F ions affects the structure of the materials. The formation of the Na-F bond can change the bonding between $O2^-$ and Na^+, and the F ions embedded in the TM layer are conducive to enhance the electrostatic attraction between the Na layer and the TM layer, thus changing the structural morphology of the material, especially for the layered and tunneled structure, and P2, O3, O2, O4 structure. Following this line of thinking, the composite structures could successfully achieve the integration of synergistic advantages stemming from the diverse structures and great improvement in electrochemical performance. Hence, it is important to reasonably prepare the complex systems composed of different structures [76]. The biphase synergy of P2-O3, P2-P3, P2-tunnel, and other intergrowth systems will be able to make full use of the advantages of various phases to develop high-performance materials, transforming the harmful irreversible phase transition into a beneficial reversible phase transition. Some new methods could integrate the merits of multiple phases to realize performance breakthroughs.

Turning to large scale applications, there have been two schools favoring either layered or polyanionic compounds such as $NaFePO_4$ [77]. Polyanionic cathode materials have been industrialized on a small scale, but polyanionic compounds inherently suffer from low stacking density [78], and it is not a perfect choice. Nowadays, the energy density of layered oxide cathode materials is similar to that of $LiFePO_4$, and the trend seems to be in favor of layered oxides due to the improvement in the composition of the material, which has led to a reduction in cost. Although there are still many problems and challenges in the industrialization, low cost and safety will be the biggest advantages of SIBs [79]. Hence, we believe that the large-scale production and commercial application of high-performance cathode materials will become possible through the F-doped strategy or multiple ions co-doped strategy. In addition, close communication between academic and industrial organizations combined with best endeavors is needed to launch full-cell SIBs into practical application. As shown in Figure 9, SIBs, as low-cost batteries, are able to integrate discontinuous energy flows from renewable sources [80]. Grid storage, wind power, and solar photovoltaic power act as the main renewable energy sources for the distributed energy system (①, ②, and ③), while the main scenarios and forms of energy use in society, such as commercial electricity, residential electricity, and industrial electricity (④, ⑤, and ⑥) will promote the development of the low-cost SIBs in the field of grid energy storage systems, renewable energy conversion, and utilization. The utilization of layered oxide cathode materials serving as rechargeable SIBs at the industrial and commercial level is anticipated to commence in the foreseeable future.

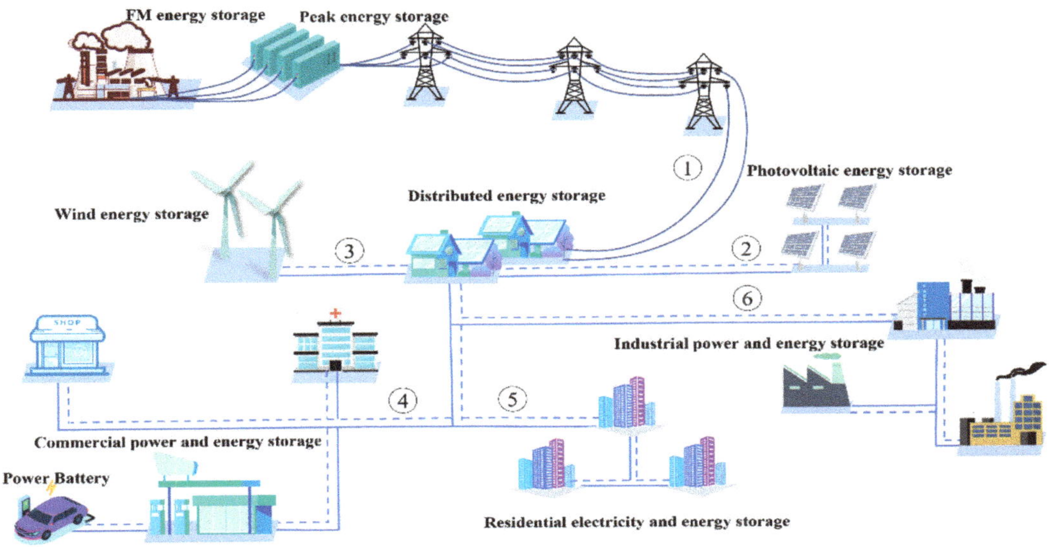

Figure 9. Possible scenarios for SIB in the context of new energy applications.

Author Contributions: Conceptualization and original draft preparation, H.W.; methodology, L.Z.; validation, Z.C.; review, L.L. and Y.W.; funding acquisition, T.D. All authors have read and agreed to the published version of the manuscript.

Funding: This work was supported by the National Natural Science Foundation of China (No. 52270177), the Natural Science Foundation of Shenyang (No. 22-315-6-13).

Data Availability Statement: All the datasets analyzed throughout the present study are available from the corresponding author on reasonable request. The data are not publicly available due to the limited content of the article.

Conflicts of Interest: The authors declare no conflict of interest.

References

1. Liu, D.; Fan, X.; Li, Z.; Liu, T.; Sun, M.; Qian, C.; Ling, M.; Liu, Y.; Liang, C. A Cation/Anion Co-Doped $Li_{1.12}Na_{0.08}Ni_{0.2}Mn_{0.6}O_{1.95}F_{0.05}$ Cathode for Lithium Ion Batteries. *Nano Energy* **2019**, *58*, 786–796. [CrossRef]
2. Han, M.H.; Gonzalo, E.; Singh, G.; Rojo, T. A Comprehensive Review of Sodium Layered Oxides: Powerful Cathodes for Na-Ion Batteries. *Energy Environ. Sci.* **2015**, *8*, 81–102. [CrossRef]
3. Vanaphuti, P.; Chen, J.; Cao, J.; Bigham, K.; Chen, B.; Yang, L.; Chen, H.; Wang, Y. Enhanced Electrochemical Performance of the Lithium-Manganese-Rich Cathode for Li-Ion Batteries with Na and F Co Doping. *ACS Appl. Mater. Inter.* **2019**, *11*, 37842–37849. [CrossRef] [PubMed]
4. Rastgoo-Deylami, M.; Javanbakht, M.; Omidvar, H. Enhanced Performance of Layered $Li_{1.2}Mn_{0.54}Ni_{0.13}Co_{0.13}O_2$ Cathode Material in Li-Ion Batteries Using Nanoscale Surface Coating with Fluorine-Doped Anatase TiO_2. *Solid State Ion.* **2019**, *331*, 74–88. [CrossRef]
5. Kim, H.; Kim, S.-B.; Park, D.-H.; Park, K.-W. Fluorine-Doped $LiNi_{0.8}Mn_{0.1}Co_{0.1}O_2$ Cathode for High-Performance Lithium-Ion Batteries. *Energies* **2020**, *13*, 4808. [CrossRef]
6. Yuan, A.; Tang, H.; Liu, L.; Ying, J.; Tan, L.; Tan, L.; Sun, R. High Performance of Phosphorus and Fluorine Co-Doped $LiNi_{0.8}Co_{0.1}Mn_{0.1}O_2$ as a Cathode Material for Lithium Ion Batteries. *J. Alloys Compd.* **2020**, *844*, 156210. [CrossRef]
7. Li, L.; Song, B.H.; Chang, Y.L.; Xia, H.; Yang, J.R.; Lee, K.S.; Lu, L. Retarded Phase Transition by Fluorine Doping in Li-Rich Layered $Li_{1.2}Mn_{0.54}Ni_{0.13}Co_{0.13}O_2$ Cathode Material. *J. Alloys Compd.* **2015**, *283*, 162–170. [CrossRef]
8. Guyomard, D.; Tarascon, J.M. Li Metal-Free Rechargeable $LiMn_2O_4$ / Carbon Cells: Their Understanding and Optimization. *J. Electrochem. Soc.* **1992**, *139*, 937–948. [CrossRef]
9. Chen, Y.; Xu, Y.; He, S.; Zhang, B.; Guo, L. A New High-Voltage Plateau of $Na_3V_2(PO_4)_3$ for Sodium Ion Batteries: A Promising Cathode with High Energy Density. *Ceram. Int.* **2021**, *47*, 26579–26583. [CrossRef]

10. Tieu, A.J.K.; Mahayoni, E.; Li, Y.; Deng, Z.; Fauth, F.; Chotard, J.-N.; Seznec, V.; Adams, S.; Masquelier, C.; Canepa, P. Zirconia-Free NaSICON Solid Electrolyte Materials for Sodium All-Solid-State Batteries. *J. Mater. Chem. A* **2023**, *11*, 23233–23242. [CrossRef]
11. Zhang, P.; Zhai, X.; Huang, H.; Zhou, J.; Li, X.; He, Y.; Guo, Z. Synergistic Na^+ and F^- Co-Doping Modification Strategy to Improve the Electrochemical Performance of Li-Rich $Li_{1.2}Mn_{0.54}Ni_{0.13}Co_{0.13}O_2$ Cathode. *Ceram. Int.* **2020**, *46*, 24723–24736. [CrossRef]
12. Zhang, J.; Wang, W.; Wang, W.; Wang, S.; Li, B. Comprehensive Review of P2-Type $Na_{2/3}Ni_{1/3}Mn_{2/3}O_2$, a Potential Cathode for Practical Application of Na-Ion Batteries. *ACS Appl. Mater. Inter.* **2019**, *11*, 22051–22066. [CrossRef] [PubMed]
13. Wei, H.; Tang, L.; Huang, Y.; Wang, Z.; Luo, Y.; He, Z.; Yan, C.; Mao, J.; Dai, K.; Zheng, J. Comprehensive Understanding of Li/Ni Intermixing in Layered Transition Metal Oxides. *Mater. Today* **2021**, *51*, 365–392. [CrossRef]
14. Chen, Y.; Cheng, J.; Sun, S.; Tian, Z.; Jiang, X.; Wang, Y.; He, Z.; Liu, C.; Huang, Q.; Guo, L. Constructing Hierarchical Porous Fe/F-Codoped $Na_3V_2(PO_4)_3$/C Composite Enwrapped with Carbon Nanotubes as High-Performance Cathode for Symmetric Sodium Ion Batteries. *J. Power Sources* **2021**, *513*, 230545. [CrossRef]
15. Liu, J.; Yang, Z.; Liu, Q.; Liu, Y. Heightening the Long Cycling Stability and Reducing Oxygen Loss of Li-Rich Cathode Materials by Mo–F Co-Doping for Half/Full Lithium-Ion Batteries. *Ceram. Int.* **2023**, *49*, 6580–6593. [CrossRef]
16. Liu, Y.; Shen, Q.; Zhao, X.; Zhang, J.; Liu, X.; Wang, T.; Zhang, N.; Jiao, L.; Chen, J.; Fan, L. Hierarchical Engineering of Porous P2-$Na_{2/3}Ni_{1/3}Mn_{2/3}O_2$ Nanofibers Assembled by Nanoparticles Enables Superior Sodium-Ion Storage Cathodes. *Adv. Funct. Mater.* **2020**, *30*, 1907837. [CrossRef]
17. Kim, S.-B.; Kim, H.; Park, D.-H.; Kim, J.-H.; Shin, J.-H.; Jang, J.-S.; Moon, S.-H.; Choi, J.-H.; Park, K.-W. Li-Ion Diffusivity and Electrochemical Performance of Ni-Rich Cathode Material Doped with Fluoride Ions. *J. Power Sources* **2021**, *506*, 230219. [CrossRef]
18. James Abraham, J.; Arro, C.R.A.; Tariq, H.A.; Kahraman, R.; Al-Qaradawi, S.; Al tahtamouni, T.M.; Shakoor, R.A. Sodium and Lithium Incorporated Cathode Materials for Energy Storage Applications—A Focused Review. *J. Power Sources* **2021**, *506*, 230098. [CrossRef]
19. Gabriel, E.; Hou, D.; Lee, E.; Xiong, H. Multiphase Layered Transition Metal Oxide Positive Electrodes for Sodium Ion Batteries. *Energy Sci. Eng.* **2022**, *10*, 1672–1705. [CrossRef]
20. Xiao, L.; Ji, F.; Zhang, J.; Chen, X.; Fang, Y. Doping Regulation in Polyanionic Compounds for Advanced Sodium-Ion Batteries. *Small* **2023**, *19*, 2205732. [CrossRef]
21. Gu, J.; Cui, K.; Niu, S.; Ge, Y.; Liu, Y.; Ma, Z.; Wang, C.; Li, X.; Wang, X. Smart Configuration of Cobalt Hexacyanoferrate Assembled on Carbon Fiber Cloths for Fast Aqueous Flexible Sodium Ion Pseudocapacitor. *J. Colloid. Interf. Sci.* **2021**, *594*, 522–530. [CrossRef] [PubMed]
22. Shimizu, T.; Mameuda, T.; Toshima, H.; Akiyoshi, R.; Kamakura, Y.; Wakamatsu, K.; Tanaka, D.; Yoshikawa, H. Application of Porous Coordination Polymer Containing Aromatic Azo Linkers as Cathode-Active Materials in Sodium-Ion Batteries. *ACS Appl. Energy Mater.* **2022**, *5*, 5191–5198. [CrossRef]
23. Wang, P.-F.; You, Y.; Yin, Y.-X.; Guo, Y.-G. Layered Oxide Cathodes for Sodium-Ion Batteries: Phase Transition, Air Stability, and Performance. *Adv. Energy Mater.* **2018**, *8*, 1701912. [CrossRef]
24. Zhang, D.; Shi, W.; Yan, Y.; Xu, S.; Chen, L.; Wang, X.; Liu, S. Fast and Scalable Synthesis of Durable $Na_{0.44}MnO_2$ Cathode Material via an Oxalate Precursor Method for Na-Ion Batteries. *Electrochim. Acta* **2017**, *258*, 1035–1043. [CrossRef]
25. Chen, H.; Wu, Z.; Zhong, Y.; Chen, T.; Liu, X.; Qu, J.; Xiang, W.; Li, J.; Chen, X.; Guo, X.; et al. Boosting the Reactivity of Ni^{2+}/Ni^{3+} Redox Couple via Fluorine Doping of High Performance $Na_{0.6}Mn_{0.95}Ni_{0.05}O_2$-F Cathode. *Electrochim. Acta* **2019**, *308*, 64–73. [CrossRef]
26. Huang, X.; Zhang, Z.; He, J.; Bai, Z.; Lu, L.; Li, J. Effects of Chromium/Fluorine Co-Doping on the Electrochemical Performance of $Li_{1.2}Ni_{0.13}Co_{0.13}Mn_{0.54}O_2$ Cathode Material for Lithium-Ion Batteries. *J. Mater. Sci.* **2021**, *56*, 9836–9851. [CrossRef]
27. Zhang, Q.; Huang, Y.; Liu, Y.; Sun, S.; Wang, K.; Li, Y.; Li, X.; Han, J.; Huang, Y. F-Doped O3-$NaNi_{1/3}Fe_{1/3}Mn_{1/3}O_2$ as High-Performance Cathode Materials for Sodium-Ion Batteries. *Sci. China Mater.* **2017**, *60*, 629–636. [CrossRef]
28. Shi, W.-J.; Yan, Y.-W.; Chi, C.; Ma, X.-T.; Zhang, D.; Xu, S.-D.; Chen, L.; Wang, X.-M.; Liu, S.-B. Fluorine Anion Doped $Na_{0.44}MnO_2$ with Layer-Tunnel Hybrid Structure as Advanced Cathode for Sodium Ion Batteries. *J. Power Sources* **2019**, *427*, 129–137. [CrossRef]
29. Li, W.; Chen, Q.; Zhang, D.; Fang, C.; Nian, S.; Wang, W.; Xu, C.; Chang, C. High Stability of Mo-F Dual-Doped O3-Type $NaNi_{1/3}Fe_{1/3}Mn_{1/3}O_2$ Cathode Material for Sodium-Ion Battery. *Mater. Today Commun.* **2022**, *32*, 103839. [CrossRef]
30. Tang, K.; Wang, Y.; Zhang, X.; Jamil, S.; Huang, Y.; Cao, S.; Xie, X.; Bai, Y.; Wang, X.; Luo, Z.; et al. High-Performance P2-Type Fe/Mn-Based Oxide Cathode Materials for Sodium-Ion Batteries. *Electrochim. Acta* **2019**, *312*, 45–53. [CrossRef]
31. Qiu, Z.; Liu, Z.; Fu, X.; Liu, J.; Zeng, Q. Improving the Cycling Performance of $LiNi_{0.8}Co_{0.15}Al_{0.05}O_2$ Cathode Materials via Zirconium and Fluorine Co-Substitution. *J. Alloy Compd.* **2019**, *806*, 136–145. [CrossRef]
32. Liu, K.; Tan, S.; Moon, J.; Jafta, C.J.; Li, C.; Kobayashi, T.; Lyu, H.; Bridges, C.A.; Men, S.; Guo, W.; et al. Insights into the Enhanced Cycle and Rate Performances of the F-Substituted P2-Type Oxide Cathodes for Sodium-Ion Batteries. *Adv. Energy Mater.* **2020**, *10*, 2000135. [CrossRef]
33. Zhou, P.; Zhang, J.; Che, Z.; Quan, Z.; Duan, J.; Wu, X.; Weng, J.; Zhao, J.; Zhou, J. Insights into the Enhanced Structure Stability and Electrochemical Performance of Ti^{4+}/F^- Co-Doped P2-$Na_{0.67}Ni_{0.33}Mn_{0.67}O_2$ Cathodes for Sodium Ion Batteries at High Voltage. *J. Energy Chem.* **2022**, *67*, 655–662. [CrossRef]

34. Cui, X.; Wang, S.; Ye, X.; Fan, X.; Gao, C.; Quan, Y.; Wen, S.; Cai, X.; Huang, J.; Li, S. Insights into the Improved Cycle and Rate Performance by Ex-Situ F and in-Situ Mg Dual Doping of Layered Oxide Cathodes for Sodium-Ion Batteries. *Energy Storage Mater.* **2022**, *45*, 1153–1164. [CrossRef]
35. Mao, Q.; Yu, Y.; Wang, J.; Zheng, L.; Wang, Z.; Qiu, Y.; Hao, Y.; Liu, X. Mitigating the P2–O2 Transition and Na^+ /Vacancy Ordering in $Na_{2/3}Ni_{1/3}Mn_{2/3}O_2$ by Anion/Cation Dual-Doping for Fast and Stable Na^+ Insertion/Extraction. *J. Mater. Chem. A* **2021**, *9*, 10803–10811. [CrossRef]
36. Hu, G.; Zhang, Y.; Zeng, J.; Fang, Z.; Cao, Y.; Peng, Z.; Tan, X.; Du, K. Mo–F Co-Doping $LiNi_{0.83}Co_{0.11}Mn_{0.06}O_2$ Stabilizes the Structure and Induces Compact Primary Particle To Improve the Electrochemical Performance. *ACS Appl. Energy Mater.* **2023**, *6*, 3834–3843. [CrossRef]
37. Hou, A.; Liu, Y.; Ma, L.; Li, T.; Ren, B.; Zhang, P. NH4F and Carbon Nanotubes Co-Modified $LiNi_{0.88}Co_{0.09}Al_{0.03}O_2$ Cathode Material with Enhanced Electrochemical Properties for Li-Ion Batteries. *J. Mater. Sci. Mater. Electron.* **2019**, *30*, 4128–4136. [CrossRef]
38. Ganesan, B.K.; Lee, Y.-S. Phase Transition Induced Enhanced Performance of Sodium-Rich $Na_{1.2}Mn_{0.8}O_{2-y}F_y$ (y = 0–0.5) Cathodes. *ACS Appl. Energy Mater.* **2023**, *6*, 960–968. [CrossRef]
39. Wang, J.; Liu, C.; Xu, G.; Miao, C.; Wen, M.; Xu, M.; Wang, C.; Xiao, W. Strengthened the Structural Stability of In-Situ F^- Doping Ni-Rich $LiNi_{0.8}Co_{0.15}Al_{0.05}O_2$ Cathode Materials for Lithium-Ion Batteries. *Chem. Eng. J.* **2022**, *438*, 135537. [CrossRef]
40. Chen, X.; Guo, W.; Li, R.; Du, P.; Zhan, X.; Gao, S. Structure, Electrochemical, and Transport Properties of Li- and F-Modified P2-$Na_{2/3}Ni_{1/3}Mn_{2/3}O_2$ Cathode Materials for Na-Ion Batteries. *Coatings* **2023**, *13*, 626. [CrossRef]
41. Zhang, Y.; Liu, G.; Su, C.; Liu, G.; Sun, H.; Qiao, D.; Wen, L. Study on the Influence of Cu/F Dual-Doping on the Fe–Mn Based Compound as Cathode Material for Sodium Ion Batteries. *J. Power Sources* **2022**, *536*, 231511. [CrossRef]
42. Si, Z.; Shi, B.; Huang, J.; Yu, Y.; Han, Y.; Zhang, J.; Li, W. Titanium and Fluorine Synergetic Modification Improves the Electrochemical Performance of $Li(Ni_{0.8}Co_{0.1}Mn_{0.1})O_2$. *J. Mater. Chem. A* **2021**, *9*, 9354–9363. [CrossRef]
43. Kang, W.; Ma, P.; Liu, Z.; Wang, Y.; Wang, X.; Chen, H.; He, T.; Luo, W.; Sun, D. Tunable Electrochemical Activity of P2–$Na_{0.6}Mn_{0.7}Ni_{0.3}O_{2-x}F_x$ Microspheres as High-Rate Cathodes for High-Performance Sodium Ion Batteries. *ACS Appl. Mater. Inter.* **2021**, *13*, 15333–15343. [CrossRef] [PubMed]
44. Fan, Y.; Ye, X.; Yang, X.; Guan, L.; Chen, C.; Wang, H.; Ding, X. Zn/Ti/F Synergetic-Doped $Na_{0.67}Ni_{0.33}Mn_{0.67}O_2$ for Sodium-Ion Batteries with High Energy Density. *J. Mater. Chem. A* **2023**, *11*, 3608–3615. [CrossRef]
45. Wen, P.; Shi, H.; Guo, D.; Wang, A.; Yu, Y.; Wu, Z.-S. Ti, F Codoped Sodium Manganate of Layered P2-$Na_{0.7}MnO_{2.05}$ Cathode for High Capacity and Long-Life Sodium-Ion Battery. *Renewables* **2023**, *1*, 81–89. [CrossRef]
46. Wang, Q.; Chu, S.; Guo, S. Progress on Multiphase Layered Transition Metal Oxide Cathodes of Sodium Ion Batteries. *Chin. Chem. Lett.* **2020**, *31*, 2167–2176. [CrossRef]
47. Palaniyandy, N. Recent Developments on Layered 3d-Transtition Metal Oxide Cathode Materials for Sodium-Ion Batteries. *Curr. Opin. Electrochem.* **2020**, *21*, 319–326. [CrossRef]
48. Huo, J.; Zhang, Y.; Kang, W.; Shen, Y.; Li, X.; Yan, Z.; Pan, Y.; Sun, W. Synthesis of F-Doped Materials and Applications in Catalysis and Rechargeable Batteries. *Nanoscale Adv.* **2023**, *5*, 2846–2864. [CrossRef] [PubMed]
49. Delmas, C.; Fouassier, C.; Hagenmuller, P. Structural Classification and Properties of the Layered Oxides. *Physica B+C* **1980**, *99*, 81–85. [CrossRef]
50. Takeda, Y.; Nakahara, K.; Nishijima, M.; Imanishi, N.; Yamamoto, O.; Takano, M.; Kanno, R. Sodium Deintercalation from Sodium Iron Oxide. *Mater. Res. Bull.* **1994**, *29*, 659–666. [CrossRef]
51. Caballero, A.; Hernán, L.; Morales, J.; Sánchez, L.; Santos Peña, J.; Aranda, M.A.G. Synthesis and Characterization of High-Temperature Hexagonal P2-$Na_{0.6}MnO_2$ and Its Electrochemical Behaviour as Cathode in Sodium Cells. *J. Mater. Chem.* **2002**, *12*, 1142–1147. [CrossRef]
52. Komaba, S.; Yabuuchi, N.; Nakayama, T.; Ogata, A.; Ishikawa, T.; Nakai, I. Study on the Reversible Electrode Reaction of $Na_{1-x}Ni_{0.5}Mn_{0.5}O_2$ for a Rechargeable Sodium-Ion Battery. *Inorg. Chem.* **2012**, *51*, 6211–6220. [CrossRef] [PubMed]
53. Kumakura, S.; Tahara, Y.; Sato, S.; Kubota, K.; Komaba, S. P′2-$Na_{2/3}Mn_{0.9}Me_{0.1}O_2$ (Me = Mg, Ti, Co, Ni, Cu, and Zn): Correlation between Orthorhombic Distortion and Electrochemical Property. *Chem. Mater.* **2017**, *29*, 8958–8962. [CrossRef]
54. Chae, M.S.; Kim, H.J.; Lyoo, J.; Attias, R.; Gofer, Y.; Hong, S.; Aurbach, D. Anomalous Sodium Storage Behavior in Al/F Dual-Doped P2-Type Sodium Manganese Oxide Cathode for Sodium-Ion Batteries. *Adv. Energy Mater.* **2020**, *10*, 2002205. [CrossRef]
55. Stoyanova, R.; Carlier, D.; Sendova-Vassileva, M.; Yoncheva, M.; Zhecheva, E.; Nihtianova, D.; Delmas, C. Stabilization of Over-Stoichiometric Mn^{4+} in Layered $Na_{2/3}MnO_2$. *J. Solid State Chem.* **2010**, *183*, 1372–1379. [CrossRef]
56. Kumakura, S.; Tahara, Y.; Kubota, K.; Chihara, K.; Komaba, S. Sodium and Manganese Stoichiometry of P2-Type $Na_{2/3}MnO_2$. *Angew. Chem. Int. Ed.* **2016**, *55*, 12760–12763. [CrossRef] [PubMed]
57. Wang, P.F.; You, Y.; Yin, Y.X.; Wang, Y.S.; Wan, L.J.; Gu, L.; Guo, Y.G. Suppressing the P2–O2 Phase Transition of $Na_{0.67}Mn_{0.67}Ni_{0.33}O_2$ by Magnesium Substitution for Improved Sodium-Ion Batteries. *Angew. Chem.* **2016**, *128*, 7571–7575. [CrossRef]
58. Choi, J.U.; Park, Y.J.; Jo, J.H.; Kuo, L.-Y.; Kaghazchi, P.; Myung, S.-T. Unraveling the Role of Earth-Abundant Fe in the Suppression of Jahn–Teller Distortion of P′2-Type $Na_{2/3}MnO_2$: Experimental and Theoretical Studies. *ACS Appl. Mater. Interfaces* **2018**, *10*, 40978–40984. [CrossRef]

59. Qiu, K.; Zhang, C.; Yan, M.; Zhao, S.; Fan, H.; An, S.; Qiu, X.; Jia, G. Structural Transformation and Electrochemical Properties of a Nanosized Flower-like R-MnO_2 Cathode in a Sodium Battery. *Phys. Chem. Chem. Phys.* **2021**, *24*, 551–559. [CrossRef]
60. Li, X.; Shang, Y.; Yan, D.; Guo, L.; Huang, S.; Yang, H.Y. Topotactic Epitaxy Self-Assembly of Potassium Manganese Hexacyanoferrate Superstructures for Highly Reversible Sodium-Ion Batteries. *ACS Nano* **2022**, *16*, 453–461. [CrossRef]
61. Tan, L.; Wu, Q.; Liu, Z.; Chen, Q.; Yi, H.; Zhao, Z.; Song, L.; Zhong, S.; Wu, X.; Li, L. Ti-Substituted O3-Type Layered Oxide Cathode Material with High-Voltage Stability for Sodium-Ion Batteries. *J. Colloid. Interf. Sci.* **2022**, *622*, 1037–1044. [CrossRef]
62. Kalapsazova, M.; Markov, P.; Kostov, K.; Zhecheva, E.; Nihtianova, D.; Stoyanova, R. Controlling at Elevated Temperature the Sodium Intercalation Capacity and Rate Capability of P3-$Na_{2/3}Ni_{1/2}Mn_{1/2}O_2$ through the Selective Substitution of Nickel with Magnesium. *Batter. Supercaps* **2020**, *3*, 1329–1340. [CrossRef]
63. Ortiz-Vitoriano, N.; Drewett, N.E.; Gonzalo, E.; Rojo, T. High Performance Manganese-Based Layered Oxide Cathodes: Overcoming the Challenges of Sodium Ion Batteries. *Energy Environ. Sci.* **2017**, *10*, 1051–1074. [CrossRef]
64. Yoda, Y.; Kubota, K.; Isozumi, H.; Horiba, T.; Komaba, S. Poly-γ-Glutamate Binder To Enhance Electrode Performances of P2-$Na_{2/3}Ni_{1/3}Mn_{2/3}O_2$ for Na-Ion Batteries. *ACS Appl. Mater. Interf.* **2018**, *10*, 10986–10997. [CrossRef] [PubMed]
65. Zhang, X.; Qiao, Y.; Guo, S.; Jiang, K.; Xu, S.; Xu, H.; Wang, P.; He, P.; Zhou, H. Manganese-Based Na-Rich Materials Boost Anionic Redox in High-Performance Layered Cathodes for Sodium-Ion Batteries. *Adv. Mater.* **2019**, *31*, 201807770. [CrossRef] [PubMed]
66. Laurita, A.; Zhu, L.; Cabelguen, P.-E.; Auvergniot, J.; Hamon, J.; Guyomard, D.; Dupré, N.; Moreau, P. Pristine Surface of Ni-Rich Layered Transition Metal Oxides as a Premise of Surface Reactivity. *ACS Appl. Mater. Interfaces* **2022**, *14*, 41945–41956. [CrossRef] [PubMed]
67. Min, K. Dual Doping with Cations and Anions for Enhancing the Structural Stability of the Sodium-Ion Layered Cathode. *Phys. Chem. Chem. Phys.* **2022**, *24*, 13006–13014. [CrossRef] [PubMed]
68. Zhao, Y.; Li, L.; Wu, Y.; Fang, Y.; Xie, H. Progress of the Elements Doped $NaFeO_2$ Cathode Materials for High Performance Sodium-ion Batteries. *ChemistrySelect* **2021**, *6*, 9701–9708. [CrossRef]
69. Hu, H.; He, H.-C.; Xie, R.-K.; Cheng, C.; Yan, T.; Chen, C.; Sun, D.; Chan, T.-S.; Wu, J.; Zhang, L. Achieving Reversible Mn^{2+}/Mn^{4+} Double Redox Couple through Anionic Substitution in a P2-Type Layered Oxide Cathode. *Nano Energy* **2022**, *99*, 107390. [CrossRef]
70. Li, G.; Zhu, W. First-Principles Investigation on the Crystal, Electronic Structures and Diffusion Barriers of F-Doped $NaMO_2$ (M=V, Cr, Co and Ni) for Rechargeable Na-Ion Batteries. *J. Solid State Chem.* **2021**, *302*, 122440. [CrossRef]
71. Wei, X.; Ma, M.; Zhang, H.; Wang, L.; Dong, C. Density Functional Theory Study of Ultra-Thin InSe Electrodes for Na and Mg Ion Storage and Transport. *Mater. Lett.* **2021**, *285*, 129091. [CrossRef]
72. Song, J.H.; Kapylou, A.; Choi, H.S.; Yu, B.Y.; Matulevich, E.; Kang, S.H. Suppression of Irreversible Capacity Loss in Li-Rich Layered Oxide by Fluorine Doping. *J. Power Sources* **2016**, *313*, 65–72. [CrossRef]
73. He, Y.-S.; Pei, L.; Liao, X.-Z.; Ma, Z.-F. Synthesis of $LiNi_{1/3}Co_{1/3}Mn_{1/3}O_{2-z}F_z$ Cathode Material from Oxalate Precursors for Lithium Ion Battery. *J. Fluor. Chem.* **2007**, *128*, 139–143. [CrossRef]
74. Sathiya, M.; Rousse, G.; Ramesha, K.; Laisa, C.P.; Vezin, H.; Sougrati, M.T.; Doublet, M.-L.; Foix, D.; Gonbeau, D.; Walker, W.; et al. Reversible Anionic Redox Chemistry in High-Capacity Layered-Oxide Electrodes. *Nat. Mater.* **2013**, *12*, 827–835. [CrossRef] [PubMed]
75. Goikolea, E.; Palomares, V.; Wang, S.; De Larramendi, I.R.; Guo, X.; Wang, G.; Rojo, T. Na-Ion Batteries—Approaching Old and New Challenges. *Adv. Energy Mater.* **2020**, *10*, 2002055. [CrossRef]
76. Fang, S.; Bresser, D.; Passerini, S. Transition Metal Oxide Anodes for Electrochemical Energy Storage in Lithium- and Sodium-Ion Batteries. *Adv. Energy Mater.* **2020**, *10*, 1902485. [CrossRef]
77. Pablos, C.; Olchowka, J.; Petit, E.; Minart, G.; Duttine, M.; Weill, F.; Masquelier, C.; Carlier, D.; Croguennec, L. Thermal Stability of $Na_{3-x}V_2(PO_4)_2F_{3-y}O_y$: Influence of F^- for O^{2-} Substitution and Degradation Mechanisms. *Chem. Mater.* **2023**, *35*, 4078–4088. [CrossRef]
78. Rozier, P.; Tarascon, J.M. Review—Li-Rich Layered Oxide Cathodes for Next-Generation Li-Ion Batteries: Chances and Challenges. *J. Electrochem. Soc.* **2015**, *162*, A2490–A2499. [CrossRef]
79. Vaalma, C.; Buchholz, D.; Weil, M.; Passerini, S. A Cost and Resource Analysis of Sodium-Ion Batteries. *Nat. Rev. Mater.* **2018**, *3*, 18013. [CrossRef]
80. Palomares, V.; Serras, P.; Villaluenga, I.; Hueso, K.B.; Carretero-González, J.; Rojo, T. Na-Ion Batteries, Recent Advances and Present Challenges to Become Low Cost Energy Storage Systems. *Energy Environ. Sci.* **2012**, *5*, 5884. [CrossRef]

molecules

MDPI

Review

Research Progress of Hydrogen Production Technology and Related Catalysts by Electrolysis of Water

Haiyao Li [1,2], Jun Guo [1,*], Zhishan Li [1] and Jinsong Wang [2]

1 Faculty of Metallugical and Energy Engineering, Kunming University of Science and Technology, Kunming 650093, China; hailiyao@outlook.com (H.L.); zhishanli@kust.edu.cn (Z.L.)
2 Faculty of Materials Science and Engineering, Kunming University of Science and Technology, Kunming 650093, China; jswang512@kust.edu.cn
* Correspondence: 1038gj@kust.edu.cn

Abstract: As a clean and renewable energy source for sustainable development, hydrogen energy has gained a lot of attention from the general public and researchers. Hydrogen production by electrolysis of water is the most important approach to producing hydrogen, and it is also the main way to realize carbon neutrality. In this paper, the main technologies of hydrogen production by electrolysis of water are discussed in detail; their characteristics, advantages, and disadvantages are analyzed; and the selection criteria and design criteria of catalysts are presented. The catalysts used in various hydrogen production technologies and their characteristics are emphatically expounded, aiming at optimizing the existing catalyst system and developing new high-performance, high-stability, and low-cost catalysts. Finally, the problems and solutions in the practical design of catalysts are discussed and explored.

Keywords: water electrolysis technique; hydrogen production; alkaline electrolysis; catalyst

1. Introduction

Hydrogen, as a clean energy carrier, plays an irreplaceable role in the future development of the energy industry. Compared with traditional fossil fuels, such as coal, natural gas, and oil, hydrogen has natural advantages of high calorific value, high energy density, and zero pollution, and is an ideal "clean energy" under the goal of carbon neutrality in China [1–3]. Furthermore, hydrogen has a wide range of sources, and water can be used as an important source of hydrogen due to the abundant hydrogen element in water. The calorific value of hydrogen is about three times as high as that of diesel. With the continuous development of hydrogen energy technology and the gradual maturity of hydrogen storage technology, the application of hydrogen energy becomes more and more extensive. Hydrogen energy has been widely used in electric power, heat energy, fuel cells, chemical synthesis, petroleum refining, metallurgical industry, and other fields. Among them, hydrogen energy has a high specific energy in thermal power generation, and the combustion products are pollution-free. Injecting hydrogen into the existing natural gas pipeline for subsequent co-combustion is an efficient hydrogen utilization method, which can make full use of the existing natural gas pipeline network infrastructure and greatly reduce storage and transportation costs. After adding hydrogen, the gas flow in the pipeline is increased, and the hydrocarbon mass ratio of natural gas, the precipitation of carbon particles, and the brightness of flame are decreased. During the combustion process of a hydrogen-doped natural gas boiler, the introduction of hydrogen can inhibit the generation of nitrogen oxides and soot and reduce the emission of CO_2. Potential applications of hydrogen fuel cells are as follows: (1) They can be used in the automobile field. The product of hydrogen fuel cells is only water. Using hydrogen fuel cells instead of fossil fuel as driving energy for vehicles can effectively reduce environmental pollution. (2) Hydrogen fuel cells can be applied to the field of ships. They play an important role in the composition of new energy structures. Compared with diesel-driven ships, ships with hydrogen fuel cells have longer endurance and meet the auxiliary energy demand of large ships. (3) Hydrogen fuel cells can be used in small unmanned aerial vehicle power systems (UVAs). Compared with the existing UAV battery technology, the endurance of UAVs driven by fuel cells has improved by 4~5 times. In the metallurgical industry, traditional coke steelmaking will produce a large amount of CO_2. Using hydrogen instead of coke as a reducing agent, the diffusion capacity of hydrogen is higher than that of CO gas obtained by coke oxidation, and the by-product of steelmaking is water, which will significantly reduce carbon emissions and promote the change in the metallurgical industry toward low-carbon and clean production. Additionally, as clean energy with zero carbon emission, hydrogen energy can help people decarbonize, fix carbon emissions, and even achieve negative carbon. Hydrogen energy used as the terminal energy can reduce or even eliminate CO_2 emissions in the industrial field through the complementary system of hydrogen and electricity.

To satisfy the increasing demand for hydrogen, researchers are constantly exploring existing hydrogen production technologies and developing new hydrogen production technologies. Hydrogen is classified into three categories according to the principle of production mode and different raw materials: chemical reforming hydrogen production, biological hydrogen production, and water electrolysis hydrogen production. Hydrogen production by chemical reforming [4,5] mainly involves high-temperature reforming or partial oxidation reforming of organic matter and fossil fuels, such as coal, natural gas, or petroleum by chemical methods. For example, methane, the main component in natural gas, is decomposed into H_2, CO_2, and CO. This kind of hydrogen production technology is quite mature. This route accounts for more than 85% of the current industrial methods, and its hydrogen production rate is 70–90%. The hydrogen prepared by this method belongs to grey hydrogen and blue hydrogen, and blue hydrogen reduces the carbon emission in grey hydrogen through carbon capture technology (CCS). However, the carbon recovery cost in the manufacturing process of blue hydrogen is very expensive, which may increase the fuel consumption by 10% compared with grey hydrogen. The biological hydrogen production method [6,7] uses organic wastes as raw materials to produce hydrogen through photosynthesis or bacterial fermentation. The key point is to cultivate high-efficiency and high-selectivity biological strains, but there are still many problems to be investigated in hydrogen production mechanism, such as strain cultivation, and bacterial metabolic pathway. Water electrolysis hydrogen production [8–10] is a completely clean way of hydrogen production, which can be used for peak shaving and energy storage of power stations—that is, the surplus electric energy of power stations is used for hydrogen production by electrolysis of water during the low power consumption period. This kind of technology has been relatively mature internationally, and some domestic hydropower stations and photovoltaic power stations have also applied this technology. The hydrogen obtained by this method possesses high purity and belongs to a class of green hydrogen. Green hydrogen is prepared by electrolytic water with renewable electricity, which not only effectively reduces the carbon emissions in the preparation process but also uses renewable raw materials, which is in line with sustainable development. However, the energy consumption of this method is high, and the on-site hydrogen production at the power station is still limited; the cost is also relatively high, with hydrogen production accounting for about 10%.

According to the prediction of the China Hydrogen Energy Alliance Research Institute, the proportion of hydrogen production from electrolytic water from renewable energy in China will reach 15%, 45%, and 70% in 2030, 2040, and 2050, respectively. With the discovery of renewable energy sources, such as photovoltaic and wind power, and the development of renewable technology, the production cost of electric energy conversion is decreasing, which provides a favorable opportunity for the development and application of hydrogen production by electrolysis of water. With the increasing energy demand, the technology of hydrogen production by electrolysis of water has undergone significant developments. Electrolytic water reaction is mainly composed of two half-reactions: cathodic hydrogen evolution reaction (HER) and anodic oxygen evolution reaction (OER). Due to the slow reaction kinetics, its energy conversion efficiency is low. Generally, noble metal catalysts (such as Pt, Ru, Ir, etc.) are needed to improve the HER/OER reaction kinetics of electrolyzed water to improve the energy conversion efficiency of the system [11–14]. However, due to the scarcity of resources and high use cost, the application of precious metal-based catalysts in large-scale hydrogen production by electrolysis of water is still limited. Based on this, the current research is focused on developing transition metal-based HER/OER electrocatalysts with high activity, high stability, and low cost. Among many transition metal-based materials, transition metal catalysts based on Ni, Co, and Mo are widely regarded by researchers because of their changeable composition and structure, abundant resources, low cost, and high catalytic activity and stability [15–20].

At present, the mainstream hydrogen production technologies include alkaline water electrolysis AWE [21,22], proton exchange membrane water electrolysis PEM [23,24], and solid oxide water electrolysis SOE [25,26], and their corresponding advantages and disadvantages are shown in Table 1. With the continuous development of electrolytic water technology, the efficiency of electrolytic hydrogen production is gradually improving and the energy consumption is decreasing. The decreased energy consumption mainly depends on optimization to improve the electrocatalytic activity of the catalyst. When selecting the catalyst, the following points should be considered: (1) Catalysts should possess high catalytic activity, long service life, and the ability to inhibit catalyst poisoning and complete the electrolytic water reaction. (2) Catalysts should possess high electrical conductivity, which promotes the charge transfer at the interface between the electrode and the electrolyte, thus improving the electrolysis efficiency. (3) Catalysts should possess a large catalytic activity area to promote more reactions. (4) The constituent elements of catalysts are earth-abundant with low cost of large-scale application. (5) Catalysts should possess good electrochemical stability, preventing them from falling off in an acid-based solution or from being poisoned or losing their activity directly because of other impurities and intermediate products of the reaction. Generally, precious metals and their alloy oxides have the best electrocatalytic activity and chemical stability due to their rich empty d orbitals, but their output is scarce and expensive, and the cost of large-scale application is high. Therefore, reducing the loading of precious metals and developing catalysts with low cost, high electrocatalytic activity (with empty d orbitals or not filled with d electrons) and high stability are the keys to improving the hydrogen production efficiency.

The design and development of high-performance catalysts can be regulated by the following two approaches: First, designing new nanomaterials /electrode structures, developing catalytic materials/electrodes with various morphologies/structures, improving the specific surface area and conductivity of catalytic electrodes, constructing open-pore structures, improving the exposure of active sites, and promoting the transport of substances will improve the electrocatalytic activity of HER/OER. This includes the preparation of heterostructure catalysts, rearranging electrons on heterostructure interfaces to modify the properties of active sites, and using the synergy of different active sites to promote the reaction kinetics [27]. Second, the intrinsic activity of electrocatalysts is regulated, and the electronic structure is optimized efficiently by strategies such as heteroatom doping, defect engineering, valence state regulation, and interface engineering, and the adsorption energy of intermediate species in the reaction is improved by regulating the electronic structure of active sites, thus improving the intrinsic activity of the catalysts. At present, there are various limitations to a single optimization method. Combining the advantages of strategies such as the optimization of nanomorphology and microstructure of materials with the regulation of intrinsic defects and compositing with metal or nonmetal carbon materials can effectively improve the catalytic activity of materials and reduce the application cost of materials.

Table 1. Type of water-electrolysis hydrogen and its characteristic.

Technologies	Diaphragm	Catalyst		Electrolyte	$T/°C$	Efficiency/%	Advantage	Disadvantage
		Anode	Cathode					
AWE	Porous materials	Ni, Co, Fe, $LaCoO_3$, $LaNiO_3$, $NiCo_2O_4$	Ni alloy, $NiMoO_4$	Alkaline water	20~90	59~70	Low cost, long service life, mature technology	Electrode corrosion, poor dynamic performance
	AEM	Ni-based materials	Ni, NiFe, $NiFe_2O_4$, PtRu/C	Pure or alkaline water	20~200	60~78	Has the advantages of alkaline electrolysis and PEM electrolysis	Low OH^- conductivity and poor high-temperature stability
PEM	Perfluorosulfonic acid membrane	RuO_2, IrO_2, $Ir_{1-x}Ru_xO_2$	Pt/C, MoS_2	Polymer, acidic electrolyte	20~200	65~82	Compact design, high responsiveness	High-cost, precious-metal catalyst
	PCC	Perovskite ceramic	Ni ceramic	Ceramic	500~1000	Up to 100%	Low cost, low energy demand, and high electrochemical reaction rate	High cost, poor mechanical stability of ceramics, difficult sealing; easy to cause hydrogen leakage
SOE	ceramic	$La_xSr_{1-x}MnO_3$, LSM-YSZ	Ni-YSZ, Ni-based ceramic, perovskite	Vapor, ceramic (oxygen ion conductor)	500~1000 °C	Up to 100%		

2. Brief Introduction of Hydrogen Production Technologies

2.1. Alkaline Water Electrolysis Hydrogen Production

Alkaline water electrolysis hydrogen production technology is the earliest industrialized production and the most mature and economical one at present. Its working efficiency is relatively high, generally 42% to 78%. After years of research and development, alkaline water electrolysis technology has made progress in two aspects: first, the efficiency of the electrode has improved, and the operating cost related to electricity consumption has been significantly reduced; second, the operating current density has increased and the investment cost has decreased [28]. The working principle of alkaline water electrolysis is shown in Figure 1 [29]. The electrolytic cell consists of two electrodes, which are separated by an airtight diaphragm. The electrolyzer has a simple structure, low requirements for raw material quality, long service life, generally reaching 10~20 years, and low cost [30]. The electrolytes used in alkaline electrolytic water technology are mainly alkaline electrolytes, such as KOH and NaOH, with a concentration of 20% to 30%. Among them, the KOH solution has high ionic conductivity and a wide application range. The diaphragm is mainly made of porous materials, such as asbestos, ceramics, and nylon. When the electrolysis temperature is 20~90 °C, water is reduced to produce hydrogen in the cathode, and OH^- passes through the diaphragm to reach the anode oxidation to produce oxygen. Although alkaline water electrolysis technology has the characteristics of low cost, long service life, mature technology, and stable operation, there are still many shortcomings in its engineering application, such as low current density, poor dynamic response, diaphragm gas leakage, alkali corrosion, and so on. To solve the above problems, researchers have developed an anion exchange membrane technology AEM, which is expected to become an improved scheme of traditional alkaline electrolysis technology and play a technical and cost advantage in large-scale hydrogen production [31].

To some extent, the AEM technology for producing hydrogen by electrolyzing water with an anion exchange membrane is a technique that combines traditional diaphragm alkaline electrolysis and proton exchange membrane electrolysis, which has the advantages of both alkaline electrolysis and PEM electrolysis [32]. AEM technology diaphragm adopts an anion exchange membrane with pure water or weak alkaline solution as an electrolyte, which can realize the circulation of OH^- from cathode to anode. The main challenges faced by AEM technology are the lack of AEM with high conductivity and alkaline corrosion resistance and increased cost due to the use of precious metal catalysts. At the same time, CO_2 entering the contact film will reduce the membrane resistance and electrode resistance, thus reducing the electrolytic performance. Therefore, the main development directions of AEM electrolyzer in the future are as follows: first, develop an AEM diaphragm with high conductivity, ion selectivity, and long-term alkaline stability; second, reduce the use of noble metal catalysts or develop high-performance catalysts without noble metals; and third, reduce the CO_2 content in the electrolytic cell to improve its electrolytic performance [33].

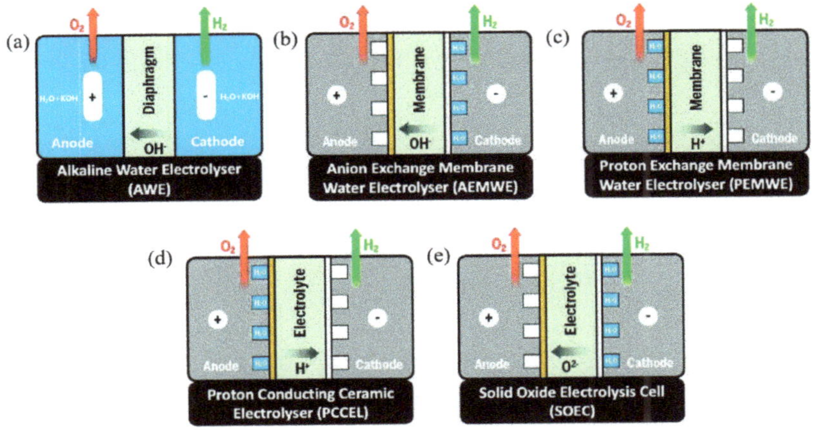

Figure 1. Schematic principle diagram of the five main types of electrolytic water hydrogen production (**a**) AWE, (**b**) AEM (**c**) PEM, (**d**) PCC, (**e**) SOE [34].

When choosing a catalyst, it is necessary to consider whether it has good electrochemical inertia, electrical conductivity, large specific surface area, catalytic activity, mechanical stability, good thermal conductivity, and strong chemical stability. Precious-based catalysts possess basically all the characteristics needed for high-performance catalysts, but their costs for application and preparation are high, and their scarce reserves and expensive prices greatly limit their wide-range use. In the industrial production of hydrogen from alkaline water electrolysis, the precious metal PtRu catalyst is used in AEM hydrogen production technology, and the widely used catalysts are mainly transition metal catalysts. Transition metals (TMs) have good catalytic activity, low price, and abundant crustal reserves, and are an ideal material to replace precious-based catalysts. TMs catalysts mainly include the following species, such as TMs and their alloys [34–36], TM oxides [37,38], TM sulfides [39,40], and TM phosphides [41,42], all of which have excellent hydrogen evolution reaction (HER) and oxygen evolution reaction (OER) activity and stability. After continuous optimization, the performance of some of them can now be comparable to that of precious-based catalysts, which have a very broad application prospect, as shown in Figure 2. The corresponding performance of the typical transition metals-based catalysts is summarized in Table 2.

Table 2. Electrochemical properties of typical transition metal-based electrocatalysts.

Transition Metal-Based Electrocatalysts	Catalyst	Overpotential/mV (10 mA cm^{-2})	Tafel Slope/mV dec^{-1}	Rct/Ω
alloys	Fe_3Co_7@PCNs for HER [35]	220	65.5	47
	Fe_3Co_7@PCNs for HER [35]	260	53.16	16
	B-Ti_2Cu_3 for HER [36]	155	103.89	14.2
	Mo-Ti_2Cu_3 for HER [36]	133	97.37	–
oxides	CoP@Co_3O_4@CC for HER [37]	73	85	70
	δ-MnO_2/SGS for HER [38]	80	42	140
sulfides	V-Ni_3S_2/NiOOH for HER [39]	154	94	5.1
	N-$NiMoO_4$/NiS_2 for HER [40]	99	74.2	–
	N-$NiMoO_4$/NiS_2 for OER [40]	283	44.3	–
phosphides	NiFeP for OER [41]	313	44	4.3
	V-CoP/Ni_2P/NF for HER [42]	20	54.2	2.6

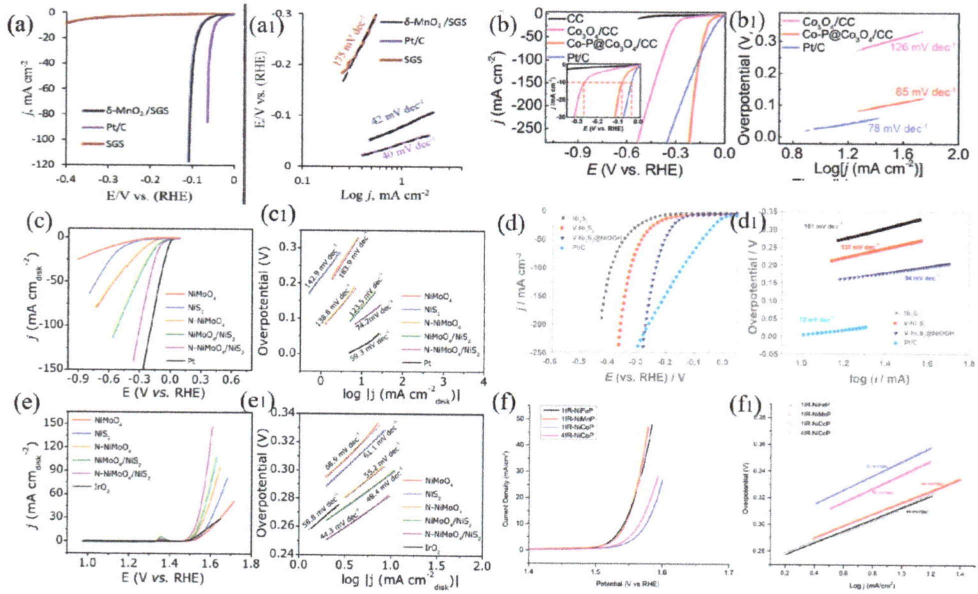

Figure 2. The HER polarization curves and Tafel plots of typical transition metal oxides (**a**–**b1**) [37,38], sulfides (**c**–**d1**) [39,40], phosphide [41] (**f**,**f1**), and OER polarization curves and Tafel plots of transition metal sulfides (**e**,**e1**) [40].

2.1.1. Transition Metal and Alloy Catalysts

Low-cost transition metals, such as Zr, Nb, Fe, Co, and Ni, possess high electrocatalytic performance in alkaline electrolytes, among which Ni is the most widely investigated and applied because of its good stability in alkaline electrolytes. At the same time, Ni is abundant in the earth's crust, easy to mine, and relatively low in preparation and use cost. Compared with other metal elements, Ni has a lower oxygen evolution overpotential and a higher oxygen evolution efficiency, so metallic Ni is widely used as an anode material for hydrogen production by electrolysis of water in alkaline electrolytes. The electrocatalytic performance of anode Ni can be optimized by increasing its specific surface area, such as porous Raney Ni anode. On the other hand, the metals on the left side of the transition system in the Periodic Table, such as Fe, Co, and Ni with incomplete or semi-full d orbitals, will have a synergistic catalytic effect on the hydrogen evolution reaction when they form alloy materials with the metals on the right side of the transition system, such as transition metals Zr, W, and Mo with paired d electrons, which will greatly reduce the hydrogen evolution overpotential of the materials. Alloy cathode electrode materials formed by transition metals can be generally divided into three categories [43]: (1) porous Raney Ni alloy materials; (2) Ni-based alloy materials, such as Ni-Mo alloy, Ni-Mo-Fe alloy, and Ni-S alloy; (3) other transition element alloys, such as Fe-R alloy and Fe-Zn-R alloy [44,45]. The typical binary alloys with high catalytic activity are anode alloy electrode materials, such as Ni-Co alloy and Ni-Fe alloy, in which Ni-Co alloy will form a dense inverse spinel $NiCo_2O_4$ oxide film on the electrode surface before oxygen evolution by electrolysis of water, which will greatly reduce the anodic oxygen evolution overpotential due to its high electrocatalytic activity. Additionally, Mn, Fe, and Cu are abundant in the crust and have no pollution effect on the environment. They are also ideal electrode materials. Ternary alloys formed by them and Ni-Co alloys, such as Ni-Co-Fe, Ni-Co-Mn, and Ni-Co-Cu alloys, have improved their electrocatalytic activity to varying degrees. Generally, transition metal and alloy electrodes are prepared by electroplating.

2.1.2. Transition Metal Oxide Catalyst

The most common perovskite oxides are oxides with ABO_3 structure formed by lanthanide metal La and transition metals Ni and Co, such as $LaNiO_3$ and $LaCoO_3$, as shown in Figure 3 [46], which have very high catalytic activity for oxygen evolution. At present, $LaNiO_3$ with a pseudo-cubic perovskite structure is the most widely used oxide. $LaNiO_3$ is the only metallic compound in perovskite-type oxides, and it is a nonstoichiometric compound. Ni ions exist in bivalent and trivalent forms; there are high-density oxygen holes in the oxides, and there is no forbidden band. The conduction band is formed by the interaction between the d electrons of Ni^{3+} and the p electrons of O^{2-}, showing the metal conductivity characteristics [47,48] Therefore, $LaNiO_3$ has high metal conductivity and catalytic activity for oxygen evolution. By introducing a small amount of Sr and Mn doping, the catalytic activity of $LaNiO_3$ can be greatly improved. Additionally, as shown in Figure 3, $LaNiO_3$ and $LaCoO_3$ composited with FeOOH possess the highest performance with a low overpotential of 264 mV and 334 mV, respectively; a low Tafel slope of 66 mV dec^{-1}, which originates from the high electrical conductivity of $LaNiO_3$ and $LaCoO_3$; and more active sites due to the existence of FeOOH.

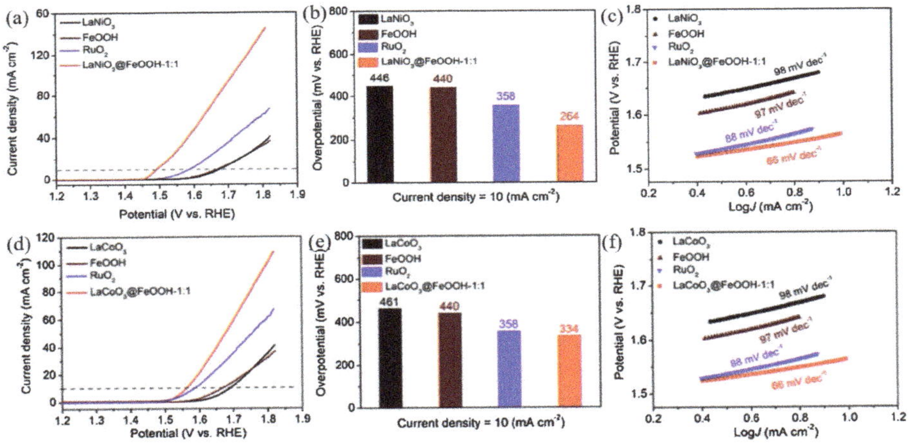

Figure 3. $LaNiO_3$, FeOOH, $LaNiO_3$@FeOOH−1:1, and RuO_2 samples: (**a**) LSV curves; (**b**) the corresponding comparison of overpotentials (η); (**c**) the corresponding Tafel plots. $LaCoO_3$, FeOOH, LaCoO3@FeOOH−1:1, and RuO_2 samples: (**d**) LSV curves; (**e**) the corresponding comparison of overpotentials (η); (**f**) the corresponding Tafel plots [47].

Spinel oxides, typically Co_3O_4 and $NiCo_2O_4$, have excellent corrosion resistance in alkaline medium, low cost, and high catalytic activity for oxygen evolution, making them the most promising anode materials for alkaline electrolytic water. Co_3O_4 has an inverse spinel structure, which is composed of cobaltous oxide (CoO) and cobaltic oxide (Co_2O_3) [49]. Cobalt ions with a valence of +2 occupy tetrahedral gaps, while cobalt ions with a valence of +3 occupy octahedral gaps, which makes the resistivity of Co_3O_4 material relatively high (40 Ω m^{-1}). By doping and compositing, the electronic interaction among the components in the oxide can produce a synergistic effect, as shown in Figure 2. $NiCo_2O_4$ is a kind of oxide with a spinel structure, in which bivalent Ni ions occupy tetrahedral gaps as A sites, while B sites are trivalent Ni ions occupying oxygen octahedral gaps. This composition with bivalent and trivalent mixed valence can improve its catalytic performance by doping bivalent Ni ions at position A with other transition metal ions [50]. $NiCo_2O_4$ has excellent electrocatalytic performance in OER and HER as shown in Figures 4 and 5 [40]. Its electrical conductivity and chemical stability are optimized by doping, and this makes oxygen adsorbing and desorbing easier. This unique structure gives it high electronic conductivity and chemical stability, as well as good adsorption and desorption performance for oxygen. Generally, methods for preparing transition metal oxide catalysts include solution impregnation (spraying) pyrolysis [51], high-temperature solid-state reaction [52], anodic electrodeposition [53], citric acid sol-gel method [54,55], coprecipitation method [47] and so on.

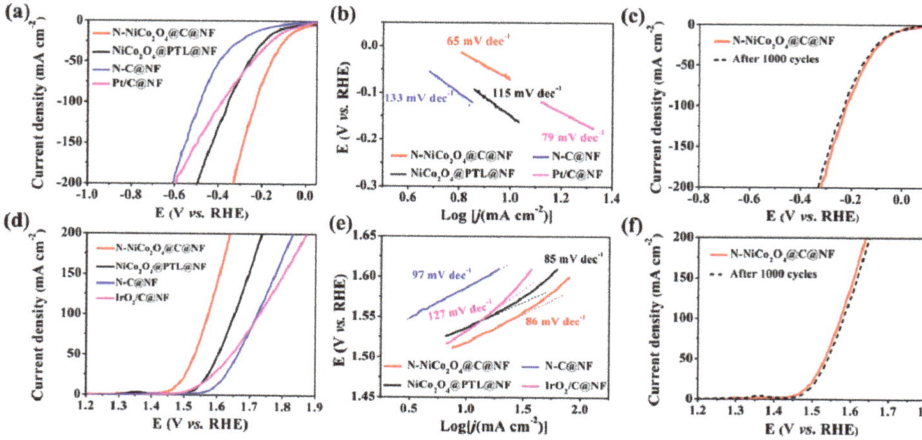

Figure 4. HER and OER performance of N−$NiCo_2O_4$@C, $NiCo_2O_4$@PTL, N−C, and Pt/C catalysts: (**a**) HER LSV curves, (**b**) Tafel plots, and (**c**) LSV curves recorded for N−$NiCo_2O_4$@C before and after 1000 cycles. (**d**) OER LSV curves, (**e**) Tafel plots, and (**f**) LSV curves received for N$NiCo_2O_4$@C before and after 1000 cycles [40].

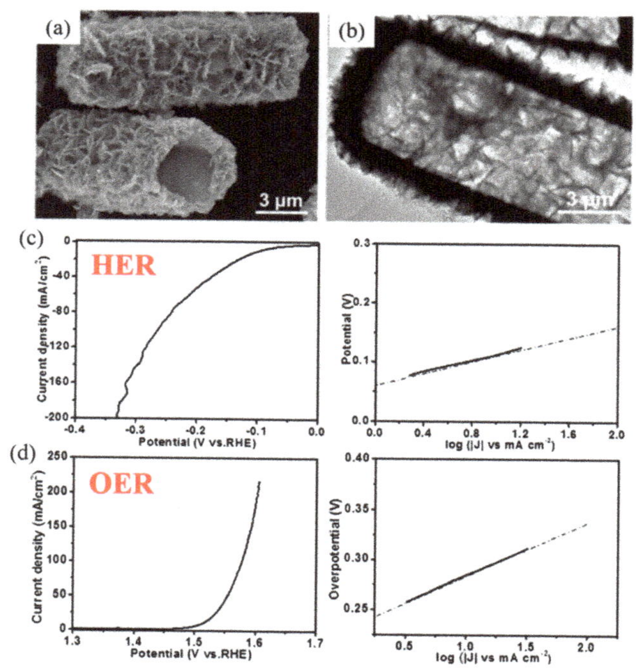

Figure 5. The microstructure and morphology of $NiCo_2O_4$ (**a**,**b**); HER of $NiCo_2O_4$ (**c**) LSV curves and corresponding Tafel plots; OER of $NiCo_2O_4$ (**d**) LSV curves and corresponding Tafel plots [40].

2.1.3. Transition Metal Sulfide Catalyst

Transition metal sulfides (TMSs) have attracted much attention in the domain of electrocatalysis because of their unique structure, composition, and excellent HER activity. The representative TMSs include MoS_2, Ni_3S_2, and WS_2 [51,56]. Nickel sulfide with different compositions and structures has excellent catalytic activity in HER [57] and OER [58] after doping, as shown in Figure 6. At present, there are different explanations for the reaction mechanism of the S atom in TMSs. Some researchers think that the high electronegativity of the S atom causes it to directly act as the adsorption and analysis site of the H atom. Other researchers believe that sulfide can adjust the electron density by producing S vacancies or constructing an S_δ-TMn+-H_2O network to improve the hydrolysis separation process [59]. Plenty of investigations show that the catalytic performance of TMSs can be improved in two ways: one is to increase the exposed number of active sites by increasing the specific surface area of the catalyst, thus improving the catalytic activity; the second is to optimize the catalytic performance by improving the inherent activity of the original active site.

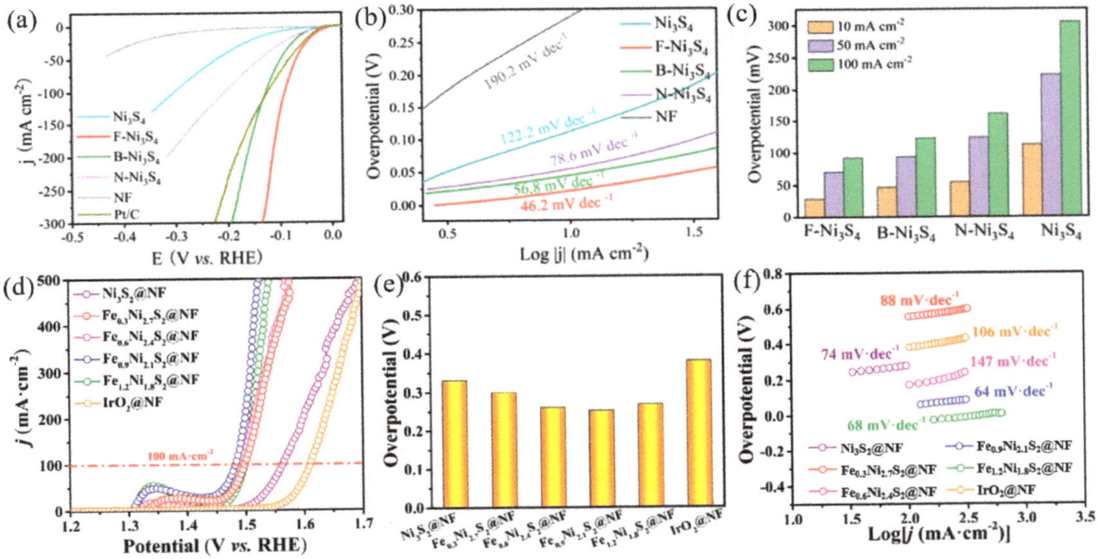

Figure 6. The HER performance of the B−, N−, F−doped Ni_3S_4 [58] (**a**) LSV polarization curves, (**b**) Tafel plot, (**c**) overpotential. The OER performance of the Fe−doped Ni_3S_2 [59] (**d**) LSV polarization curves, (**e**) overpotential, (**f**) Tafel plot.

2.1.4. Transition Metal Phosphide Catalyst

Transition metal phosphides (TMPs) are one of the most widely investigated electrocatalytic materials for hydrogen evolution in HER. P atoms with high electronegativity can extract electrons from neighboring transition metals and capture positively charged protons, so they can be used as active sites to stabilize reaction intermediates [60]. TMPs have stable structure, ceramic and metal characteristics, good thermal and electrical conductivity, and thermodynamic stability, etc., and have always been known as the "quasi-platinum catalyst". Metal phosphides tend to form more isotropic structures rather than layered structures. Therefore, they can have more coordinative unsaturated surface atoms and are not constrained by the stacking load on the electrode. Furthermore, they have metallic properties, which are beneficial to their application in the electrochemical direction. Typical monometallic TMPs include Ni-P, Co-P, and Fe-P. Compared with monometallic TMPs, bimetallic TMPs, such as Ni-Co-P [61–63] and Fe-Co-P [64], can significantly reduce the catalytic overpotential and have more excellent catalytic performance in the field of electrolytic hydrogen production as shown in Figure 7. Brewer-Engel valence bond theory shows that the transition metals with empty or semi-empty d orbitals (d electrons are less than d orbitals) and transition metals with paired d electrons (d electrons are greater than d orbitals) will have a synergistic effect, which is beneficial to the cathodic hydrogen evolution reaction.

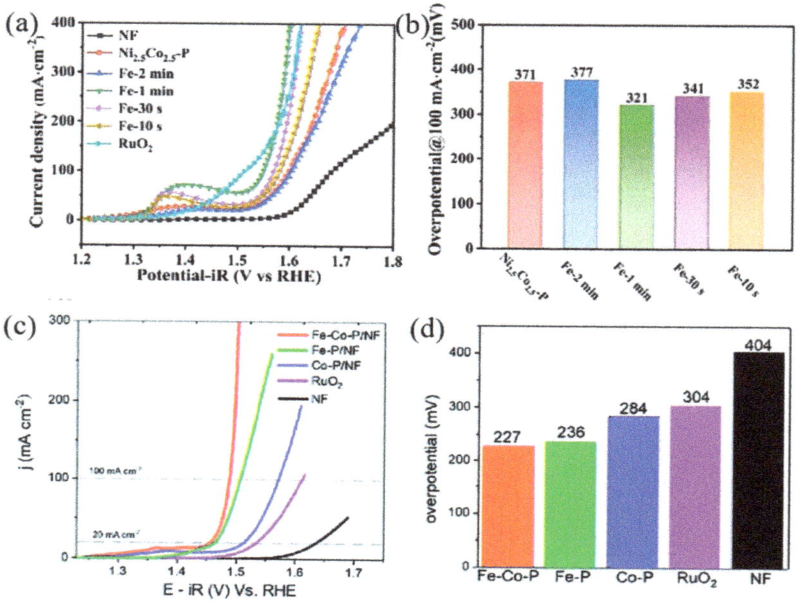

Figure 7. Transition metal phosphides (**a**) the polarization curves for Ni−Co−P after Fe addition (**b**) corresponding Tafel plots [64]; (**c**) OER polarization curves for Multicomponent alloy phosphide, (**d**) corresponding overpotentials [65].

2.2. Proton Exchange Membrane Water Electrolysis Hydrogen Production

Hydrogen production by electrolysis of water with a proton exchange membrane is a technology developed from the 1960s to the 1970s. After a long period of development, its equipment has had a high degree of integration, but there are bottlenecks in many technical aspects. At present, it is mainly imported from developed countries, such as Europe and America. The principle of hydrogen production by PEM is shown in Figure 8. The PEM electrolyzer adopts a bipolar structure, and the electrical connection between cells is carried out by a bipolar plate, which plays an important role in discharging the generated gas [9]. The anode, the cathode, and the membrane group form a membrane electrode assembly. At the anode end, water is catalytically oxidized by the catalyst on the membrane to generate oxygen, electrons, and protons, and the protons generated at the anode are circulated to the cathode end through the membrane and reduced to generate hydrogen. PEM technology uses a perfluorosulfonic acid proton exchange membrane as an electrolytic membrane. Compared with traditional membranes, PEM membrane has the advantages of stable chemical properties, high proton conductivity, nonporous gas isolation, and so on, and can be integrated with electrodes to reduce the extra resistance and power loss caused by the distance between the two electrodes [65]. Therefore, this technology can improve the purity of hydrogen production, and at the same time obtain large current density and rapid response, which is suitable for renewable energy power generation systems with large fluctuations.

However, the investment cost of PEM membrane is high, and almost all the catalysts used depend on platinum-based precious metals and their alloys, as well as precious metal oxides, such as PtO_2, RuO_2, and IrO_2 [66], which greatly increases the use cost. At present, the research on noble metal catalysts mainly focuses on the control of microstructure [67] and composition [68]. Figure 8 shows several methods of membrane electrode assembly in PEM electrolytic cells. When the catalyst layers are directly deposited onto membranes, protonic transport is most likely improved, due to the geometrically shortest protonic pathway for a given PEM thickness, as indicated by the orange arrow. At the anode side, as shown in Figure 8A, only a few titanium fibers directly connected electric contact between the catalyst and membrane layer. When depositing the anode catalysts directly on the titanium diffusion layer, the deposition of large parts of the catalyst layer up to multiple hundreds of micrometers into the pores of the titanium diffusion layer (Figure 8B) is expected. In this case, fewer parts of the catalyst layer are directly in contact with the membrane. Therefore, the reduced protonic interface between the catalyst layer and the PEM leads to an increased ohmic resistance, but the increased interface between the titanium fibers and catalysts makes the electron transport improve, as indicated by the blue arrow. For the anodic porous transport electrodes (aPTE), the carbon-based surface is more planar compared to the titanium substrates. Therefore, one would assume less poorly connected catalyst areas (Figure 8D). For the ai PTE, as shown in Figure 8C, we expected a behavior based on the aPTE and cathode porous transport electrode base cases, but an improvement of the ionic transport and consequently an improved high-frequency resistance. By changing the microstructure of a noble metal catalyst, its electrochemical performance is greatly improved, and the reaction overpotential can be further reduced, or it can be combined with other non-noble metal materials, such as PtCu alloy, to reduce the loading and cost while ensuring its catalytic activity. Furthermore, the typical synthesized electrocatalysts with high electrochemical performance are shown in Table 3. However, compared with AWE technology, PEM technology requires higher water quality [22], which increases not only the production cost but also the supply cost of raw materials. Moreover, due to the imperfect manufacturing process and lack of practical engineering experience, the voltage fluctuates greatly during operation, and the theoretical electrolysis effect cannot be achieved [69]. Although the purity of hydrogen prepared by PEM technology is higher, the preparation cost and service life of membrane and catalysts are difficult to compare with AWE.

Table 3. Electrochemical properties of typical electrocatalysts for PEM electrolyzer.

Electrocatalysts for PEM Electrolyzer	Catalyst	Overpotential/mV (10 mA cm^{-2})	Tafel Slope/mV dec^{-1}	Rct/Ω
oxides	$Ir_{0.6}Mn_{0.4}O_x$ for OER [70]	212	40	5.2
	Ir/ATO 70% for OER [71]	256	-	-
sulfides	GDL/(CNTs+FeMoS) for HER [72]	180	57	-
	FeS_2 for HER [73]	870 mV at 1 A cm^{-2}	-	-
phosphides	$Ni_{78}P_{22}$ for HER [74]	105	38	1.16
	$Ni_{71.5}Mo_{26.5}P_2$ for HER [75]	28	29	
	20% FeP/CB for HER [76]	51	101	-

Proton conductive ceramic water electrolysis hydrogen production technology PCC uses solid oxide ceramics as electrolytes, which reduces the cost of hydrogen production by PEM [77]. The electrolyte and catalyst used are similar to solid oxide fuel cell SOFC [34], mainly including zirconate, cobaltate, and ferrite perovskite ceramic materials [78]. Among them, the typical perovskite strontium zirconate ceramics show excellent proton conductivity at the temperature of about 700 °C, which is beneficial to the production of high-purity hydrogen at the cathode and can simplify the steam electrolysis device. The main working principle diagram is shown in Figure 1d. The disadvantage of hydrogen production by electrolysis of water with proton conductive ceramics is the overvoltage caused by high ohmic polarization loss at the connection between batteries. Contrary to the tubular battery, the flat battery makes the structure more compact, thus improving hydrogen production efficiency. At present, the main obstacle to the industrial application of hydrogen production technology by electrolytic water with proton conductive ceramics is the long-term stability of electrolyzer in high-temperature and high-humidity environments [79], and there are also problems of electrode aging and deactivation. It works at high temperatures. Ceramic materials are generally used as catalysts, such as perovskite material with proton conduction as an anode and nickel-based ceramic material as a cathode. Ceramic oxide catalysts have excellent high-temperature stability, and their catalytic performance can be optimized by doping and introducing oxygen vacancies.

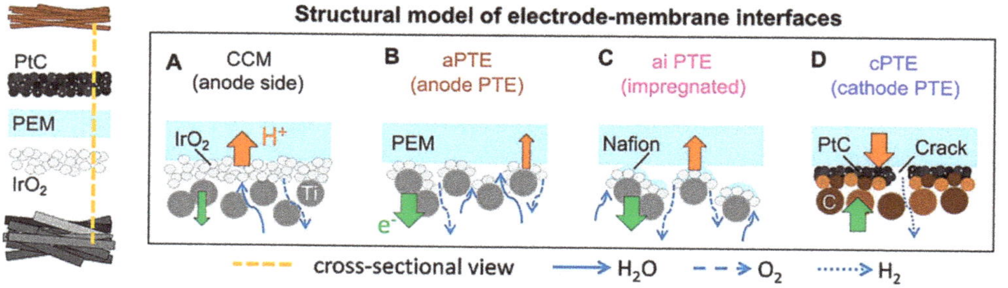

Figure 8. Model of the membrane-electrode interfaces and transport processes in different MEA configurations [68].

2.3. Solid Oxide Water Electrolysis Hydrogen Production

Solid oxide ceramic is used as an electrolyte in solid oxide electrolytic water technology SOE. Contrary to SOFC, the working temperature is between 500~1000 °C. The high working temperature leads to higher efficiency than AWE and PEM technology, and the highest efficiency can be close to 100% [80]. Hydrogen is generated by water electrolysis under high temperature [81,82]. In the solid oxide electrolyzer, water is converted into water vapor at high temperature, and the current electrolyzes the water molecules adsorbed on the cathode catalytic layer into H^+ and O^{2-}; the free electrons of H^+ transmitted through the external circuit are reduced to H_2, and O^{2-} passes through the solid electrolyte layer to reach the anode catalytic layer. At the same time, the lost electrons are converted into oxygen, and the free electrons enter the external circuit [9]. It is required that the electrolyte has high oxygen ion conductivity, so that O^{2-} can pass through the electrolyte layer, and its electronic conductivity is very low to prevent short circuits. The anode and cathode are porous structures, which are beneficial for gas diffusion and constructing the interfaces of three-phase catalytic.

The biggest advantage of SOE technology lies in its high energy conversion efficiency, which can effectively reduce the energy consumption required in the electrolytic procedure [82]. As the PCC technology, its working temperature is higher, and the catalyst used is high-temperature ceramic material. The anode is composed of perovskite or perovskite ceramic composite, and the cathode is made of nickel-based ceramic composite material, so a precious metal catalyst is not needed, and the preparation cost of the catalyst is low [83]. Additionally, SOE technology can also be used for the electrolytic reduction of CO_2 to realize CO_2 conversion and emission reduction [82]. Although the catalytic efficiency of SOE technology is increasing with the increase of working temperature, high working temperature also creates some problems, such as difficulty in sealing, and higher requirements for the chemical and mechanical stability of electrodes and catalytic materials in high-temperature and high-humidity environments, which limits the development of SOE technology to some extent. Furthermore, the gas produced by the cathode is a mixture of hydrogen and water vapor, which needs to be further separated and purified, which, in turn, increases the cost compared with conventional liquid water electrolysis. At present, due to the short life of the battery stack, the need for auxiliary components in the process of electrolytic hydrogen production and the high temperature and other safety issues, SOE technology is still in the laboratory research and development stage, and it cannot be commercialized in the short term [8].

After years of research and development, the catalytic performance and chemical stability of electrolyzed water hydrogen production technology have been greatly improved, but there are still some problems with the actual industrial application. First, water electrolysis OER is a slow four-electron transfer process. To drive the reaction, it is necessary to apply a very high overpotential, resulting in high energy consumption. The oxygen evolution performance of the catalyst can be improved by structural optimization. Second, the economic competitiveness of hydrogen production technology by electrocatalytic decomposition of water still needs to be strengthened, and the development of electrocatalysts with low cost and rich resources is still the main research direction in the future. Moreover, the hydrogen evolution reaction in the semi-reaction of electrolytic water is a relatively fast two-electron transfer process, and it is still an effective measure to reduce the energy consumption in the process of electrolytic water by using electrocatalyst to reduce the overpotential of OER and reduce the energy loss in the process of water decomposition. Finally, because acidic or alkaline electrolytes are corrosive, long-term operation will corrode industrial equipment, which further hinders the industrial application of electrolytic water. Therefore, it will be the future development direction to investigate high-performance electrocatalysts in neutral electrolytes. In neutral electrolytes, more materials will be used as suitable electrocatalysts to realize water electrolysis. However, the long-term stability of the catalyst itself is affected by many factors. The researchers found that the current density, electrolyte pH, electrolyte temperature, and the decomposition of the catalytic substrate have a great influence on the long-term stability of the catalytic materials. The optimal parameters could be obtained by comparing the effects of various parameters on the cathodic catalytic hydrogen evolution [84–86].

The design of the catalytic electrode structure and the optimization of the catalytic material structure are very important to improve the efficiency of hydrogen production by electrolysis of water. The conventional powder material spraying method uses a binder in electrode preparation, which increases additional resistance, hinders the contact between active sites and electrolyte, hinders the transfer of substances, and leads to the decrease of apparent activity of the catalyst [87]. In situ construction of an integrated electrode on a three-dimensional conductive substrate can effectively improve the conductivity of the electrode, increase the number and utilization rate of active sites, and promote substance transfer. Common three-dimensional electrode substrate materials include carbon-based and metal-based materials, and different substrate materials have different charge transfer capabilities, so their catalytic activities are also different. In addition, different microstructures of catalytic materials have different specific surface areas, which leads to different active sites. Through the microstructure design of the catalyst, the number of active sites exposed on the catalyst surface can be increased, the stability of the catalytic electrode can be improved, and the charge transfer, electrolyte diffusion, and timely release of H_2 can be improved. By combining the control of different electrode substrates and nanostructures, the catalytic limit of electrodes can be broken, and the material activity and electrolytic efficiency can be effectively improved [88,89]. Therefore, it is necessary to consider a multi-scale design strategy and explore the mechanism of various factors on catalyst activity and stability. For the practical application process of the catalysts, the catalytic activity, and chemical and mechanical stability of the catalytic material need to be considered to ensure that the catalytic electrode material has high catalytic efficiency and service life. To realize large-scale commercial applications, the catalyst's preparation and use costs need to be thoroughly analyzed. Noble metal catalysts are essential for some noble metals in special catalytic systems due to their excellent catalytic activity and stability. Based on this, it is necessary to consider how to reduce the usage of noble metals on the premise of ensuring the catalytic activity of materials and recycling the subsequent noble metals.

Based on the finding of this study, the future research directions of hydrogen production by water electrolysis mainly include the following: (1) considering the high catalytic activity and electrochemical stability of the noble metal catalyst, especially its excellent stability in acidic electrolysis system, to develop an economical, efficient, and stable electrocatalyst composed of noble metal and base metal [90]; (2) based on the high catalytic activity of the obtained catalyst, further improving the structural stability of the catalyst and preventing large deformation during the reaction; (3) designing an electrolyzer with zero gap structure between anode and cathode to improve the current density and optimize electrolysis efficiency.

3. Conclusions

Hydrogen production technology by water electrolysis has the advantages of environmental friendliness, high purity of hydrogen production, and simple preparation process, which has broad development potential under the background of global energy shortage and successive transformation of high-carbon energy. At present, the main technologies for hydrogen production by electrolysis of water can be divided into three categories, including alkaline water electrolysis, proton exchange membrane water electrolysis, and solid oxide water electrolysis. The key to the development of hydrogen production technology by electrolysis of water lies in the selection of catalysts and the optimization of their performance. Optimizing the existing catalyst systems and developing high-performance catalysts, reducing the use cost, and improving the stability of catalysts are of great significance for improving the efficiency of OER and HER, reducing the overpotential of electrodes, and reducing energy consumption. Precious metals, precious metal alloys, and oxides are still the main catalysts with the best performance, but the use cost of precious metal catalysts is high, so it is very important to develop high-performance and low-cost catalysts. Transition metal catalysts are widely used because of their low preparation cost and many optimized electrocatalytic properties comparable to those of precious metal catalysts. Catalysts can be optimized using alloying, element doping, multi-structure compounding, and micro-morphology regulation, but the single method of optimization has limitations, and the combination of multi-scale methods can realize the diversified promotion of catalysts from composition and structure to performance. With the increasing demand for clean energy, the technology of hydrogen production from electrolyzed water will be further developed and is expected to boost energy transformation and environmental optimization.

Author Contributions: Conceptualization, H.L. and J.G.; validation, J.G., Z.L. and J.W.; writing—original draft preparation, H.L. and J.G.; writing—review and editing, J.G. and J.W.; supervision, J.G. and J.W.; funding acquisition, J.W. All authors have read and agreed to the published version of the manuscript.

Funding: This research was funded by the National Natural Science Foundation of China (Grant No. 52101258), the Yunnan Major Scientific and Technological Projects (grant NO. 202202AG050017-2).

Conflicts of Interest: The authors declare no conflict of interest.

References

1. Muradov, N.Z.; Veziroğlu, T.N. "Green" path from fossil-based to hydrogen economy: An overview of carbon-neutral technologies. *Int. J. Hydrogen Energy* **2008**, *33*, 6804–6839. [CrossRef]
2. Chen, X.; Lin, B. Towards carbon neutrality by implementing carbon emissions trading scheme: Policy evaluation in China. *Energy Policy* **2021**, *157*, 112510. [CrossRef]
3. Hou, Z.M.; Xiong, Y.; Luo, J.S.; Fang, Y.L.; Haris, M.; Chen, Q.J.; Yue, Y.; Wu, L.; Wang, Q.C.; Huang, L.C.; et al. International experience of carbon neutrality and prospects of key technologies: Lessons for China. *Pet. Sci.* **2023**, *20*, 893–909. [CrossRef]
4. Taherian, Z.; Khataee, A.; Han, N.; Orooji, Y. Hydrogen production through methane reforming processes using promoted-Ni/mesoporous silica: A review. *J. Ind. Eng. Chem.* **2022**, *107*, 20–30. [CrossRef]

5. Gao, Y.; Jiang, J.; Meng, Y.; Yan, F.; Aihemaiti, A. A review of recent developments in hydrogen production via biogas dry reforming. *Energy Convers. Manag.* **2018**, *171*, 133–155. [CrossRef]
6. Kumar, G.; Mathimani, T.; Rene, E.R.; Pugazhendhi, A. Application of nanotechnology in dark fermentation for enhanced biohydrogen production using inorganic nanoparticles. *Int. J. Hydrogen Energy* **2019**, *44*, 13106–13113. [CrossRef]
7. Jayachandran, V.; Basak, N.; De Philippis, R.; Adessi, A. Novel strategies towards efficient molecular biohydrogen production by dark fermentative mechanism: Present progress and future perspective. *Bioprocess. Biosyst. Eng.* **2022**, *45*, 1595–1624. [CrossRef]
8. Wang, S.; Lu, A.; Zhong, C.J. Hydrogen production from water electrolysis: Role of catalysts. *Nano Converg.* **2021**, *8*, 1–23. [CrossRef]
9. Ahmed, S.F.; Mofijur, M.; Nuzhat, S.; Rafa, N.; Musharrat, A.; Lam, S.S.; Boretti, A. Sustainable hydrogen production: Technological advancements and economic analysis. *Int. J. Hydrogen Energy* **2022**, *47*, 37227–37255. [CrossRef]
10. Zhang, B.; Zhang, S.X.; Yao, R.; Wu, Y.H.; Qiu, J.S. Progress and prospects of hydrogen production: Opportunities and challenges. *J. Electron. Sci. Technol.* **2021**, *19*, 100080. [CrossRef]
11. Tahir, M.; Pan, L.; Idrees, F.; Zhang, X.; Wang, L.; Zou, J.-J.; Wang, Z.L. Electrocatalytic oxygen evolution reaction for energy conversion and storage: A comprehensive review. *Nano Energy* **2017**, *37*, 136–157. [CrossRef]
12. Peng, J.; Tao, P.; Song, C.; Shang, W.; Deng, T.; Wu, J. Structural evolution of Pt-based oxygen reduction reaction electrocatalysts. *Chin. J. Catal.* **2022**, *43*, 47–58. [CrossRef]
13. Yang, W.; Zhang, W.; Liu, R.; Lv, F.; Chao, Y.; Wang, Z.; Guo, S. Amorphous Ru nanoclusters onto Co-doped 1D carbon nanocages enables efficient hydrogen evolution catalysis. *Chin. J. Catal.* **2022**, *43*, 110–115. [CrossRef]
14. Liu, Y.; Liang, X.; Chen, H.; Gao, R.; Shi, L.; Yang, L.; Zou, X. Iridium-containing water-oxidation catalysts in acidic electrolyte. *Chin. J. Catal.* **2021**, *42*, 1054–1077. [CrossRef]
15. Anantharaj, S.; Ede, S.R.; Sakthikumar, K.; Karthick, K.; Mishra, S.; Kundu, S. Recent trends and perspectives in electrochemical water splitting with an emphasis on sulfide, selenide, and phosphide catalysts of Fe, Co, and Ni: A review. *ACS Catal.* **2016**, *6*, 8069–8097. [CrossRef]
16. Anantharaj, S.; Kundu, S.; Noda, S. Progress in nickel chalcogenide electrocatalyzed hydrogen evolution reaction. *J. Mater. Chem. A* **2020**, *8*, 4174–4192. [CrossRef]
17. Yin, J.; Jin, J.; Lin, H.; Yin, Z.; Li, J.; Lu, M.; Guo, L.; Xi, P.; Tang, Y.; Yan, C.-H. Optimized metal chalcogenides for boosting water splitting. *Adv. Sci.* **2020**, *7*, 1903070. [CrossRef]
18. Peng, L.; Shoaib, S.; Shah, A.; Wei, Z. Recent developments in metal phosphide and sulfide electrocatalysts for oxygen evolution reaction. *Chin. J. Catal.* **2018**, *39*, 1575–1593. [CrossRef]
19. Lv, Z.; Mahmood, N.; Tahir, M.; Pan, L.; Zhang, X.; Zou, J.-J. Fabrication of zero to three dimensional nanostructured molybdenum sulfides and their electrochemical and photocatalytic applications. *Nanoscale* **2016**, *8*, 18250–18269. [CrossRef]
20. Li, Y.; Nakamura, R. Structural change of molybdenum sulfide facilitates the electrocatalytic hydrogen evolution reaction at neutral pH as revealed by in situ Raman spectroscopy. *Chin. J. Catal.* **2018**, *39*, 401–406. [CrossRef]
21. Brauns, J.; Turek, T. Alkaline water electrolysis powered by renewable energy: A review. *Processes* **2020**, *8*, 248. [CrossRef]
22. Zeng, K.; Zhang, D.K. Recent progress in alkaline water electrolysis for hydrogen production and application. *Prog. Energy Combust. Sci.* **2010**, *36*, 307–326. [CrossRef]
23. Falcão, D.S.; Pinto, A. A review on PEM electrolyzer modelling: Guidelines for beginners. *J. Clean. Prod.* **2020**, *261*, 121184. [CrossRef]
24. Reza, A.; Brian, P.; Lin, S.; Wang, J.; Zhao, Y.; Xu, H.; Pivovar, B.; Tian, B.; Chen, X. A roadmap to low-cost hydrogen with hydroxide exchange membrane electrolyzers. *Adv. Mater.* **2019**, *31*, 1805876.
25. Anwar, S.; Khan, F.; Zhang, Y.H.; Djire, A. Recent development in electrocatalysts for hydrogen production through water electrolysis. *Int. J. Hydrogen Energy* **2021**, *46*, 32284–32317. [CrossRef]
26. Ursua, A.; Gandia, L.M.; Sanchis, P. Hydrogen production from water electrolysis: Current status and future trends. *Proc. IEEE* **2012**, *100*, 410–426. [CrossRef]
27. Li, Z.; Hu, M.; Wang, P.; Liu, J.; Yao, J.; Li, C. Heterojunction catalyst in electrocatalytic water splitting. *Coord. Chem. Rev.* **2021**, *439*, 213953. [CrossRef]
28. Yang, E.; Mohamed, H.O.; Park, S.G.; Obaid, M.; Al-Qaradawi, S.Y.; Castaño, P.; Chon, K.; Chae, K.J. A review on self-sustainable microbial electrolysis cells for electro-biohydrogen production via coupling with carbon-neutral renewable energy technologies. *Bioresour. Technol.* **2021**, *320*, 124363. [CrossRef]
29. Khatib, F.N.; Wilberforce, T.; Ijaodol, O.; Ogungbemi, E.; El-Hassana, Z.; Durranta, A.; Thompsona, J.; Olabi, A. Material degradation of components in polymer electrolyte membrane (PEM) electrolytic cell and mitigation mechanisms: A review. *Renew. Sustain. Energy Rev.* **2019**, *111*, 1–14. [CrossRef]
30. Dawood, F.; Anda, M.; Shafiullah, G.M. Hydrogen production for energy: An overview. *Int. J. Hydrogen Energy* **2020**, *45*, 3847–3869. [CrossRef]
31. Park, J.E.; Kang, S.Y.; Oh, S.H. High-performance anion exchange membrane water electrolysis. *Electrochim. Acta* **2019**, *295*, 99–106. [CrossRef]
32. Vincent, I.; Kruger, A.; Bessarabov, D. Hydrogen production by water electrolysis with an ultrathin anion-exchange membrane(AEM). *Int. J. Electrochem. Sci.* **2018**, *13*, 11347–11358. [CrossRef]

33. Chen, P.J.; Hu, X.L. High-efficiency anion exchange membrane water electrolysis employing non-noble metal catalysts. *Adv. Energy Mater.* **2020**, *10*, 2002285. [CrossRef]
34. Chatenet, M.; Pollet, B.G.; Dekel, D.R. Water electrolysis: From textbook knowledge to the latest scientific strategies and industrial developments. *Chem. Soc. Rev.* **2022**, *51*, 4583–4762. [CrossRef] [PubMed]
35. Liu, T.; Li, M.; Bo, X.; Zhou, M. Designing transition metal alloy nanoparticles embedded hierarchically porous carbon nanosheets as high-efficiency electrocatalysts toward full water splitting. *J. Colloid Interface Sci.* **2019**, *537*, 280–294. [CrossRef]
36. Dong, S.; Li, Y.; Zhao, Z.; Hu, X.; Li, R.; Sun, S.; Zhang, X. Reparation of porous Ti-Cu alloy by one-step sintering method and application of hydrogen evolution reaction. *J. Electroanal. Chem.* **2022**, *918*, 116448. [CrossRef]
37. Yang, L.; Zhou, H.; Qin, X.; Guo, X.; Cui, G.; Asiri, A.M.; Sun, X. Cathodic electrochemical activation of Co_3O_4 nanoarrays: A smart strategy to significantly boost the hydrogen evolution activity. *Chem. Commun.* **2018**, *54*, 2150–2153. [CrossRef]
38. Begum, H.; Ahmed, M.S.; Jeon, S.; Jeon, S. δ-MnO_2 nanoflowers on sulfonated graphene sheets for stable oxygen reduction and hydrogen evolution reaction. *Electrochim. Acta* **2019**, *296*, 235–242. [CrossRef]
39. Wang, Y.; Liu, J.C.; Liao, Y.F. Hetero-structured V-Ni_3S_2@NiOOH core-shell nanorods from an electrochemical anodization for water splitting. *J. Alloys Compd.* **2021**, *856*, 158219. [CrossRef]
40. Ha, Y.; Shi, L.X.; Yan, X.X.; Chen, Z.L.; Li, Y.P.; Xu, W.; Wu, R.B. Multifunctional Electrocatalysis on a Porous N-doped $NiCo_2O_4$@C Nanonetwork. *ACS Appl. Mater. Interfaces* **2019**, *11*, 45546–45553. [CrossRef]
41. Köse, K.Ö.; Aydınol, M.K. Development of activated carbon/bimetallic transition metal phosphide composite materials for electrochemical capacitors and oxygen evolution reaction catalysis. *Int. J. Energy Res.* **2022**, *46*, 22078–22088. [CrossRef]
42. Wang, L.; Gong, N.; Zhou, Z.; Peng, W.; Li, Y.; Zhang, F.; Fan, X. Electronic modulation of multi-element transition metal phosphide by V-doping for high-efficiency and pH-universal hydrogen evolution reaction. *Int. J. Hydrogen Energy* **2022**, *47*, 18305–18313. [CrossRef]
43. Kibsgaard, J.; Tsai, C.; Chan, K.; Benck, J.D.; Nørskov, J.K.; Pedersen, F.A.; Jaramillo, T.F. Designing an improved transition metal phosphide catalyst for hydrogen evolution using experimental and theoretical trends. *Energy Environ. Sci.* **2015**, *8*, 3022–3029. [CrossRef]
44. Bodner, M.; Hofer, A.; Hacker, V. H_2 generation from alkaline electrolyzer. *Wiley Energy Environ.* **2015**, *4*, 365–381. [CrossRef]
45. Rosalbino, F.; Macciò, D.; Angelini, E.; Saccone, A.; Delfino, D. Electrocatalytic properties of Fe-R (R = rare earth metal) crystalline alloys as hydrogen electrodes in alkaline water electrolysis. *J. Alloys Compd.* **2005**, *403*, 275–282. [CrossRef]
46. Rosalbino, F.; Macciò, D.; Angelini, E.; Saccone, A.; Delfino, S. Characterization of Fe-Zn-R (R = rare earth metal) crystalline alloys as electrocatalysts for hydrogen evolution. *J. Hydrogen Energy* **2008**, *33*, 2660–2667. [CrossRef]
47. Li, Z.; Lv, L.; Ao, X.; Li, A.; Sun, H.; An, P.; Xue, X.; Li, Y.; Liu, M.; Wang, C.; et al. An effective method for enhancing oxygen evolution kinetics of $LaMO_3$ (M = Ni, Co, Mn) perovskite catalysts and its application to a rechargeable zinc–air battery. *Appl. Catal. B Environ.* **2020**, *262*, 118291. [CrossRef]
48. Matsumoto, Y.; Manade, H.; Sato, E. Oxygen Evolution on $La_{1-x}Sr_xCoO_3$ Electrodes in Alkaline Solutions. *J. Electrochem. Soc.* **1980**, *4*, 811–817. [CrossRef]
49. Mizusaki, J.; Mima, Y.; Yamauchis, S.; Fueki, K.; Tagawa, H. Nonstoichiometry of the perovskite-type oxides $La_{1-x}Sr_xCoO_{3-\delta}$. *J. Solid State Chem.* **1989**, *80*, 102–111. [CrossRef]
50. Svegl, F.; Orel, B.; Svegl, I.G.; Kaucic, V. Characterization of spinel Co_3O_4 and Li-doped Co_3O_4 thin film electrocatalysts prepared by the Sol-Gel route. *J. Electrochim. Acta* **2000**, *45*, 4359–4371. [CrossRef]
51. Gao, X.; Zhang, H.; Li, Q.; Yu, X.; Hong, Z.; Zhang, X.; Liang, C.; Lin, Z. Hierarchical $NiCo_2O_4$ hollow microcuboids as bifunctional electrocatalysts for overall water-splitting. *Angew. Chem. Int. Ed.* **2016**, *55*, 6290–6294. [CrossRef] [PubMed]
52. Lankhost, M.H.R.; Bouwmeester, H.J.M.; Verweij, H. Importance of electronic band structure to nonstoichiometric behaviour of $La_{0.8}Sr_{0.2}CoO_3$. *Solid State Ion.* **1997**, *96*, 21–27. [CrossRef]
53. Otagawa, T.; Bockris, J.O.M. Lanthanum Nickelate as Electrocatalyst Oxygen Evolution. *J. Electrochem. Soc.* **1982**, *10*, 2391–2392. [CrossRef]
54. Rogachev, A.S.; Fourmont, A.; Kovalev, D.Y.; Vadchenko, S.G.; Kochetov, N.A.; Shkodich, N.F.; Baras, F.; Politano, O. Mechanical alloying in the Co-Fe-Ni powder mixture: Experimental study and molecular dynamics simulation. *Powder Technol.* **2022**, *399*, 117187. [CrossRef]
55. Atanova, A.V.; Seregin, D.S.; Zhigalina, O.M.; Khmelenin, D.N.; Orlov, G.A.; Turkina, D.I.; Sigov, A.S.; Vorotilov, K.A. Control of Columnar Grain Microstructure in CSD $LaNiO_3$ Films. *Molecules* **2023**, *28*, 1938. [CrossRef]
56. Chang, H.; Liang, Z.; Wang, L.; Wang, C. Research progress in improving the oxygen evolution reaction by adjusting the 3d electronic structure of transition metal catalysts. *Nanoscale* **2022**, *14*, 5639–5656. [CrossRef]
57. Wang, M.; Zhang, L.; He, Y.; Zhu, H. Recent advances in transition-metal-sulfide-based bifunctional electrocatalysts for overall water splitting. *J. Mater. Chem. A* **2021**, *9*, 5320–5363. [CrossRef]
58. Wang, J.; Zhang, Z.; Song, H.; Zhang, B.; Liu, J.; Shai, X.; Miao, L. Water dissociation kinetic-oriented design of nickel sulfides via tailored dual sites for efficient alkaline hydrogen evolution. *Adv. Funct. Mater.* **2020**, *31*, 2008578. [CrossRef]
59. Fei, B.; Chen, Z.; Liu, J.; Xu, H.; Yan, X.; Qing, H.; Chen, M.; Wu, R. Ultrathinning nickel sulfide with modulated electron density for efficient water splitting. *Adv. Energy Mater.* **2020**, *10*, 2001963. [CrossRef]
60. Joo, J.; Kim, T.; Lee, J.; Choi, A.I.; Lee, K.Y. Morphology-controlled metal sulfides and phosphides for electrochemical water splitting. *Adv. Mater.* **2019**, *31*, 1806682. [CrossRef] [PubMed]

61. Pei, Y.; Cheng, Y.; Chen, J.; Smith, W.; Dong, P.; Ajayan, P.M.; Ye, M.; Shen, J. Recent developments of transition metal phosphides as catalysts in the energy conversion field. *J. Mater. Chem. A* **2018**, *6*, 23220–23243. [CrossRef]
62. Li, Y.; Li, R.; Wang, D.; Xu, H.; Meng, F.; Dong, D.; Jiang, J.; Zhang, J.; An, M.; Yang, P. A review: Target-oriented transition metal phosphide design and synthesis for water splitting. *Int. J. Hydrogen Energy* **2021**, *46*, 5131–5149. [CrossRef]
63. Shi, Y.; Zhang, B. Recent advances in transition metal phosphide nanomaterials: Synthesis and applications in hydrogen evolution reaction. *Chem. Soc. Rev.* **2016**, *45*, 1529–1541. [CrossRef]
64. Su, W.; Wang, D.; Zhou, Q.; Zheng, X. Preparation of 3D Fe-Co-Ni-OH/NiCoP electrode as a highly efficient electrocatalyst in the oxygen evolution reactions. *J. Alloys Compd.* **2023**, *941*, 168578. [CrossRef]
65. Xu, H.; Zhu, J.; Wang, P.; Chen, D.; Zhang, C.; Xiao, M.; Ma, Q.; Bai, H.; Qin, R.; Ma, J.; et al. Fe–Co–P multi-heterostructure arrays for efficient electrocatalytic water splitting. *J. Mater. Chem. A* **2021**, *9*, 24677–24685. [CrossRef]
66. Barbir, F. PEM electrolysis for production of hydrogen from renewable energy sources. *Sol. Energy* **2005**, *78*, 661–669. [CrossRef]
67. Carmo, M.; Fritz, D.L.; Merge, J.; Stolten, D. A comprehensive review on PEM water electrolysis. *Int. J. Hydrogen Energy* **2013**, *38*, 4901–4934. [CrossRef]
68. Bühler, M.; Holzapfel, P.; McLaughlin, D.; Thiele, S. From catalyst coated membranes to porous transport electrode based configurations in PEM water electrolyzers. *J. Electrochem. Soc.* **2019**, *166*, F1070. [CrossRef]
69. Kumar, S.S.; Himabindu, V. Hydrogen production by PEM water electrolysis—A review. *Mat. Sci. Energy Technol.* **2019**, *2*, 442–454.
70. Zhang, S.; Ma, X.; He, Y.; Zhu, Y.; Wang, Z. Amorphous mixed Ir-Mn oxide catalysts for the oxygen evolution reaction in PEM water electrolysis for H_2 production. *Int. J. Hydrogen Energy* **2023**, *48*, 10532–10544. [CrossRef]
71. Han, S.-B.; Mo, Y.-H.; Lee, Y.-S.; Lee, S.-G.; Park, D.-H.; Park, K.-W. Mesoporous iridium oxide/Sb-doped SnO_2 nanostructured electrodes for polymer electrolyte membrane water electrolysis. *Int. J. Hydrogen Energy* **2020**, *45*, 1409–1416. [CrossRef]
72. Morozan, A.; Johnson, H.; Roiron, C.; Genay, G.; Aldakov, D.; Ghedjatti, A.; Nguyen, C.T.; Tran, P.D.; Kinge, S.; Artero, V. Nonprecious Bimetallic Iron−Molybdenum Sulfide Electrocatalysts for the Hydrogen Evolution Reaction in Proton Exchange Membrane Electrolyzers. *ACS Catal.* **2020**, *10*, 14336–14348. [CrossRef]
73. Giovanni, C.D.; Reyes-Carmona, Á.; Coursier, A.; Nowak, S.; Greneche, J.-M.; Lecoq, H.; Mouton, L.; Roziere, J.; Jones, D.; Peron, J.; et al. Low-Cost Nanostructured Iron Sulfide Electrocatalysts for PEM Water Electrolysis. *ACS Catal.* **2016**, *6*, 2626–2631. [CrossRef]
74. Kim, H.; Park, H.; Kim, D.-K.; Choi, I.; Kim, S.-K. Pulse-electrodeposited nickel phosphide for high-performance proton exchange membrane water electrolysis. *J. Alloys Compd.* **2019**, *785*, 296–304. [CrossRef]
75. Bang, H.T.; Yeo, K.-R.; Choi, K.J.; Jung, W.S.; Kim, S.-K. Ternary Ni-Mo-P catalysts for enhanced activity and durability in proton exchange membrane water electrolysis. *Int. J. Energy Res.* **2022**, *46*, 13023–13034. [CrossRef]
76. Brito, J.; Restivo, J.; Sousa, J.P.S.; Spera, N.C.M.; Falcão, D.S.; Rocha, A.; Pinto, A.M.F.R.; Pereira, M.F.R.; Soares, O.S.G.P. Implementation of Transition Metal Phosphides as Pt-Free Catalysts for PEM Water Electrolysis. *Energies* **2022**, *15*, 1821. [CrossRef]
77. Rashid, M.D.; Al Mesfer, M.K.; Naseem, H.; Danish, M. Hydrogen production by water electrolysis: A review of alkaline water electrolysis, PEM water electrolysis and high temperature water electrolysis. *Int. J. Eng. Adv. Technol.* **2015**, *4*, 80–93.
78. Revankar, S.; Bindra, H. *Storage and Hybridization of Nuclear Energy*; Academica Press: Cambridge, MA, USA, 2018; pp. 49–117.
79. Medvedev, D. Trends in research and development of protonic ceramic electrolysis cells. *Int. J. Hydrogen Energy* **2019**, *44*, 26711–26740. [CrossRef]
80. Herring, J.S.; O'Brien, J.E.; Stoots, C.M.; Hawkes, G.L.; Hartvigsen, J.J.; Shahnam, M. Progress in high-temperature electrolysis for hydrogen production using planar SOFC technology. *Int. J. Hydrogen Energy* **2006**, *32*, 440–450. [CrossRef]
81. Younas, M.; Shafique, S.; Hafeez, A.; Javed, F.; Rehman, F. An overview of hydrogen production: Current status, potential, and challenges. *Fuel* **2022**, *316*, 123317. [CrossRef]
82. Kim, J.; Jun, A.; Gwon, O.; Yoo, S.; Liu, M.; Shin, J.; Lim, T.H.; Kim, G. Hybrid-solid oxide electrolysis cell: A new strategy for efficient hydrogen production. *Nano Energy* **2018**, *44*, 121–126. [CrossRef]
83. Meng, X.; Chen, M.; Gu, A.; Wu, X.; Liu, B.; Zhou, J.; Mao, Z. China's hydrogen development strategy in the context of double carbon targets. *Nat. Gas Ind. B* **2022**, *9*, 521–547. [CrossRef]
84. Zhang, S.; Yang, C.; Jiang, Y.; Li, P.; Xia, C. A robust fluorine-containing ceramic cathode for direct CO_2 electrolysis in solid oxide electrolysis cells. *J. Energy Chem.* **2023**, *77*, 300–309. [CrossRef]
85. Vidales, A.G.; Choi, K.; Omanovic, S. Nickel-cobalt-oxide cathodes for hydrogen production by water electrolysis in acidic and alkaline media. *Int. J. Hydrogen Energy* **2018**, *43*, 12917–12928. [CrossRef]
86. Vidalesa, A.G.; Millanb, N.C.; Bock, C. Modeling of anion exchange membrane water electrolyzers: The influence of operating parameters. *Chem. Eng. Res. Des.* **2023**, *194*, 636–648. [CrossRef]
87. Vidales, A.G.; Omanovic, S. Evaluation of nickel-molybdenum-oxides as cathodes for hydrogen evolution by water electrolysis in acidic, alkaline, and neutral media. *Electrochim. Acta* **2018**, *262*, 115–123. [CrossRef]
88. Yu, Z.; Xu, J.; Li, Y.; Wei, B.; Zhang, N.; Li, Y.; Bondarchuk, O.; Miao, H.; Araujo, A.; Wang, Z.; et al. Ultrafine oxygen-defective iridium oxide nanoclusters for efficient and durable water oxidation at high current densities in acidic media. *J. Mater. Chem. A* **2020**, *8*, 24743–24751. [CrossRef]

89. Vidales, A.G.; Sridhar, D.; Meunier, J.-L.; Omanovic, S. Nickel oxide on directly grown carbon nanofibers for energy storage applications. *J. Appl. Electrochem.* **2020**, *50*, 1217–1229. [CrossRef]
90. Vidales, A.G.; Dam-Quang, L.; Hong, A.; Omanovic, S. The influence of addition of iridium-oxide to nickel-molybdenumoxide cathodes on the electrocatalytic activity towards hydrogen evolution in acidic medium and on the cathode deactivation resistance. *Electrochim. Acta* **2019**, *302*, 198–206. [CrossRef]

molecules MDPI

Review

The Progress of Hard Carbon as an Anode Material in Sodium-Ion Batteries

Suchong Tan [1], Han Yang [1], Zhen Zhang [1], Xiangyu Xu [1], Yuanyuan Xu [1], Jian Zhou [1], Xinchi Zhou [1], Zhengdao Pan [1], Xingyou Rao [1], Yudong Gu [2], Zhoulu Wang [1], Yutong Wu [1,*], Xiang Liu [1,*] and Yi Zhang [1,*]

[1] School of Energy Sciences and Engineering, Nanjing Tech University, Nanjing 211816, China
[2] Jiangsu Svace Intelligent Technology Co., Ltd., Nanjing 210023, China
[*] Correspondence: yw19@njtech.edu.cn (Y.W.); iamxliu@njtech.edu.cn (X.L.); zhangy@njtech.edu.cn (Y.Z.)

Abstract: When compared to expensive lithium metal, the metal sodium resources on Earth are abundant and evenly distributed. Therefore, low-cost sodium-ion batteries are expected to replace lithium-ion batteries and become the most likely energy storage system for large-scale applications. Among the many anode materials for sodium-ion batteries, hard carbon has obvious advantages and great commercial potential. In this review, the adsorption behavior of sodium ions at the active sites on the surface of hard carbon, the process of entering the graphite lamellar, and their sequence in the discharge process are analyzed. The controversial storage mechanism of sodium ions is discussed, and four storage mechanisms for sodium ions are summarized. Not only is the storage mechanism of sodium ions (in hard carbon) analyzed in depth, but also the relationships between their morphology and structure regulation and between heteroatom doping and electrolyte optimization are further discussed, as well as the electrochemical performance of hard carbon anodes in sodium-ion batteries. It is expected that the sodium-ion batteries with hard carbon anodes will have excellent electrochemical performance, and lower costs will be required for large-scale energy storage systems.

Keywords: sodium-ion battery; electrochemical performance; sodium-ion storage; hard carbon anode

1. Introduction

Lithium-ion batteries are the main energy storage systems for mobile electronic devices such as electric vehicles, mobile phones, and laptops [1,2]. The lithium-ion batteries are the most mature energy storage devices at present, but their high cost and poor safety stagnate their commercial application [3–5]. Researchers are committed to finding products with a capacity and service life comparable to lithium-ion batteries, with lower costs and better safety performance, to satisfy the energy storage requirements of large-scale power grids in the country [6–16]. After long-term efforts, scientists finally found that sodium-ion batteries are the most promising alternatives to lithium-ion batteries [17–23]. Through investigation, it was found that the crustal abundance of sodium metal (2.74%) is more than 420 times that of lithium metal (0.0065%), which is rich in resources and evenly distributed worldwide [24–26]. Moreover, the two elements are of the same main group and have similar physical and chemical properties and similar electrochemical behaviors, and many technologies applied to lithium-ion batteries can also be applied to sodium-ion batteries [27–33]. Sodium-ion batteries have attracted wide attention as a substitute for lithium-ion batteries [34–38]. Anode materials play an important role in facilitating sodium-ion batteries with outstanding electrochemical performance. The design of novel anode materials with excellent performance and low cost can accelerate the commercialization of sodium-ion batteries [39–48]. Among the many anode electrode materials of sodium-ion batteries, hard carbon materials have the superiority of high capacity, low price, and low working voltage, and their unique structure is conducive to sodium-ion adsorption and reversible embedding/removal, showing excellent sodium storage performance, making them the most likely anode materials to be commercialized [49–57]. When commercializing hard carbon materials, troubles such as low first-cycle coulombic efficiency, terrible rate performance, and poor cycle stability are also faced [58–60]. In order to achieve an in-depth understanding of the sodium storage behavior of hard carbon anodes, this review analyzes in detail the mechanism of sodium-ion storage in hard carbon materials with different structures and then proposes three kinds of methods to improve the storage capacity of hard carbon materials, including morphology and structure construction, heteroatom doping, and electrolyte optimization. It is hoped that it can guide the synthesis of hard carbon materials with excellent properties and realize the application of sodium-ion batteries with hard carbon as the anode in large-scale national power grids as soon as possible.

2. Mechanism of Sodium Storage by Hard Carbon

Hard carbon is a kind of carbon material that cannot be graphitized at high temperatures, while graphite is a long-range ordered carbon material with a crystal structure [61–68]. The microstructure of hard carbon is obviously distinctive regarding graphite, which shows a short-range ordered structure with local graphite regions inside. Hard carbon prepared with various precursors has distinguishing characteristics, morphology, and structure, which makes the storage capacity of sodium ions vary greatly among hard carbon materials with distinguishable structures [69–73]. The difficulty in understanding the mechanism of sodium-ion storage in hard carbon has hindered the design of anode materials with excellent performance for sodium-ion batteries. Scientists have made many efforts to solve the controversial problems in developing sodium-ion batteries, and enormous achievements have been made during this period [74–76]. This review summarizes four different sodium-storage models for hard carbon anodes [77,78] (Figure 1), including the "embedding-adsorption" model [79,80], the "adsorption-embedding" model [3,81,82], the "three-stage model" [83], and the "adsorption-filling" model [84]. The following section describes the storage of sodium ions in hard carbon.

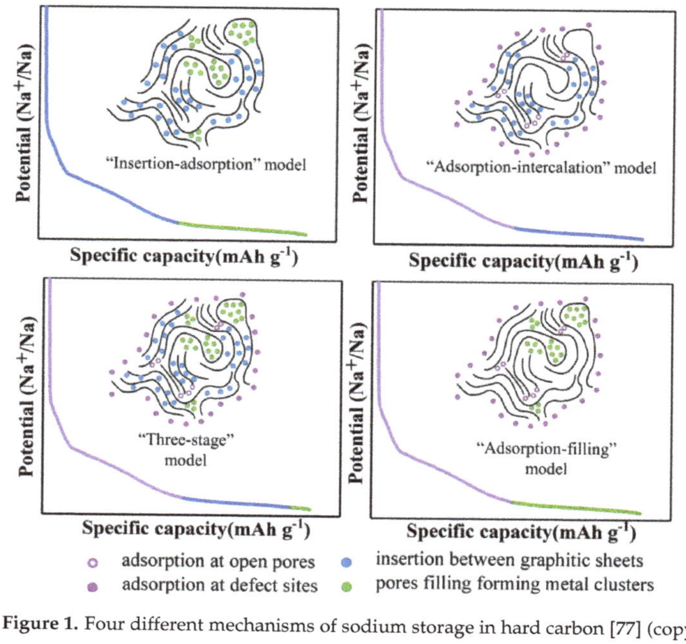

Figure 1. Four different mechanisms of sodium storage in hard carbon [77] (copyright Wiley, 2022).

2.1. "*Embedding-Adsorption" Model*

This sodium storage model was first proposed by Stevens et al. in 2000. As a glucose precursor, hard carbon materials were synthesized by pyrolysis, which served as the anode of sodium-ion batteries [85]. The storage mechanism of sodium ions in a hard carbon negative electrode was proposed, with the sodium ions embedded into the carbon layer under a high-voltage range, then absorbed into the pores at a low-voltage range [86] (Figure 2a). Subsequently, Stevens et al. found that sodium ions could enter the layers of hard carbon materials through wide-angle in-situ X-ray diffraction studies, and the layer spacing expanded with the insertion process. They also conducted small-angle diffraction studies and found sodium ions in the nanopores of hard carbon materials [87]. Ilic et al. processed commercial hard carbon by ball milling at different lengths and studied its sodium ion storage mechanism through various characterization methods [79]. It was found that by extending the milling time, the amount of gas adsorption and the pore volume increased through physical nitrogen adsorption and small-angle X-ray diffraction.

On the basis of the quenched solid density functional theory model, the pore and specific surface area of the sample increases with an increase in ball milling time, confirming the presence of large quantities of closed pores in the sample. Electrochemical tests have confirmed that hard carbon materials with high closed porosity can store more charges in the platform region because of the existence of quasi-metallic sodium particles in the closed pores, which is consistent with the embedding-adsorption mechanism. Xu et al. used bamboo as a precursor and combined acidizing pretreatments and a two-step carbonization method to introduce the carbonyl group and closed micropores into bamboo-sourced hard carbon materials [88]. Through testing, it was found that the carbonyl group provided additional active sites, improving the reversible adsorption capacity of sodium ions at the slope areas, while the closed micropores promoted the storage capacity of the sodium ions in the platform range. There is an existing similar mechanism of sodium ion storage of the platform region in the embedding-adsorption model.

2.2."*Adsorption-Embedding" Model*

The "embedding-adsorption" mechanism has been recognized by some people but has also been opposed by others. The core idea of the "embedding-adsorption" mechanism is that the capacity of the low-voltage platform region is positively correlated with the porosity of the anode material. By increasing the carbonization temperature, the sodium storage capacity of the low voltage platform region can be increased while the specific surface area of material is reduced, which is contrary to the "embedding-adsorption" mechanism. On this basis, the opposite mechanism of "adsorption-embedding" sodium storage was proposed [88] (Figure 2b). Cao et al. summarized the "adsorption-embedding" model in 2012. In the high-voltage region, sodium ions were adsorbed by the pores of the materials, while in the low-voltage range, they were inserted between the carbon layers [89]. Jin et al. used situ X-ray diffraction (situ XRD) to prove that the peak (002) was offset during the discharge-charge process [82] (Figure 2c). Figure 2d is the Raman image of the hard carbon near spheres (HCNSs). There are a couple of peaks at 1340 and 1580 cm^{-1} that indicate the existence of defects. By using the galvanostatic intermittent titration technique (GITT) test (Figure 2e), it was verified that the diffusion coefficient of sodium ions in the high-potential region is higher than in the low-voltage region. By using the $i = av^b$ equation, it was proved that the diffusion process of sodium ions is dominant near 0.1 V, which further confirms the "adsorption-embedding" mechanism in hard carbon. Lu et al. synthesized hard carbon materials by a simple ball-milling method and confirmed the influence of the microstructure on the sodium storage capacity of hard anodes [90]. By increasing the milling time, the graphite-like nanoregions in hard carbon become smaller and thinner, and the specific surface area and micropore volume increased along with the degree of structural disorder, while the capacity and initial coulombic efficiency (ICE) of the platform region decreased. The capacity of the high-voltage region is derived from the activity and defect sites of the sodium ions adsorbed on the surface of the material. In contrast, the low-voltage region corresponds to the insertion of sodium ions into graphite layers, which is consistent with the "adsorption-embedding" mechanism.

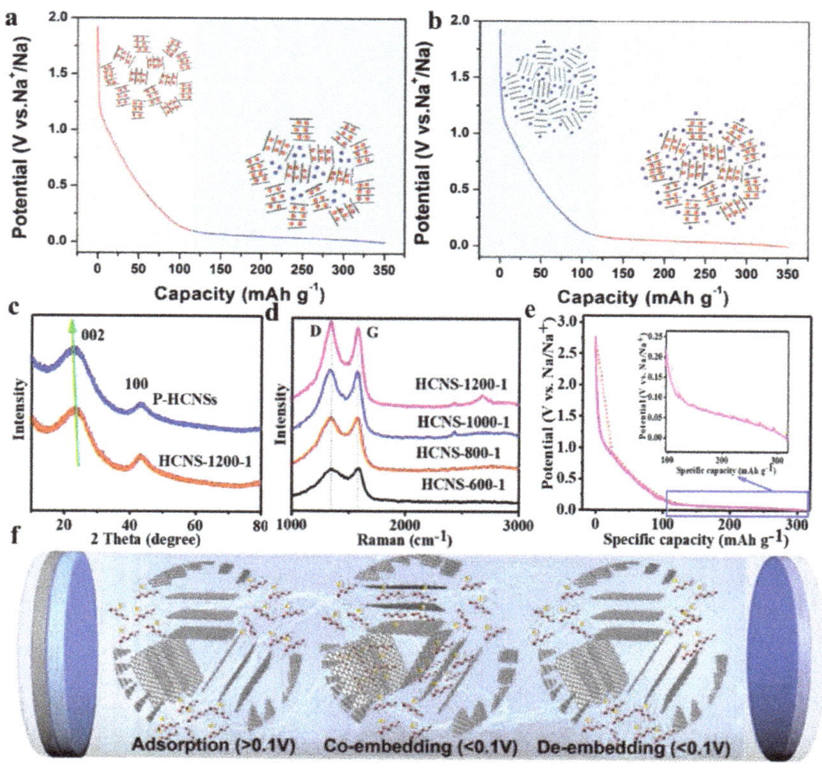

Figure 2. Model of sodium-storage mechanism using hard carbon; (**a**) embedding-adsorption mechanism; (**b**) adsorption-embedding mechanism [86] (copyright Wiley, 2017); (**c**) XRD of P-HCNS and CNS-1200-1; (**d**) Raman of HCNSs; (**e**) the GITT Curves of CNS-1200-1 [82] (copyright Elsevier, 2021); (**f**) schematic diagram of adsorption-co-embedding mechanism [91] (copyright Wiley, 2022).

Using asphalt as a precursor, Yuan et al. synthesized a battery using a carbon anode with diverse crystallinity and pore structures under various carbonization temperatures and finally proposed a universal sodium storage mechanism by combining it with electrochemical performance [92]. In a high-voltage region of greater than 0.3 V, the sodium ions adsorbed in the holes, defects, and impurity sites in the presence of the hard carbon. A low-voltage region of less than 0.3 V for the sodium ions is inserted into the middle of the carbon layer. Only hard carbon has a capacity in the plateau region of less than 0.1 V, resulting from the filling of sodium ions in closed pores. Jiang et al. proposed an "adsorption-coembedding" mechanism, that is, in a high-voltage region of greater than 0.1 V, the sodium ions adsorb in the amorphous region of hard carbon, and with the further progress of a discharge reaction, the sodium ions and ether-based solvent chelates co-embed in the platform region of less than 0.1 V [91] (Figure 2f). The "adsorption-embedding" mechanism is consistent with the Na^+ storage mechanism of most hard carbon.

2.3. "*Adsorption-Insertion-Hole Filling" Model*

Although there exist many experimental data from the two mechanisms of sodium storage demonstrated above, with the deepening of research, there are still many experimental phenomena that the above two mechanisms cannot explain. With the improvement of experimental methods and characterization techniques, researchers continue to discover new mechanisms of sodium-ion storage, such as the "adsorption-insertion-hole-filling" sodium storage mechanism. Ren et al. synthesized a series of independent, flexible microfiber carbon papers (MFCPs) as sodium-ion battery anode materials at a pyrolysis temperature of 900–1500 °C using filter paper as a precursor and a simple graphite plate-assisted method [93]. MFCP-1300 exhibited the most excellent electrochemical performance, including up to 96.3% ICE and a high platform capacity, attributed to a large graphite-like structure and low defect and porosity. The mechanism of sodium storage in MFCP-1300 can be divided into three parts: adsorption of sodium ions in edge and surface defects (>0.1 V), implantation of exposed sodium ions into the interlayer (0.1–0.03 V), and the filling of closed pores (<0.03 V) (Figure 3a–c). Song et al. used esterified starch as a carbon source to prepare hard carbon materials by quantitatively adjusting the oxygen content through low-temperature hydrogen reduction [94]. The storage mechanism of sodium ions in hard carbon materials was confirmed by cyclic voltammetry (CV) situ Raman (Figure 3d): In the first stage (open-circuit voltage (OCV) of 0.6 V), the sodium ions were adsorbed on the defect sites. Through low-temperature carbonization, the defect sites in hard carbon materials increase. The defects provide additional sodium-ion storage sites and improve the capacity retention rate at high rates. In the second stage (0.6–0.1 V), the sodium ions entered the graphite-like region. Thanks to a high proportion of graphite-like regions, hard carbon materials provide a suitable microstructure for sodium-ion insertion and high capacity in the inclined region. In the third stage (0.1–0.001 V), the sodium ions are filled in the closed micropore. When the insertion site was saturated, closed-hole filling was the main sodium storage behavior in the platform region.

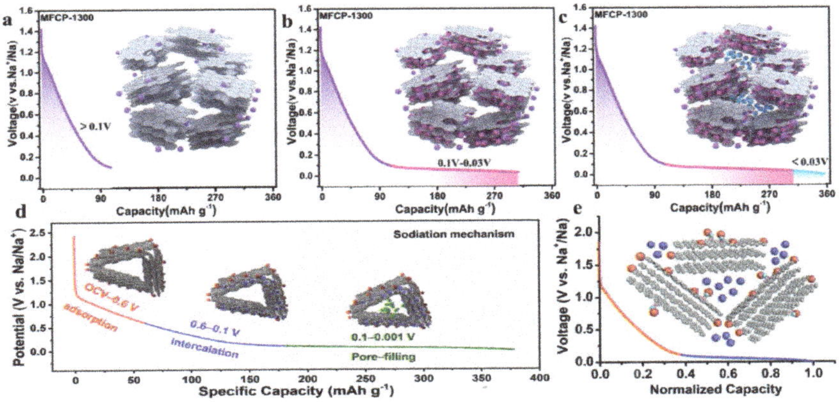

Figure 3. (**a–c**), schematic diagram of Na^+ storage mechanism of MFCP-1300 [93] (**d**), schematic diagram of Na^+ storage [94] (copyright Elsevier, 2022) (**e**), schematic diagram of adsorption-pore filling mechanism [95] (copyright Wiley, 2018).

2.4. "Adsorption-Hole Filling" Model

Bai et al. synthesized composite materials by injecting sulfur elements into hard carbon and confirmed the two-step storage behavior of sodium ions in anodes through experiments [95] (Figure 3e), showing the voltage curve change of hard carbon materials with a low carbonation temperature have higher defects and heteroatom concentration and are amorphous structures.

When increasing the carbonization temperature, the defects and heteroatom concentration in the materials can be reduced, and the graphite structure of materials can be improved. In the Na^+ insertion process, Na^+ adsorption at the defect or heteroatom site leads to a change in binding energy, which makes the voltage curve appear inclined. The pores of the hard carbon material were filled with Na^+ and verified by a series of experiments, including the use of sulfur to fill the pores, varying the heating temperature to adjust the pore structure, the use of different electrolyte systems, the comparison of fuzzy and conflict spectral analyses in the literature, and the pores filled with Na^+ with weak binding energy, similar to the deposit of sodium metal. The result is a platform-shaped low-pressure area. Fei et al. used cotton as a carbon source to pyrolyze hard carbon materials at 1300 °C [96]. Ex situ X-ray photoelectron spectroscopy (XPS) and GITT confirmed that the sodium storage mechanism of hard carbon materials was adsorption and pore filling. The inclined region above 0.12 V corresponds to the adsorption of Na^+ between disordered graphite layers. Na^+ fills the nanopore in the platform region near 0 V. Due to the different precursors and carbonization temperatures, hard carbon materials have a complex morphology and structure, and it is very difficult to determine the storage mechanism of sodium ions in them. In particular, the sodium storage mechanism in the platform area is controversial, which also puts forward clear requirements for future research.

3. Challenges and Solutions of Hard Carbon

3.1. Problems with Hard Carbon

As an anode of sodium-ion batteries, hard carbon has excellent performance, low cost, and uniform distribution, which is conducive to commercialization, but this also has the problem of a wasted rate performance and low ICE. Hard carbon materials have abundant internal pores and surface defects. The first charge and discharge process cause serious irreversible reactions, including electrolyte that is deposited at the electrode surface to form a solid electrolyte interphase (SEI) membrane, which consumes many sodium ions, surface defects, and internal pores in the cycle and will also cause irreversible reactions, which are the main reason for the low ICE of hard carbon materials. The capacity of the whole battery is influenced by active matter loss, so the specific surface area and defects of hard carbon can be reduced, and some pores can be closed to reduce the irreversible reaction to achieve the purpose of improving ICE and obtaining the anode materials of sodium-ion batteries.

3.2. Improvement Strategy

3.2.1. Morphology and Structure Regulation

Yang et al. used walnut shell as a carbon source to prepare anode material through high-temperature carbonization and then hydrothermally treated hard carbon with cetyltrimethyl-ammonium bromide (CTAB)/KOH [97] (Figure 4a). Using a scanning electron microscope (SEM) to analyze CWS-CK (Figure 4b), it was shown that the surface presents a massive structure with micropores, which is conducive to electrolyte diffusion in the hard carbon main body. High-resolution transmission electron microscopy (HRTEM) images of CWS and CVS-CK can be seen in Figure 4c,d, respectively. Through comparison, it was found that the micropores of CVS-CK were transformed into mesoporous ones, which slow down the diffusion of sodium ions or even have difficulty containing Na^+. On the contrary, mesoporous pores are beneficial to sodium-ion diffusion. The specific surface area and pore volume of the samples are plotted vs. the pore size, respectively (Figure 4 e,f). The pores of CWS concentrate at 0.6–2 nm. Besides, the mesopore volume of CWS-CK is larger than CWS. The CWS-CK contains abundant mesopores that accelerate Na^+ intercalation. The maximum capacity of CWS-CK under 0.02 A g^{-1} can attain 283.7 mAh g^{-1}, 83% higher than that of hard carbon that is rich in micropores, which still has a capacity of 189.4 mAh g^{-1} at 0.2 A g^{-1} after 320 cycles.

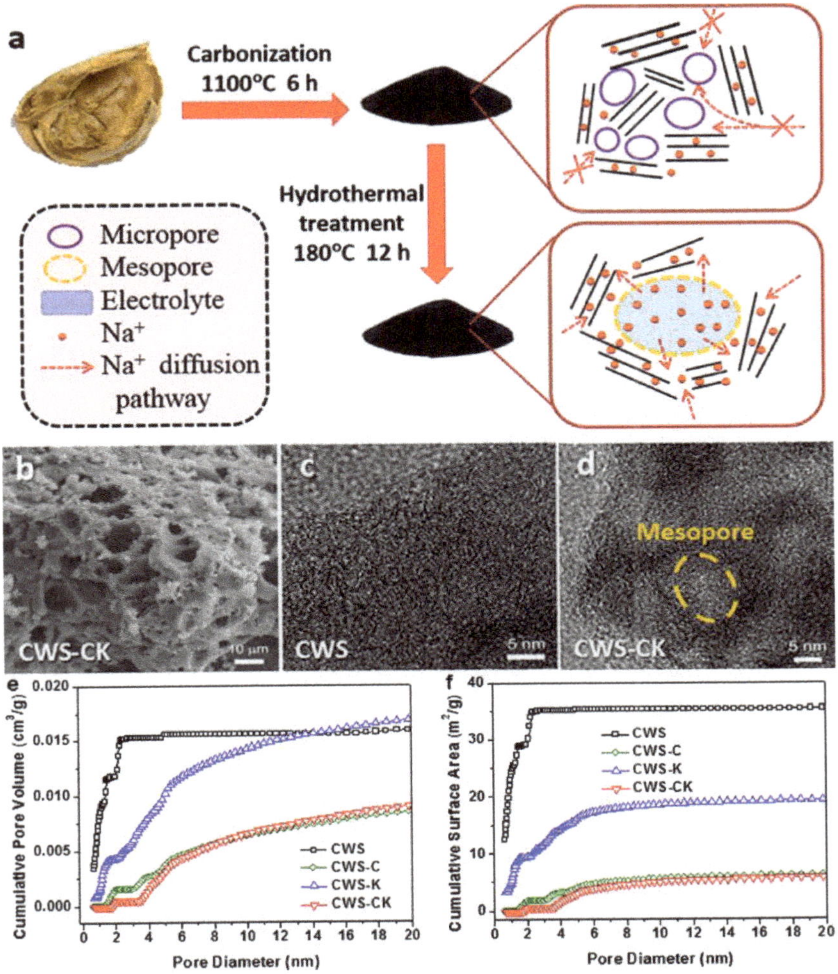

Figure 4. (**a**), Synthesis process of the hard carbon using walnut shell as a precursor. (**b**) SEM image of CWS-CK, (**c**,**d**) HRTEM images of CWS and CWS-CK, (**e**,**f**) specific pore volume and specific surface area of CWS-CK, CWS-C, CWS, and CWS-K [97] (copyright Elsevier 2020).

Jin et al. synthesized several spheroidal hard carbons under different carbonization conditions, and the effect of microstructure on Na^+ storage behavior was investigated [82]. It was confirmed by experimental data that the overall sodium-ion storage capacity could be improved by increasing the material order and platform capacity while maintaining the appropriate layer spacing (>0.364 nm). The "adsorption capacity" of the high-voltage region could be markedly increased by increasing the specific surface area. The electrode material had a high capacity of 305 mAh g^{-1}. The platform area had a capacity of 170 mAh g^{-1} and 210 mAh g^{-1} at 0.02 A g^{-1} and 1 A g^{-1}, respectively. Yin et al. synthesized different hard carbon materials (HMM-1300-ZBE) through the ZnO auxiliary etching strategy, which can achieve extremely fast sodium-ion storage with a voltage between 0.01 V and 2 V, which increases the diffusion rate of sodium ions by two orders of magnitude [98]. Through characterization, it was proved that the short graphite layer could be made thicker and longer by increasing the oxygen content via a volume-etching strategy, which improves the ion transport rate. In the electrochemistry test, the discharge capacity of HCM-1300-ZBE at the first cycle is 501 mAh g^{-1} at 0.05 A g^{-1} (Figure 5a), and HCM-1300-ZBE shows excellent rate performance. The capacity is 230.4 mAh g^{-1} and 107 mAh g^{-1} at 20 A g^{-1} and 50 A g^{-1} (Figure 5b,c), and the outstanding cycle stability is 344 mAh g^{-1} at a current density of 2 A g^{-1}. The capacity retention rate is more than 99.99% after 3000 cycles (Figure 5d). The electrodes were used at a wide range of temperatures, with a capacity of 426 mAh g^{-1} under −40 °C, a mere 14.9% reduction from 25 °C, which is the lowest reported in the literature. Figure 5f shows the diffusion and storage of sodium ions of HCM-1500/1300 and HCM1300/1500-ZBE. Finally, the whole battery was tested using the negative electrode of HCM-1300-ZBE and the positive electrode of $Na_3V_2(PO_4)_3$(NVP). It was found that the capacity reached 39.95 mAh g^{-1}, and the average voltage of the whole battery reached 3.31 V at 10 A g^{-1}. The power densities and energy densities are 6.73 kW kg^{-1} and 294.6 Wh kg^{-1}, respectively.

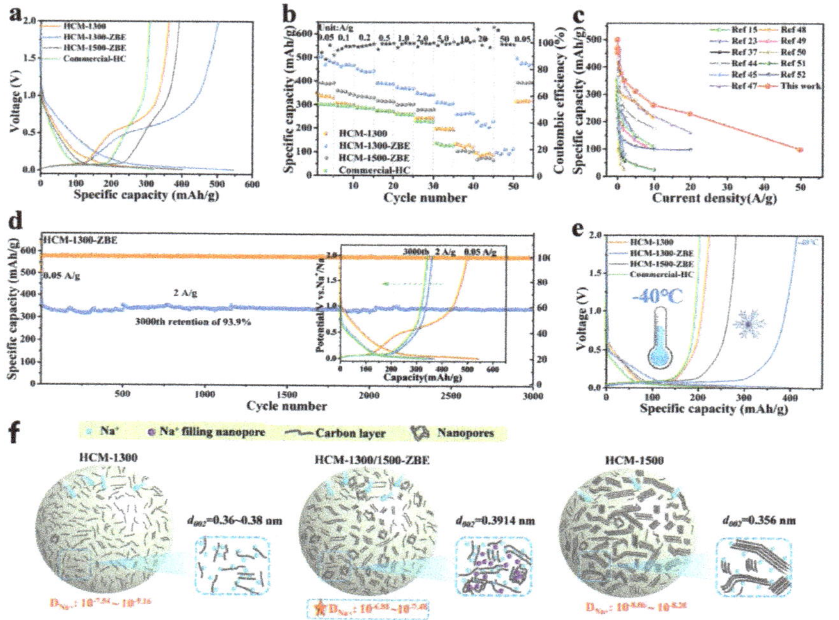

Figure 5. Electrochemical characterizations. (**a**) Galvanostatic charge/discharge (GCD) curves of HCM-1300, HCM-1300-ZBE, HCM-1500-ZBE, and commercial HC with the voltage range from 0.01 V to 2 V at 50 mAh g^{-1} of the second cycle. (**b**) Rate performances at 0.05, 0.1, 0.2, 0.5, 1.0, 2.0, 5.0, 10, 20, and 50 A g^{-1}, respectively. (**c**) The electrochemical performance of different HC materials [15,23,37,44,45,47–52]. (**d**) The long-term cycling performance of HCM-1300-ZBE at 2 A g^{-1}. (**e**) The second cycle of the GCD curve with −40 °C, (**f**) showing sodium-ion diffusion in HCM-1500, HCM-1300, and HCM-1300/1500-ZBE [98] (copyright Wiley 2022).

Liu et al. synthesized many permeable channel-like hard carbons with interconnections through an ingenious controlled phase-transition method, improving ionic diffusion and electrode and electrolyte interface and affinity [99]. The experimental results show that porous hard carbon materials with a crosslinked mesoporous structure can improve the capacity of sodium-ion storage. When compared with other initial hard carbon materials with lower pore content, the ICE of the synthesized porous hard carbon anode increased from 51.5% to 68.3%, with a high capacity of 332.7 mAh g^{-1} at 50 mA g^{-1}, as well as a capacity retention rate enhanced from 46.5% to 67.4% at 2 A g^{-1}. The capacity retention rate was enhanced from 86.4% to 95% after 90 cycles, with anode capacity, cycle stability, and rate performance all comprehensively enhanced. Yang et al. used melt diffusion carbonization to make the micropores inside of hard carbon become ultrafine pores [100]. In situ XRD confirmed that the ultrafine pores provided additional sodium-ion storage sites, increasing the capacity of the anode. The ICE of hard carbon increased up to 80.6%, with a capacity of 346 mAh g^{-1} at 0.03 A g^{-1}, and regions with a voltage of less than 1 V contributed more than 90% to the capacity. The surface capacity was 5.32/6.14 mAh cm^{-2} at −20 °C and 25 °C, respectively, under a high load of 19 mg cm^{-2}. Porous hard carbon with a 3D structure can be constructed by morphological and structural regulation to realize the fast diffusion of sodium ions and improve the cycle stability and rate performance of sodium-ion batteries.

3.2.2. Doping by Heteroatom

The number of defects and layer spacings can be regulated by heteroatom doping. Wu et al. adopted a simple boron-doping strategy to optimize the multiscale carbon-based structure in terms of spherical morphology and crystal parameters while improving the storage capacity of the sodium ions and the battery's electrochemical performance [101]. By using boric acid as a boron source and glucose as a precursor, hard carbon nanospheres were synthesized through high-temperature carbonization, presenting a polydisperse nanosphere shape and expanded layer spacing. These structural features can improve sodium-ion storage and rate performance. The capacity of the hard carbon anode in the full voltage range significantly increased under boron doping, with the capacity of the high-voltage region increased by three times, while the low-voltage region increased by 67%. The reaction curve of the B-doped hard carbon spheres (BHCS-1200) negative electrode at a pulse current of 0.03 A g^{-1}, with the calculated chemical diffusion rate of the sodium ions (D_{Na+}), was calculated (Figure 6a). In Figure 6b, the D_{Na+} of the two electrodes is compared, and it is found that the diffusion mechanism of Na^+ in the two electrode materials has many similarities, indicating that doping boron atoms do not change the mechanism of sodium storage behavior. From 0.1 to 0.03 V, the D_{Na+} of BHCS-1200 was higher than HCS-1200, illustrating that heteroatom doping increased the storage capacity of the sodium ions. The capacity of the region with a voltage larger than 0.1 V comes from the sodium-ion filling of the micropores. The in-situ XRD pattern of the BHCS-1200 electrode at 30 mA g^{-1} can be seen in Figure 6c. At 21°~22°, BHCS-1200 has a strong peak, indicating that the layer spacing of BHCS-1200 becomes wider after the initial discharge process, confirming that sodium-ion embedding occurs in the low discharge region. The density functional theory (DFT) calculation of BHCS-1200 confirmed that the diffusion and intercalation kinetics of the sodium ions in doped hard carbon improved, and the intercalation capacity increased. Yu et al. used tamarind fruit as a carbon source to prepare hard carbon materials through a cost-effective and scalable self-assembly technology through the natural super accumulation and enrichment of calcium ions [102]. After further hydrothermal treatment, self-binding and evenly distributed calcium ions act as a buffer layer to expand the distance between hard carbon layers. By optimizing the calcining temperature, the natural pore structure can be retained to a large extent. The hard carbon anode rich in calcium ions shows excellent performance. The capacity was 326.7 mAh g^{-1} after 250 cycles at 0.05 A g^{-1}, and the ICE was 70.39%.

Liang et al. proved that the carbon nanoribbon array modified by metallic bismuth nanospheres coated with a carbon layer was synthesized by introducing MOF, which had a three-dimensional frame superstructure [103]. It was found that a thin and uniform SEI film was generated on the anode surface during the discharge and charging cycles, and the porous structure formed on the electrode surface not only improved the utilization rate of active substances but also reduced the diffusion distance of electrons or ions. As shown in Figure 7, they analyzed the thickness and chemical composition of the SEI film carbon nanobelt arrays decorated with carbon-layer-coated metallic Bi nanospheres (Bi@C⊂CFs) in the two matrix electrolytes of dimethyl ether-based (DME) and ethylene carbonate (EC)/diethyl carbonate (DEC)-based by ex-in situ HRTEM and XPS. Figure 7a,e show HRTEM images of the anodes after 20 cycles in the DME-based and EC/DEC-based electrolytes, respectively. In the DME-based electrolyte, the SEI film is homogeneous, and the main component is RCH_2ONa (Figure 7b–d). In the EC/DEC-based electrolyte, the SEI film is thick and uneven, and the main component is $ROCO_2Na$, which easily decomposes, resulting in the increased thickness of the SEI film (Figure 7f–h). The alloying/dealloying models of Bi in a sodium-ion battery are shown in Figure 7i. The Bi@C⊂CFs material shows excellent electrochemical properties, an outstanding performance of 305 mAh g^{-1} at 5 A g^{-1} past 5000 cycles, an outstanding capacity of 308.8 mAh g^{-1} at 80 A g^{-1}; good cycle performance provides the possibility for commercial applications.

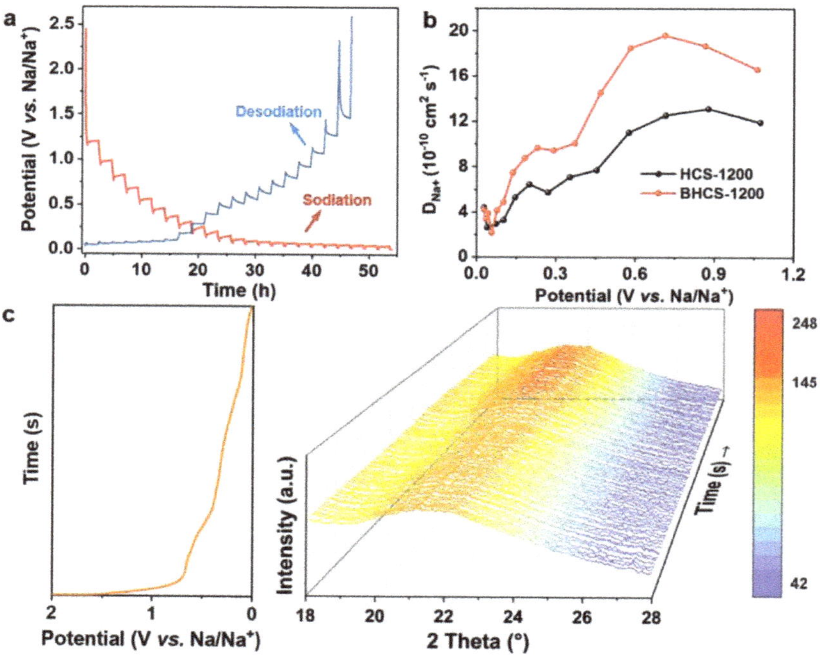

Figure 6. (**a**) GITT curve of BHCS-1200. (**b**) D_{Na+} of BHCS-1200 and HCS-1200. (**c**) Operando XRD patterns of BHCS-1200 at 30 mAh g^{-1} [101] (copyright The Royal Society of Chemistry 2022).

Wang et al. synthesized phosphorus-doped hollow carbon nanorods (P-HCNs) with a phosphorus content of 7.5% via a one-step method [104]. The P-HCNs showed a high capacity of 260 mAh g^{-1} at 1.0 A g^{-1} after 500 cycles and high ICE (73%). The mechanism of different contents of phosphorus on the structure of hard carbon was investigated by XRD, XPS, and Raman spectrum techniques. Li et al. obtained several phosphorus-doped hard carbon materials by directly calcinating phospho-solidified epoxy resin under various temperatures. They analyzed the evolution of the microstructure and groups containing phosphorus as well as the sodium-storage properties with increases in temperature [105]. By raising the carbonization temperature, the phosphorus groups of P-O and P-C become P-P bonds with higher activity, and the layer spacing presents a non-monotone upward trend. The increase in active phosphorus groups and the increase in layer spacing are conducive to providing more sodium-ion transport channels and active sites. With a high capacity of 379.3 mAh g^{-1} at 100 mAh g^{-1}, and the outstanding rate performance was as high as 158.1 mAh g^{-1}, with long cycle stability over 6500 cycles at 5.0 A g^{-1} and nearly no capacity loss. Yan et al. used ordinary PCl_3 and C_6H_{12} as the phosphorus and carbon sources to successfully establish a nitrogen-bulking anaerobic reaction system and simultaneously realized carbonization and in-situ P doping [106]. In the synthesized hard carbon material, the mass fraction of phosphorus reached 30%, which was higher than previously reported phosphor-doped carbon material, and the phosphorus atoms replaced the carbon atoms embedded in the lattice. The adsorption energy of the sodium ions and intercalation spacing increased significantly. There was a reversible capacity of 510.4 mAh g^{-1} with a rate performance of 397.1 mAh g^{-1} at 10 A g^{-1} at a working voltage of 0.54 V. Huang et al. used tannic acid (TA) and amino acid as a carbon source to synthesize a high proportion of active nitrogen-doped hard carbon (PTA-Lys-800) [107]. The electrochemical performance of the PTA-Lys-T//Na cell is shown in Figure 8. The first 3 cycles of the CV curve shown in Figure 8a is the PTA-Lys-800 at 0.2 mV s^{-1}. Under a scan speed of 0.2 mV s^{-1} for the first three cycles, the CV curve of PTA-Lys-800 can be seen in Figure 8a. A region of irreversible regions appears between 0.01 V and 0.65 V due to the formation of the SEI film. In the subsequent cycles, the curves almost coincide, showing that PTA-Lys-80 has excellent reversibility.

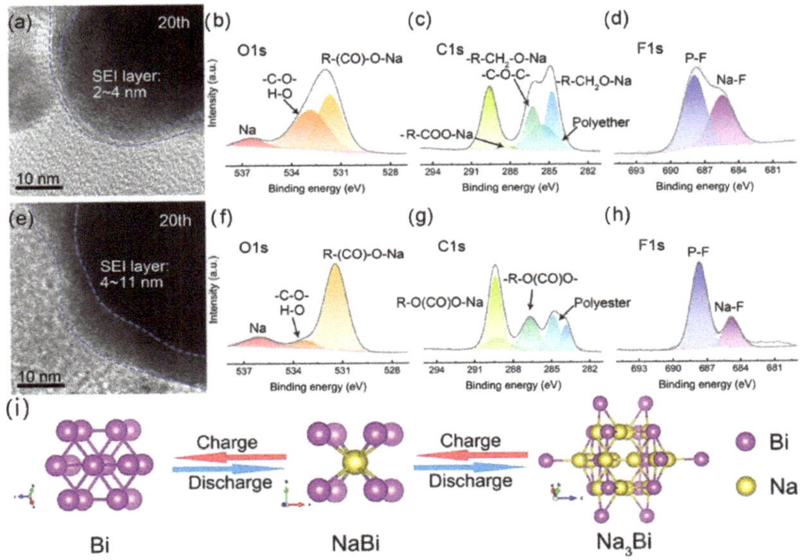

Figure 7. (**a**) HRTEM images of the Bi@C⊂CFs after 20 cycles in DME-based (**a**) and EC/DEC-based (**e**) electrolytes. (**b–d**,**f–h**) The high-resolution XPS spectra of C, O, and F for Bi@C⊂CFs with EC/DEC-based and DME-based electrolytes after cycles. (**i**) Reaction mechanism of Bi in charge and discharge process [103] (copyright Wiley 2022).

Figure 8b shows the GCD curves of the sodium-ion batteries when the voltage is 0.01–3.0 V. It can deliver 318.3, 504.0, and 715.4 mAh g^{-1} at initial discharge, while the initial charge capacities are 591.1, 504.0, and 715.4 mAh g^{-1} for PTA-Lys-900, PTA-Lys-800, and PTA-Lys-700, respectively. The ICE is 55.2, 51.3, and 53.8%. The PTA-Lys-800 delivered excellent cycling performance, with a high reversible capacity of 338.8 mAh g^{-1} at 100 mA g^{-1} after 100 cycles and a capacity retention rate of over 85% (Figure 8c). The PTA-Lys-800 electrode delivered an outstanding rate performance (Figure 8d). The PTA-Lys-800 electrode showed a high reversible capacity of 131.1 mAh g^{-1} after 5000 cycles at 4 A g^{-1}, demonstrating that the long-term cycling stability of PTA-Lys-800 is better than the TA-800 electrode (Figure 8e). The number of active sites can be increased by simple heteroatom doping to enhance the storage capacity of sodium ions. The sodium-ion battery anode materials designed by single-atom doping have high capacity and excellent cyclic stability, but further studies are needed to understand the mechanism more deeply.

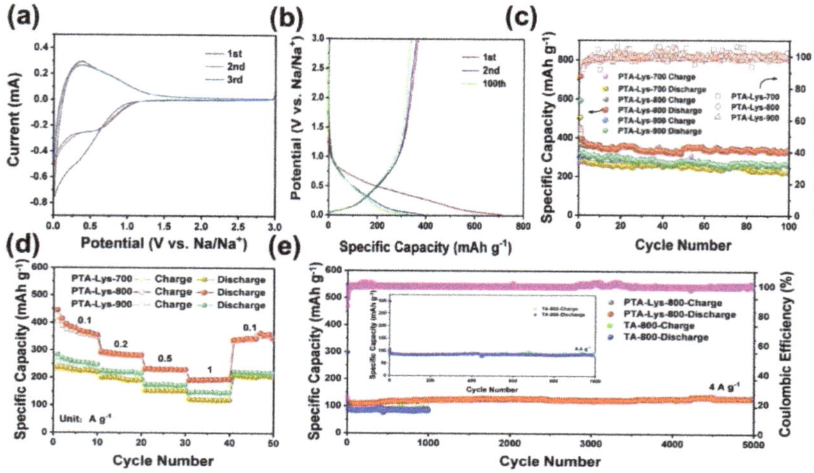

Figure 8. (**a**), CV curve of PTA-Lys-800. (**b**) The GCD curve of PTA-Lys-800 at 100 mA g^{-1}. (**c**) Electrochemical stability of PTA-Lys-800 at 100 mAh g^{-1}. (**d**) Rate performance of PTA-NHCS-T. (**e**) Cycling performance of PTA-Lys-800 and TA-800 at 4 A g^{-1} [107] (copyright Wiley 2022).

3.2.3. Electrolyte Optimization

Niitani et al. confirmed that all-solid sodium-ion batteries with a hard carbon anode and a sodium carborane electrolyte (Na $(CB_9H_{10})_{0.7}(CB_{11}H_{12})_{0.3}$) have excellent fast-charging performance and verified that (Na $(CB_9H_{10})_{0.7}(CB_{11}H_{12})_{0.3}$) has good plasticity through experiments [8]. By generating a thin oxide layer to establish a tight, stable connection between the electrolyte and electrode interface, reducing the interface impedance of the active material load of hard carbon to 5.8 mg cm^{-2} (0.68 Ω cm^{-2} at 0.1 V vs. Na^+/Na), the full battery has an area capacity over 1 mAh cm^{-2} for 5 min. Yu et al. designed a homotype heterojunction using hard carbon to generate a steady, solid electrolyte interface, effectively increasing the ICE by 16.4% [58]. A simple surface engineering strategy was used for the construction of a homogeneous amorphous Al_2O_3 layer using HC, which shielded the active site. Finally, electrolyte decomposition and adverse factors were slowed down by inhibiting the decomposition of $NaPF_6$ in ether-based electrolytes and reducing the accumulation of NaF, forming a thin and dense solid electrolyte interface to reduce interface impedance, improving not only the ICE but also the storage capacity of sodium ions. The optimized reversible capacity of the hard carbon anode occurred at 321.5 mAh g^{-1} at 50 mA g^{-1}. After 2000 cycles at 1 A g^{-1}, the capacity retention rates of the optimized hard carbon and untreated hard carbon were 86.9% and 52.6%, respectively. Ma et al. used hard carbon to study the relationship between SEI films and sodium-ion storage performance and found that a good SEI film is not directly related to its components, such as NaF and Na_2O [108]. However, the fine structure arrangement of the components of a nano SEI film is the basis for achieving fast sodium-ion storage and a stable interface for a "good" SEI film. A double-layer SEI film composed of the inner layer of inorganic matter and the outer layer of organic matter has a long cycle life and outstanding rate performance. The SEI membrane is the gateway for sodium ions from the electrolyte to the electrode. In order to transform the crystal structure of the hard carbon surface using a cylindrical solvent and play the role of "pseudo-SEI", a 1 M $NaPF_6^-$ inTEGDME electrolyte is used, which also facilitates the fast and stable storage of sodium ions, with an excellent performance of 192 mAh g^{-1} at 2 C and a cycle stability of 1100 cycles at 0.5 C. The interface stability of hard carbon materials over long-term cycles is poor. The continuous accumulation of electrolyte decomposition substances will seriously increase the interface impedance and sharply decrease the discharge capacity. Zhang et al. used n-phenyl bis (trifluoridemethansulfonimide) as an electrolyte film-forming additive, and could effectively improve the long-cycle performance of the hard carbon negative electrode of sodium-ion batteries, making the cyclic stability of half-batteries increase from 0% to 50% after 500 cycles, and the improved interphase stability makes the capacity retention rate of the whole battery increase by 52% after 100 cycles [109]. Jin et al. designed a non-flammable local high concentration electrolyte for highly reversible sodium-ion batteries [110]. Through high-resolution cryo-TEM, it was found that an extremely thin but solid interface layer was formed on the positive electrode surface, which could inhibit the transition metal dissolution and reduce the structural changes on the electrode surface. The formation of a solid electrolyte interface rich in inorganic substances on the hard carbon surface can reduce the side reactions between hard carbon and an electrolyte. These stable interfaces ensure the high coulombic efficiency and long-cycle stability of the hard carbon negative electrode, with a capacity of 247.9 mAh g^{-1} after 500 cycles and a capacity retention rate of 94.8%. Lohani et al. formed a thin, dense SEI film on the surface of hard carbon by directly contacting hard carbon that was moistened with a vinyl carbonate electrolyte with sodium metal [111]. During incubation, the hard carbon negative electrode was partially passivated by a thin SEI film. This sodium-ion cell had a first-cycle coulombic efficiency of 97% and provided a stable area capacity of 1.4 mAh cm^{-2} at 0.1 mA cm^{-2} at a current rate of 0.1 mA cm^{-2}, showing excellent rate capacity. With a high current density of 0.5 mA cm^{-2}, the initial capacity of 1.18 mAh cm^{-2} was maintained, and excellent cycling stability was achieved with 1.0 mAh cm^{-2} after 500 cycles. Dong [112] found that the sodium storage performance of a hard carbon (HC) anode in ether electrolytes is higher and more effective than conventional ester electrolytes. The experiment results confirm that the resistances originated from the charge transfer (Rct) and SEI film (RSEI) in EC/DEC-based electrolytes are both higher than those in DEGDME-based electrolytes, that is, 137.1 Ω versus 64.36 Ω for Rct and 111.6 Ω versus 23.93 Ω for RSEI, respectively. Therefore, these results imply the facilitation of Na storage kinetics in the DEGDME-based electrolyte. Zeng et al. constructed a discontinuous SEI film on a hard carbon negative electrode by regulating NaPF6 in an ether-based electrolyte according to the storage mechanism of sodium ions in the low-pressure range, including the co-embedding of sodium ions into the carbon layer and the desolating sodium ions into the hole of the carbon [113]. The discharge capacity was 459.7 mAh g^{-1} at 0.1 C, and the capacity was 357.2 mAh g^{-1} at 1 C after 500 cycles, with both being higher than other electrolytes. Hou et al. directly demonstrated the internal relationship between solvent structure and electrochemical reaction by controlling the anions in the solvent shell, which determined the storage kinetics of the sodium ions and the SEI membrane's evolution process [114]. The results showed that the weak coulomb interaction between Na^+ and PF_6^- could improve the transport volume and storage kinetics of Na^+ in liquid. PF_6^- is induced to preferentially reduce and promote the addition of the additive into the solvated structure to form a thin, compact, protective, and layered form of SEI. Excellent SEI can effectively protect the anode, inhibit solvent decomposition, and improve battery performance. As expected, the 5 Ah $Na_3V_2(PO_4)_3$ hard carbon pouch cells with a well-matched PF_6^--based electrolyte showed a high specific energy of 129 Wh kg^{-1}, a superior capacity retention rate of 90% at 5 C after 700 cycles. The electrolyte plays an important role in the transfer of sodium ions between the anode and cathode. A thin and dense SEI film can be formed via an ether-based electrolyte, which can enhance the rate capacity of a sodium-ion battery with mixed ester-based electrolytes, although efforts have been made recently to develop advanced electrodes and electrolytes. Simultaneously, one of the key components of sodium-ion batteries is the separator [115].

Kang et al. induced a thin and compact SEI by a functionalized separator with an sp^2 carbon conjugated-covalent organic framework (sp^2 c-COF). A NaF-rich SEI can be generated by using a COF separator to destroy the structure of $NaPF_6$ in the electrolyte. With a current density of 20 mA cm^{-2}, the Na | COF | Na symmetric cell exhibits stable cycling stability over 1500 h, and the cell presented excellent rate performance (50C) and outstanding cycling stability over 5000 cycles at 10 C [116]. Niu et al. prepared a separator using an ether imide (PEI) or polyvinylpyrrolidone (PVP) mixture by immersion phase separation (Figure 9a). The PEI/PVP separator exhibited high ionic conductivity (1.14 mS cm^{-1}), excellent thermal stability (over 180 °C), and outstanding flexibility and mechanical strength. The carbon/Na cell showed a high reversible capacity of 119.4 mAh g^{-1} at 0.5 C and good cycling performance [117]. Zhou et al. prepared all-cellulose composite (ACC) separators coated with $Zn_5(OH)_8Cl_2{\cdot}H_2O$ (ZHC) particles (Figure 9b). Batteries benefit from ZHC on the surface because it promotes sodium-ion passage and improves cycling performance. The hard carbon/Na cell showed outstanding cycling stability and a high reversible specific capacity of 315 mAh g^{-1} at 0.025 A g^{-1} [118]. Separators are part of sodium-ion batteries and are proposed to allow for the fast diffusion of sodium ions.

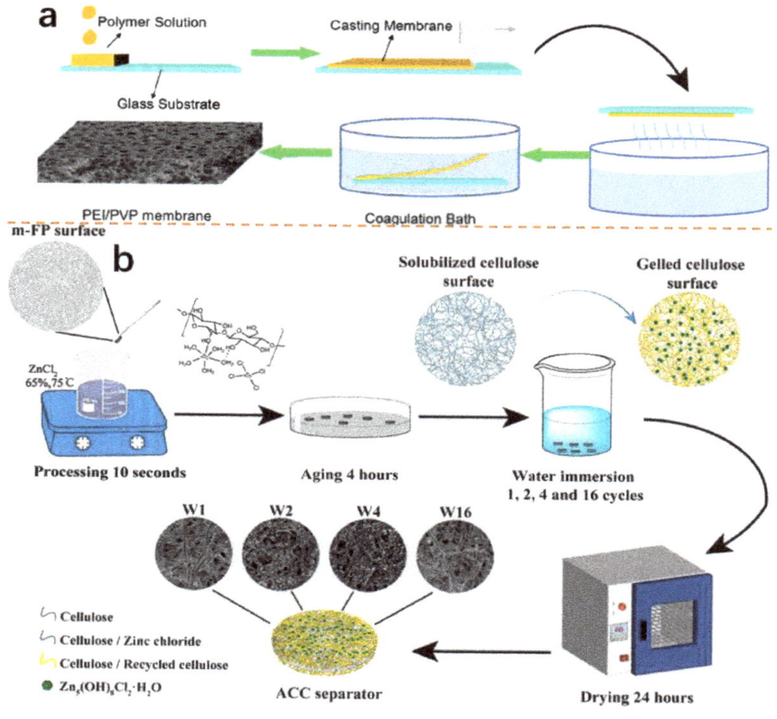

Figure 9. (**a**), Preparation process of PEI/PVP separators (Copyright, American Chemical Society 2021). (**b**) Preparation process of ACC separators (Copyright, Elsevier 2021).

Table 1 shows some of the recent keywords related to hard carbon as an anode in sodium-ion batteries.

Table 1. Summary of representative publications on hard carbon as a SIB anode.

Precursor	Temperature (°C)	Electrolyte System	Doped Heteroatom (1)/Regular(2)	Current Density (mA g^{-1})	Capacity (mAh g^{-1})	ICE (%)	Cycle Number	Capacity Retention Rate (%)	Ref
hazelnut shell	1400	1 M $NaPF_6$ in EC/DMC	2	20	342.0	91	100	91	[119]
fungus-treated basswood	1300	1 M $NaPF_6$ in DEDM	2	200	175.6	88.2	536	86.4	[120]
sucrose	1300	1 M $NaPF_6$ DEGDME	2	30	282.6	91.2	100	92	[121]
$H_2C_2O_4 \cdot 2H_2O$	1300	1 M $NaClO_4$ PC with 5% FEC	1	100	278.0	68.7	100	93.8	[49]
Epoxy phenol novolac resin	1800	1 M $NaPF_6$ in DME	2	50 500	480.0 423.2	84.6	1000	92	[122]
trisodium citrate/ hexam-ethylenete-tramine	800	1 M $NaClO_4$ PC with 5% FEC	2	5000	238.0	84.5	5000	98. 7	[123]
sugarcane waste-derived	1200	1 M $NaClO_4$ PC with 5% FEC	2	50	323.6		500	96	[110]
sucrose	1200	1 M $NaPF_6$ in EC/DEC	2	50 1000	318.0 77.90	97.1	2000	86.9	[58]
$C_6H_{12}O_6 \cdot H_2O$	600	1 M $NaPF_6$ in EC/DMC		120 1200	109.0 98.2		300 600	80.3 71.8	[124]
Hard carbon		1 M $NaPF_6$ in EC/DEC	2	400	153.2	84.49	300	94.8	[125]
citrate sodium	800	1 M $NaPF_6$ in EC/DEC with 5%FEC	1	50 1000	280 193		200 2000	99 115	[126]
(H3BTC)	600	1 M $NaPF_6$ in EC/DEC	1	5000	305		5000	89.4	[103]
sodium ligninsulfonate	1200	1 M $NaPF_6$ in DME	2	20	210		1000	99	[82]
water-soluble starch	900	1 M $NaClO_4$ in EC/DEC	2	25 100	343.1 197.8	80.1	200	87.2	[127]
3-aminophenol	1300	1 M $NaPF_6$ in DEDM	2	2000	344		3000	99.9	[50]
corn starch	1100	1 M $NaClO_4$ in EC/DEC	2	50	270		50	85.2	[124]

4. Summary and Outlook

Hard carbon, as a material for an anode in a sodium-ion battery, has the characteristics of high capacity, low cost, and wide distribution, showing great commercial value. Using hard carbon as the material for the anode of sodium-ion batteries has excellent application prospects in large-scale energy storage systems. Hard carbon has been commercialized, which has the advantage of low cost, but also has the problem of low capacity. In order to increase the value of hard carbon in practical applications, the next goal is to enhance the capacity of hard carbon. This paper first introduces four different mechanisms of sodium storage using hard carbon, including the "embedding-adsorption", "adsorption-embedding", "adsorption-insertion-pore filling", and "adsorption-pore filling" cashier models. Based on sodium storage mechanism analysis, the problems of poor cycling performance and first-cycle coulombic efficiency regarding hard carbon materials were put forward, and three optimization strategies, including structural morphology regulation, heteroatom doping, and electrolyte regulation, were introduced to solve these problems. The hard carbon anode materials were analyzed and summarized below.

There are different ways to store sodium when synthesizing hard carbon materials using different precursors, and the mechanism of storing sodium in the low-voltage region is controversial. The development of advanced characterization techniques is crucial for clarifying the storage mechanism of sodium ions in the platform region, and the detailed storage behavior of sodium can guide the design of hard carbon negative electrodes.

The number of surface defects can be increased by heteroatomic doping, changing the conductivity of the electrode materials, and expanding the layer spacing of graphite, which improves the adsorption capacity of sodium ions, but also improves the diffusion rate, and finally achieves improvements in its capacity and rate performance. Adjusting the morphology, such as increasing the proportion of closed micropores and reducing the specific surface area of hard carbon, can reduce the consumption of sodium ions and improve the capacity and ICE during the production of SEI film.

An excellent electrolyte can form a dense and thin SEI film on the anode surface to reduce the impedance of sodium ion diffusion. Simultaneously, the introduction of additives to reduce side reactions and electrolyte decomposition can improve the stability of the electrolyte and elevate the cycle stability and rate performance of sodium-ion batteries. Developing biomass-based precursors that are low-cost and abundant resources and synthesizing hard carbon with a lower defect degree and a small specific surface area are both conducive to commercializing sodium-ion batteries.

Funding: This work is supported by the National Key Research and Development Project (2020YFE0100100).

Institutional Review Board Statement: Not applicable.

Informed Consent Statement: Not applicable.

Data Availability Statement: All date that support the findings of this study are include within the article.

Conflicts of Interest: The authors declare no conflict of interest.

References

1. Ren, J.; Wang, Z.; Xu, P.; Wang, C.; Gao, F.; Zhao, D.; Liu, S.; Yang, H.; Wang, D.; Niu, C.; et al. Porous Co(2)VO(4) nanodisk as a high-energy and fast-charging anode for lithium-ion batteries. *Nanomicro Lett.* **2021**, *14*, 5. [PubMed]
2. Huang, Y.; Yang, H.; Zhang, Y.; Zhang, Y.; Wu, Y.; Tian, M.; Chen, P.; Trout, R.; Ma, Y.; Wu, T.-H.; et al. A safe and fast-charging lithium-ion battery anode using MXene supported Li_3VO_4. *J. Mater. Chem. A* **2019**, *7*, 11250–11256. [CrossRef]
3. Sun, Y.K. A rising tide of co-free chemistries for li-ion batteries. *ACS Energy Lett.* **2022**, *7*, 1774–1775. [CrossRef]
4. Zeng, L.; Fang, Y.; Xu, L.; Zheng, C.; Yang, M.Q.; He, J.; Xue, H.; Qian, Q.; Wei, M.; Chen, Q. Rational design of few-layer MoSe(2) confined within ZnSe-C hollow porous spheres for high-performance lithium-ion and sodium-ion batteries. *Nanoscale* **2019**, *11*, 6766–6775. [CrossRef]
5. Luo, F.; Feng, X.; Zeng, L.; Lin, L.; Li, X.; Kang, B.; Xiao, L.; Chen, Q.; Wei, M.; Qian, Q. In situ simultaneous encapsulation of defective MoS2 nanolayers and sulfur nanodots into SPAN fibers for high rate sodium-ion batteries. *Chem. Eng. J.* **2021**, *404*, 123430. [CrossRef]
6. Guo, X.; Wang, C.; Wang, W.; Zhou, Q.; Xu, W.; Zhang, P.; Wei, S.; Cao, Y.; Zhu, K.; Liu, Z.; et al. Vacancy manipulating of molybdenum carbide MXenes to enhance Faraday reaction for high performance lithium-ion batteries. *Nano Res. Energy* **2022**, *1*, e9120026. [CrossRef]
7. Peng, J.; Zhang, W.; Liu, Q.; Wang, J.; Chou, S.; Liu, H.; Dou, S. Prussian blue analogues for sodium-ion batteries: Past, present, and future. *Adv. Mater.* **2022**, *34*, e2108384. [CrossRef]
8. Lin, C.; Yang, X.; Xiong, P.; Lin, H.; He, L.; Yao, Q.; Wei, M.; Qian, Q.; Chen, Q.; Zeng, L. High-rate, large capacity, and long life dendrite-free zn metal anode enabled by trifunctional electrolyte additive with a wide temperature range. *Adv. Sci.* **2022**, *9*, e2201433. [CrossRef]
9. Yuan, Z.; Xiao, F.; Fang, Y.; Xiong, P.; Sun, X.; Duan, X.; Yang, X.; Fan, H.; Wei, M.; Qian, Q.; et al. Defect engineering on VO_2(B) nanoleaves/graphene oxide for the high performance of cathodes of zinc-ion batteries with a wide temperature range. *J. Power Sources* **2023**, *559*, e912002617. [CrossRef]
10. Peng, J.; Zhang, W.; Hu, Z.; Zhao, L.; Wu, C.; Peleckis, G.; Gu, Q.; Wang, J.Z.; Liu, H.K.; Dou, S.X.; et al. Ice-assisted synthesis of highly crystallized prussian blue analogues for all-climate and long-calendar-life sodium ion batteries. *Nano Lett.* **2022**, *22*, 1302–1310. [CrossRef]
11. Kong, N.; Du, H.; Li, Z.; Lu, T.; Xia, S.; Tang, Z.; Song, S. Nano heterojunction of double MOFs for improved CO_2 photocatalytic reduction performance. *Colloids Surf. A Physicochem. Eng. Asp.* **2023**, *663*, 131005. [CrossRef]
12. Zhou, W.; Cao, G.; Yuan, M.; Zhong, S.; Wang, Y.; Liu, X.; Cao, D.; Peng, W.; Liu, J.; Wang, G.; et al. Core-shell engineering of conductive fillers toward enhanced dielectric properties: A universal polarization mechanism in polymer conductor composites. *Adv. Mater.* **2023**, *35*, e2207829. [CrossRef]
13. Gao, F.; Mei, B.; Xu, X.; Ren, J.; Zhao, D.; Zhang, Z.; Wang, Z.; Wu, Y.; Liu, X.; Zhang, Y. Rational design of ZnMn2O4 nanoparticles on carbon nanotubes for high-rate and durable aqueous zinc-ion batteries. *Chem. Eng. J.* **2022**, *448*, 1377742. [CrossRef]
14. Zhang, M.; Wang, W.; Xia, G.; Wang, L.; Wang, K. Self-powered electronic skin for remote human–machine synchronization. *ACS Appl. Electron. Mater.* **2023**, *5*, 498–508. [CrossRef]
15. Wang, J.; Xin, S.; Xiao, Y.; Zhang, Z.; Li, Z.; Zhang, W.; Li, C.; Bao, R.; Peng, J.; Yi, J.; et al. Manipulating the water dissociation electrocatalytic sites of bimetallic nickel-based alloys for highly efficient alkaline hydrogen evolution. *Angew. Chem. Int. Ed. Engl.* **2022**, *61*, e202202518.

16. Tian, Y.; Chen, S.; He, Y.; Chen, Q.; Zhang, L.; Zhang, J. A highly reversible dendrite-free Zn anode via spontaneous galvanic replacement reaction for advanced zinc-iodine batteries. *Nano Res. Energy* **2022**, *1*, e9120025. [CrossRef]
17. Peng, J.; Zhang, B.; Hua, W.; Liang, Y.; Zhang, W.; Du, Y.; Peleckis, G.; Indris, S.; Gu, Q.; Cheng, Z.; et al. A disordered rubik's cube-inspired framework for sodium-ion batteries with ultralong cycle lifespan. *Angew. Chem. Int. Ed. Engl.* **2023**, *62*, e202215865. [CrossRef]
18. Zhao, D.; Zhang, Z.; Ren, J.; Xu, Y.; Xu, X.; Zhou, J.; Gao, F.; Tang, H.; Liu, S.; Wang, Z.; et al. Fe_2VO_4 nanoparticles on rGO as anode material for high-rate and durable lithium and sodium ion batteries. *Chem. Eng. J.* **2023**, *451*, 138882. [CrossRef]
19. Li, X.; Liang, H.; Qin, B.; Wang, M.; Zhang, Y.; Fan, H. Rational design of heterostructured bimetallic sulfides ($CoS_{(2)}$/NC@$VS_{(4)}$) with $VS_{(4)}$ nanodots decorated on $CoS_{(2)}$ dodecahedron for high-performance sodium and potassium ion batteries. *J. Colloid Interface Sci.* **2022**, *625*, 41–49. [CrossRef]
20. Peng, J.; Ou, M.; Yi, H.; Sun, X.; Zhang, Y.; Zhang, B.; Ding, Y.; Wang, F.; Gu, S.; López, C.A.; et al. Defect-free-induced Na^+ disordering in electrode materials. *Energy Environ. Sci.* **2021**, *14*, 3130–3140. [CrossRef]
21. Deng, Q.; Wang, M.; Liu, X.; Fan, H.; Zhang, Y.; Yang, H.Y. Ultrathin cobalt nickel selenides ($Co_{(0.5)}Ni_{(0.5)}Se_{(2)}$) nanosheet arrays anchoring on $Ti_{(3)}C_{(2)}$ MXene for high-performance $Na^{(+)}$/$K^{(+)}$ batteries. *J. Colloid Interface Sci.* **2022**, *626*, 700–709. [CrossRef] [PubMed]
22. Peng, J.; Wang, J.; Yi, H.; Hu, W.; Yu, Y.; Yin, J.; Shen, Y.; Liu, Y.; Luo, J.; Xu, Y.W.; et al. A dual-insertion type sodium-ion full cell based on high-quality ternary-metal prussian blue analogs. *Adv. Energy Mater.* **2018**, *8*, 1702856. [CrossRef]
23. Liu, X.; Liu, Z.; Yang, W.; Wang, M.; Qin, B.; Zhang, Y.; Liu, Z.; Fan, H. In situ fragmented and confined CoP nanocrystals into sandwich-structure MXene@CoP@NPC heterostructure for superior sodium-ion storage. *Mater. Today Chem.* **2022**, *26*, 101002. [CrossRef]
24. Hwang, J.Y.; Myung, S.T.; Sun, Y.K. Sodium-ion batteries: Present and future. *Chem. Soc. Rev.* **2017**, *46*, 3529–3614. [CrossRef]
25. Gong, C.; Pu, S.D.; Zhang, S.; Yuan, Y.; Ning, Z.; Yang, S.; Gao, X.; Chau, C.; Li, Z.; Liu, J.; et al. The role of an elastic interphase in suppressing gas evolution and promoting uniform electroplating in sodium metal anodes. *Energy Environ. Sci.* **2023**, *404*, 535–545. [CrossRef]
26. Zhang, W.; Peng, J.; Hua, W.; Liu, Y.; Wang, J.; Liang, Y.; Lai, W.; Jiang, Y.; Huang, Y.; Zhang, W.; et al. Architecting amorphous vanadium oxide/mxene nanohybrid via tunable anodic oxidation for high-performance sodium-ion batteries. *Adv. Energy Mater.* **2021**, *11*, 2100757. [CrossRef]
27. Lu, T.; Du, H.; Kong, N.; Li, H.; Xu, S.; Li, Z.; Du, X. U-shaped micropores induced dielectric and piezoelectric tunability in bismuth sodium titanate-based ceramics. *J. Mater. Sci. Mater. Electron.* **2023**, *34*, 797. [CrossRef]
28. Wang, J.; Guo, X.; Apostol, P.; Liu, X.; Robeyns, K.; Gence, L.; Morari, C.; Gohy, J.-F.; Vlad, A. High performance Li-, Na-, and K-ion storage in electrically conducting coordination polymers. *Energy Environ. Sci.* **2022**, *15*, 3923–3932. [CrossRef]
29. Liu, Q.; Hu, Y.; Yu, X.; Qin, Y.; Meng, T.; Hu, X. The pursuit of commercial silicon-based microparticle anodes for advanced lithium-ion batteries: A review. *Nano Res. Energy* **2022**, *1*, 9120037. [CrossRef]
30. Zhang, Y.; Du, H.; Lu, J.; Li, Z.; Lu, T.; Hu, Y.; Ma, Y. Enhanced cyclic stability of Ga_2O_3@PDA-C nanospheres as pseudocapacitive anode materials for lithium-ion batteries. *Fuel* **2023**, *334*, 126683. [CrossRef]
31. Xu, G.-L.; Gong, Y.-D.; Miao, C.; Wang, Q.; Nie, S.-Q.; Xin, Y.; Wen, M.-Y.; Liu, J.; Xiao, W. Sn nanoparticles embedded into porous hydrogel-derived pyrolytic carbon as composite anode materials for lithium-ion batteries. *Rare Met.* **2022**, *41*, 3421–3431. [CrossRef]
32. Yan, W.; Gao, X.; Jin, X.; Liang, S.; Xiong, X.; Liu, Z.; Wang, Z.; Chen, Y.; Fu, L.; Zhang, Y.; et al. Nonporous gel electrolytes enable long cycling at high current density for lithium-metal anodes. *ACS Appl. Mater. Interfaces* **2021**, *13*, 14258–14266. [CrossRef]
33. Li, C.; Liu, B.; Jiang, N.; Ding, Y. Elucidating the charge-transfer and Li-ion-migration mechanisms in commercial lithium-ion batteries with advanced electron microscopy. *Nano Res. Energy* **2022**, *1*, e9120031. [CrossRef]
34. Vaalma, C.; Buchholz, D.; Weil, M.; Passerini, S. A cost and resource analysis of sodium-ion batteries. *Nat. Rev. Mater.* **2018**, *3*, 18013. [CrossRef]
35. Hirsh, H.S.; Li, Y.; Tan, D.H.S.; Zhang, M.; Zhao, E.; Meng, Y.S. Sodium-ion batteries paving the way for grid energy storage. *Adv. Energy Mater.* **2020**, *10*, 2001274. [CrossRef]
36. Yang, C.; Xin, S.; Mai, L.; You, Y. Materials design for high-safety sodium-ion battery. *Adv. Energy Mater.* **2020**, *11*, 2000974. [CrossRef]
37. Peng, J.; Zhang, W.; Wang, J.; Li, L.; Lai, W.; Yang, Q.; Zhang, B.; Li, X.; Du, Y.; Liu, H.; et al. Processing rusty metals into versatile prussian blue for sustainable energy storage. *Adv. Energy Mater.* **2021**, *11*, 2102356. [CrossRef]
38. Peng, J.; Zhang, W.; Wang, S.; Huang, Y.; Wang, J.Z.; Liu, H.K.; Dou, S.X.; Chou, S.L. The emerging electrochemical activation tactic for aqueous energy storage: Fundamentals, applications, and future. *Adv. Funct. Mater.* **2022**, *32*, 2111720. [CrossRef]
39. Yuan, M.; Liu, H.; Ran, F. Fast-charging cathode materials for lithium & sodium ion batteries. *Mater. Today* **2023**, *2*, 1361–7021.
40. Wang, H.; Ding, F.; Wang, Y.; Han, Z.; Dang, R.; Yu, H.; Yang, Y.; Chen, Z.; Li, Y.; Xie, F.; et al. In situ plastic-crystal-coated cathode toward high-performance na-ion batteries. *Acs Energy Lett.* **2023**, *8*, 1434–1444. [CrossRef]
41. Chen, Y.; Zhao, B.; Yang, Y.; Cao, A. Toward high-areal-capacity electrodes for lithium and sodium ion batteries. *Adv. Energy Mater.* **2022**, *12*, 2201834. [CrossRef]
42. Lemoine, K.; Hémon-Ribaud, A.; Leblanc, M.; Lhoste, J.; Tarascon, J.-M.; Maisonneuve, V. Fluorinated materials as positive electrodes for li- and na-ion batteries. *Chem. Rev.* **2022**, *122*, 14405–14439. [CrossRef] [PubMed]

43. Voronina, N.; Shin, M.Y.; Kim, H.J.; Yaqoob, N.; Guillon, O.; Song, S.H.; Kim, H.; Lim, H.D.; Jung, H.G.; Kim, Y.; et al. Hysteresis-suppressed reversible oxygen-redox cathodes for sodium-ion batteries. *Adv. Energy Mater.* **2022**, *12*, 2103939. [CrossRef]
44. Yuan, Y.; Wei, Q.; Yang, S.; Zhang, X.; Jia, M.; Yuan, J.; Yan, X. Towards high-performance phosphate-based polyanion-type materials for sodium-ion batteries. *Energy Storage Mater.* **2022**, *50*, 760–782. [CrossRef]
45. Lin, C.-C.; Liu, H.-Y.; Kang, J.-W.; Yang, C.-C.; Li, C.-H.; Chen, H.-Y.T.; Huang, S.-C.; Ni, C.-S.; Chuang, Y.-C.; Chen, B.-H.; et al. In-situ X-ray studies of high-entropy layered oxide cathode for sodium-ion batteries. *Energy Storage Mater.* **2022**, *51*, 159–171. [CrossRef]
46. Liang, X.; Yu, T.Y.; Ryu, H.H.; Sun, Y.K. Hierarchical O3/P2 heterostructured cathode materials for advanced sodium-ion batteries. *Energy Storage Mater.* **2022**, *47*, 515–525. [CrossRef]
47. Yu, C.-X.; Li, Y.; Wang, Z.-H.; Wang, X.-R.; Bai, Y.; Wu, C. Surface engineering based on in situ electro-polymerization to boost the initial Coulombic efficiency of hard carbon anode for sodium-ion battery. *Rare Met.* **2022**, *41*, 1616–1625. [CrossRef]
48. Fan, M.-P.; Chen, Y.-C.; Chen, Y.-M.; Huang, Z.-X.; Wu, W.-L.; Wang, P.; Ke, X.; Sun, S.-H.; Shi, Z.-C. NiS2 nanosheet arrays on stainless steel foil as binder-free anode for high-power sodium-ion batteries. *Rare Met.* **2022**, *41*, 1294–1303. [CrossRef]
49. Niitani, K.; Ushiroda, S.; Kuwata, H.; Ohata, H.N.; Shimo, Y.; Hozumi, M.; Matsunaga, T.; Nakanishi, S. Hard carbon anode with a sodium carborane electrolyte for fast-charging all-solid-state sodium-ion batteries. *Acs Energy Lett.* **2022**, *7*, 145–149. [CrossRef]
50. Chen, F.; Di, Y.; Su, Q.; Xu, D.; Zhang, Y.; Zhou, S.; Liang, S.; Cao, X.; Pan, A. Vanadium-modified hard carbon spheres with sufficient pseudographitic domains as high-performance anode for sodium-ion batteries. *Carbon Energy* **2022**, *5*, e191. [CrossRef]
51. Li, Q.; Zhang, J.; Zhong, L.; Geng, F.; Tao, Y.; Geng, C.; Li, S.; Hu, B.; Yang, Q.H. Unraveling the key atomic interactions in determining the varying li/na/k storage mechanism of hard carbon anodes. *Adv. Energy Mater.* **2022**, *12*, 2201734. [CrossRef]
52. Zhu, Y.; Wang, Y.; Wang, Y.; Xu, T.; Chang, P. Research progress on carbon materials as negative electrodes in sodium- and potassium-ion batteries. *Carbon Energy* **2022**, *4*, 1182–1213. [CrossRef]
53. Liu, M.; Wang, Y.; Wu, F.; Bai, Y.; Li, Y.; Gong, Y.; Feng, X.; Li, Y.; Wang, X.; Wu, C. Advances in carbon materials for sodium and potassium storage. *Adv. Funct. Mater.* **2022**, *32*, 2203117. [CrossRef]
54. Peng, W.; Zhou, W.; Li, T.; Zhou, J.; Yao, T.; Wu, H.; Zhao, X.; Luo, J.; Liu, J.; Zhang, D. Towards inhibiting conductivity of Mo/PVDF composites through building MoO3 shell as an interlayer for enhanced dielectric properties. *J. Mater. Sci. Mater. Electron.* **2022**, *33*, 14735–14753. [CrossRef]
55. Yao, T.; Zhou, W.; Peng, W.; Zhou, J.; Li, T.; Wu, H.; Zheng, J.; Lin, N.; Liu, D.; Hou, C. Insights into concomitant enhancements of dielectric properties and thermal conductivity of PVDF composites filled with core@ double-shell structured Zn@ZnO@PS particles. *J. Appl. Polym. Sci.* **2022**, *139*, e53069. [CrossRef]
56. Zhang, M.; Li, Y.; Wu, F.; Bai, Y.; Wu, C. Boost sodium-ion batteries to commercialization: Strategies to enhance initial Coulombic efficiency of hard carbon anode. *Nano Energy* **2021**, *82*, 105738. [CrossRef]
57. Zhao, L.F.; Hu, Z.; Lai, W.H.; Tao, Y.; Peng, J.; Miao, Z.C.; Wang, Y.X.; Chou, S.L.; Liu, H.K.; Dou, S.X. Hard carbon anodes: Fundamental understanding and commercial perspectives for na-ion batteries beyond li-ion and k-ion counterparts. *Adv. Energy Mater.* **2020**, *11*, 2002704. [CrossRef]
58. Yu, C.; Li, Y.; Ren, H.; Qian, J.; Wang, S.; Feng, X.; Liu, M.; Bai, Y.; Wu, C. Engineering homotype heterojunctions in hard carbon to induce stable solid electrolyte interfaces for sodium-ion batteries. *Carbon Energy* **2023**, *5*, e220. [CrossRef]
59. Lu, Z.; Geng, C.; Yang, H.; He, P.; Wu, S.; Yang, Q.-H.; Zhou, H. Step-by-step desolvation enables high-rate and ultra-stable sodium storage in hard carbon anodes. *Proc. Natl. Acad. Sci. USA* **2022**, *119*, e2210203119. [CrossRef]
60. Zhao, D.; Jiang, S.; Yu, S.; Ren, J.; Zhang, Z.; Liu, S.; Liu, X.; Wang, Z.; Wu, Y.; Zhang, Y. Lychee seed-derived microporous carbon for high-performance sodium-sulfur batteries. *Carbon* **2023**, *201*, 864–870. [CrossRef]
61. Huang, S.; Li, Z.; Wang, B.; Zhang, J.; Peng, Z.; Qi, R.; Wang, J.; Zhao, Y. N-doping and defective nanographitic domain coupled hard carbon nanoshells for high performance lithium/sodium storage. *Adv. Funct. Mater.* **2018**, *28*, 1706294. [CrossRef]
62. Li, X.; Sun, J.; Zhao, W.; Lai, Y.; Yu, X.; Liu, Y. Intergrowth of graphite-like crystals in hard carbon for highly reversible na-ion storage. *Adv. Funct. Mater.* **2021**, *32*, 2106980. [CrossRef]
63. Yu, Z.-L.; Xin, S.; You, Y.; Yu, L.; Lin, Y.; Xu, D.-W.; Qiao, C.; Huang, Z.-H.; Yang, N.; Yu, S.-H.; et al. Ion-Catalyzed synthesis of microporous hard carbon embedded with expanded nanographite for enhanced lithium/sodium storage. *J. Am. Chem. Soc.* **2016**, *138*, 14915–14922. [CrossRef]
64. Chen, K.H.; Goel, V.; Namkoong, M.J.; Wied, M.; Müller, S.; Wood, V.; Sakamoto, J.; Thornton, K.; Dasgupta, N.P. Enabling 6C fast charging of li-ion batteries with graphite/hard carbon hybrid anodes. *Adv. Energy Mater.* **2020**, *11*, 2003336. [CrossRef]
65. Zhang, X.; Han, S.; Xiao, P.; Fan, C.; Zhang, W. Thermal reduction of graphene oxide mixed with hard carbon and their high performance as lithium ion battery anode. *Carbon* **2016**, *11*, 600–607. [CrossRef]
66. Alvin, S.; Chandra, C.; Kim, J. Controlling intercalation sites of hard carbon for enhancing Na and K storage performance. *Chem. Eng. J.* **2021**, *411*, 128490. [CrossRef]
67. Zhao, X.; Ding, Y.; Xu, Q.; Yu, X.; Liu, Y.; Shen, H. Low-temperature growth of hard carbon with graphite crystal for sodium-ion storage with high initial coulombic efficiency: A general method. *Adv. Energy Mater.* **2019**, *9*, 1803648. [CrossRef]
68. Zhao, D.; Ge-Zhang, S.; Zhang, Z.; Tang, H.; Xu, Y.; Gao, F.; Xu, X.; Liu, S.; Zhou, J.; Wang, Z.; et al. Three-dimensional honeycomb-like carbon as sulfur host for sodium-sulfur batteries without the shuttle effect. *ACS Appl Mater Interfaces* **2022**, *14*, 54662–54669. [CrossRef]

69. Lei, H.; Li, J.; Zhang, X.; Ma, L.; Ji, Z.; Wang, Z.; Pan, L.; Tan, S.; Mai, W. A review of hard carbon anode: Rational design and advanced characterization in potassium ion batteries. *InfoMat* **2022**, *4*, e12272. [CrossRef]
70. Gong, H.; Chen, Y.; Chen, S.; Xu, C.; Yang, Y.; Ye, Y.; Huang, Z.; Ning, R.; Cui, Y.; Bao, Z. Fast-charging of hybrid lithium-ion/lithium-metal anodes by nanostructured hard carbon host. *Acs Energy Lett.* **2022**, *7*, 4417–4426. [CrossRef]
71. Chi, C.; Liu, Z.; Lu, X.; Meng, Y.; Huangfu, C.; Yan, Y.; Qiu, Z.; Qi, B.; Wang, G.; Pang, H.; et al. Balance of sulfur doping content and conductivity of hard carbon anode for high-performance K-ion storage. *Energy Storage Mater.* **2022**, *54*, 668–679. [CrossRef]
72. Sun, C.; Zhang, X.; An, Y.; Li, C.; Wang, L.; Zhang, X.; Sun, X.; Wang, K.; Zhang, H.; Ma, Y. Low-temperature carbonized nitrogen doped hard carbon nanofiber towards high-performance sodium-ion capacitors. *Energy Environ. Mater.* **2023**, *17*, e12603. [CrossRef]
73. Liu, C.; Lu, Q.; Gorbunov, M.V.; Omar, A.; Martinez, I.G.G.; Zhao, P.; Hantusch, M.; Permana, A.D.C.; He, H.; Gaponik, N.; et al. Ultrasmall CoS nanoparticles embedded in heteroatom-doped carbon for sodium-ion batteries and mechanism explorations via synchrotron X-ray techniques. *J. Energy Chem.* **2023**, *79*, 373–381. [CrossRef]
74. Kim, H.; Hyun, J.C.; Kim, D.-H.; Kwak, J.H.; Lee, J.B.; Moon, J.H.; Choi, J.; Lim, H.-D.; Yang, S.J.; Jin, H.M.; et al. Revisiting lithium- and sodium-ion storage in hard carbon anodes. *Adv. Mater.* **2023**, *20*, 2209128. [CrossRef]
75. Ye, W.; Li, X.; Zhang, B.; Liu, W.; Cheng, Y.; Fan, X.; Zhang, H.; Liu, Y.; Dong, Q.; Wang, M.-S. Superfast mass transport of na/k via mesochannels for dendrite-free metal batteries. *Adv. Mater.* **2023**, e2210447. [CrossRef]
76. Surta, T.W.; Koh, E.; Li, Z.; Fast, D.B.; Ji, X.; Greaney, P.A.; Dolgos, M.R. Combining experimental and theoretical techniques to gain an atomic level understanding of the defect binding mechanism in hard carbon anodes for sodium ion batteries. *Adv. Energy Mater.* **2022**, *12*, 2200647. [CrossRef]
77. Chen, X.; Liu, C.; Fang, Y.; Ai, X.; Zhong, F.; Yang, H.; Cao, Y. Understanding of the sodium storage mechanism in hard carbon anodes. *Carbon Energy* **2022**, *4*, 1133–1150. [CrossRef]
78. Sun, N.; Qiu, J.; Xu, B. Understanding of sodium storage mechanism in hard carbons: Ongoing development under debate. *Adv. Energy Mater.* **2022**, *12*, 2200715. [CrossRef]
79. Ilic, I.K.; Schutjajew, K.; Zhang, W.; Oschatz, M. Changes of porosity of hard carbons during mechanical treatment and the relevance for sodium-ion anodes. *Carbon* **2022**, *186*, 55–63. [CrossRef]
80. Yu, K.; Wang, X.; Yang, H.; Bai, Y.; Wu, C. Insight to defects regulation on sugarcane waste-derived hard carbon anode for sodium-ion batteries. *J. Energy Chem.* **2021**, *55*, 499–508. [CrossRef]
81. Alvin, S.; Cahyadi, H.S.; Hwang, J.; Chang, W.; Kwak, S.K.; Kim, J. Revealing the intercalation mechanisms of lithium, sodium, and potassium in hard carbon. *Adv. Energy Mater.* **2020**, *10*, 2000283. [CrossRef]
82. Jin, Q.; Wang, K.; Li, H.; Li, W.; Feng, P.; Zhang, Z.; Wang, W.; Zhou, M.; Jiang, K. Tuning microstructures of hard carbon for high capacity and rate sodium storage. *Chem. Eng. J.* **2021**, *417*, 128104. [CrossRef]
83. Hou, L.; Liu, T.; Wang, H.; Bai, M.; Tang, X.; Wang, Z.; Zhang, M.; Li, S.; Wang, T.; Zhou, K.; et al. Boosting the reversible, high-rate Na^+ storage capability of the hard carbon anode via the synergistic structural tailoring and controlled presodiation. *Small* **2023**, *20*, 2207638. [CrossRef] [PubMed]
84. Michael, M.; Mangayarkarasi, N.; Edgardo Maximiliano Gavilán, A.; Anshika, B.; Swoyam, P.; Heather, A.; Maria, C.-R.; Maria-Magdalena, T.; Ezequiel, L.; Harry Ernst, H. Sodiation energetics in pore size controlled hard carbons determined via entropy profiling. *J. Mater. Chem. A* **2023**, *10*, 1039.
85. Stevens, D.A.; Dahn, J.R. High capacity anode materials for rechargeable sodium-ion batteries. *J. Electrochem. Soc.* **2000**, *147*, 1271. [CrossRef]
86. Qiu, S.; Xiao, L.; Sushko, M.L.; Han, K.S.; Shao, Y.; Yan, M.; Liang, X.; Mai, L.; Feng, J.; Cao, Y.; et al. Manipulating adsorption-insertion mechanisms in nanostructured carbon materials for high-efficiency sodium ion storage. *Adv. Energy Mater.* **2017**, *7*, 1700403. [CrossRef]
87. Stevens, D.A.; Dahn, J.R. The mechanisms of lithium and sodium insertion in carbon materials. *J. Electrochem. Soc.* **2001**, *148*, A803–A811. [CrossRef]
88. Xu, T.; Qiu, X.; Zhang, X.; Xia, Y. Regulation of surface oxygen functional groups and pore structure of bamboo-derived hard carbon for enhanced sodium storage performance. *Chem. Eng. J.* **2023**, *452*, 139514. [CrossRef]
89. Cao, Y.; Xiao, L.; Sushko, M.L.; Wang, W.; Schwenzer, B.; Xiao, J.; Nie, Z.; Saraf, L.V.; Yang, Z.; Liu, J. Sodium ion insertion in hollow carbon nanowires for battery applications. *Nano Lett.* **2012**, *12*, 3783–3787. [CrossRef]
90. Lu, H.; Ai, F.; Jia, Y.; Tang, C.; Zhang, X.; Huang, Y.; Yang, H.; Cao, Y. Exploring sodium-ion storage mechanism in hard carbons with different microstructure prepared by ball-milling method. *Small* **2018**, *14*, 2108092. [CrossRef]
91. Jiang, N.; Chen, L.; Jiang, H.; Hu, Y.; Li, C. Introducing the solvent co-intercalation mechanism for hard carbon with ultrafast sodium storage. *Small* **2022**, *18*, 2108092. [CrossRef]
92. Yuan, M.; Cao, B.; Liu, H.; Meng, C.; Wu, J.; Zhang, S.; Li, A.; Chen, X.; Song, H. Sodium storage mechanism of nongraphitic carbons: A general model and the function of accessible closed pores. *Chem. Mater.* **2022**, *34*, 3489–3500. [CrossRef]
93. Ren, Q.; Wang, J.; Yan, L.; Lv, W.; Zhang, F.; Zhang, L.; Liu, B.; Shi, Z. Manipulating free-standing, flexible and scalable microfiber carbon papers unlocking ultra-high initial Coulombic efficiency and storage sodium behavior. *Chem. Eng. J.* **2021**, *425*, 131656. [CrossRef]

94. Song, M.; Yi, Z.; Xu, R.; Chen, J.; Cheng, J.; Wang, Z.; Liu, Q.; Guo, Q.; Xie, L.; Chen, C. Towards enhanced sodium storage of hard carbon anodes: Regulating the oxygen content in precursor by low-temperature hydrogen reduction. *Energy Storage Mater.* **2022**, *51*, 620–629. [CrossRef]
95. Bai, P.; He, Y.; Zou, X.; Zhao, X.; Xiong, P.; Xu, Y. Elucidation of the sodium-storage mechanism in hard carbons. *Adv. Energy Mater.* **2018**, *8*, 1703217. [CrossRef]
96. Li, Y.; Hu, Y.-S.; Titirici, M.-M.; Chen, L.; Huang, X. Hard carbon microtubes made from renewable cotton as high-performance anode material for sodium-ion batteries. *Adv. Energy Mater.* **2016**, *6*, 1600659. [CrossRef]
97. Yang, L.; Hu, M.; Zhang, H.; Yang, W.; Lv, R. Pore structure regulation of hard carbon: Towards fast and high-capacity sodium-ion storage. *J. Colloid Interface Sci.* **2020**, *566*, 257–264. [CrossRef]
98. Yin, X.; Lu, Z.; Wang, J.; Feng, X.; Roy, S.; Liu, X.; Yang, Y.; Zhao, Y.; Zhang, J. Enabling fast na+ transfer kinetics in the whole-voltage-region of hard-carbon anodes for ultrahigh-rate sodium storage. *Adv. Mater.* **2022**, *34*, 2109282. [CrossRef]
99. Liu, M.; Wu, F.; Bai, Y.; Li, Y.; Ren, H.; Zhao, R.; Feng, X.; Song, T.; Wu, C. Boosting sodium storage performance of hard carbon anodes by pore architecture engineering. *Acs Appl. Mater. Interfaces* **2021**, *13*, 47671–47683. [CrossRef]
100. Yang, J.; Wang, X.; Dai, W.; Lian, X.; Cui, X.; Zhang, W.; Zhang, K.; Lin, M.; Zou, R.; Loh, K.P.; et al. From micropores to ultra-micropores inside hard carbon: Toward enhanced capacity in room-/low-temperature sodium-ion storage. *Nano-Micro Lett.* **2021**, *13*, 98. [CrossRef]
101. Wu, D.; Sun, F.; Qu, Z.; Wang, H.; Lou, Z.; Wu, B.; Zhao, G. Multi-scale structure optimization of boron-doped hard carbon nanospheres boosting the plateau capacity for high performance sodium ion batteries. *J. Mater. Chem. A* **2022**, *10*, 17225–17236. [CrossRef]
102. Yu, K.; Zhao, H.; Wang, X.; Zhang, M.; Dong, R.; Li, Y.; Bai, Y.; Xu, H.; Wu, C. Hyperaccumulation route to ca-rich hard carbon materials with cation self-incorporation and interlayer spacing optimization for high-performance sodium-ion batteries. *Acs Appl. Mater. Interfaces* **2020**, *12*, 10544–10553. [CrossRef] [PubMed]
103. Liang, Y.; Song, N.; Zhang, Z.; Chen, W.; Feng, J.; Xi, B.; Xiong, S. Integrating Bi@C nanospheres in porous hard carbon frameworks for ultrafast sodium storage. *Adv. Mater.* **2022**, *34*, 2202673. [CrossRef] [PubMed]
104. Wang, X.; Hou, M.; Shi, Z.; Liu, X.; Mizota, I.; Lou, H.; Wang, B.; Hou, X. Regulate phosphorus configuration in high p-doped hard carbon as a superanode for sodium storage. *ACS Appl. Mater. Interfaces* **2021**, *13*, 12059–12068. [CrossRef] [PubMed]
105. Li, N.; Yang, Q.; Wei, Y.; Rao, R.; Wang, Y.; Sha, M.; Ma, X.; Wang, L.; Qian, Y. Phosphorus-doped hard carbon with controlled active groups and microstructure for high-performance sodium-ion batteries. *J. Mater. Chem. A* **2020**, *8*, 20486–20492. [CrossRef]
106. Yan, J.; Li, H.; Wang, K.; Jin, Q.; Lai, C.; Wang, R.; Cao, S.; Han, J.; Zhang, Z.; Su, J.; et al. Ultrahigh phosphorus doping of carbon for high-rate sodium ion batteries anode. *Adv. Energy Mater.* **2021**, *11*, 2003911. [CrossRef]
107. Huang, G.; Kong, Q.; Yao, W.; Wang, Q. High Proportion of Active Nitrogen-doped hard carbon based on mannich reaction as anode material for high-performance sodium-ion batteries. *ChemSusChem* **2023**, *20*, e202202070. [CrossRef]
108. Ma, M.; Cai, H.; Xu, C.; Huang, R.; Wang, S.; Pan, H.; Hu, Y.-S. Engineering solid electrolyte interface at nano-scale for high-performance hard carbon in sodium-ion batteries. *Adv. Funct. Mater.* **2021**, *31*, 2100278. [CrossRef]
109. Zhang, W.; Zeng, F.; Huang, H.; Yu, Y.; Xu, M.; Xing, L.; Li, W. Enhanced interphasial stability of hard carbon for sodium-ion battery via film-forming electrolyte additive. *Nano Res.* **2022**, *11*, 5974. [CrossRef]
110. Jin, Y.; Xu, Y.; Le, P.M.L.; Vo, T.D.; Zhou, Q.; Qi, X.; Engelhard, M.H.; Matthews, B.E.; Jia, H.; Nie, Z.; et al. Highly reversible sodium ion batteries enabled by stable electrolyte-electrode interphases. *Acs Energy Lett* **2020**, *5*, 3212–3220. [CrossRef]
111. Lohani, H.; Kumar, A.; Kumari, P.; Ahuja, A.; Gautam, M.; Sengupta, A.; Mitra, S. Artificial organo-fluoro-rich anode electrolyte interface and partially sodiated hard carbon anode for improved cycle life and practical sodium-ion batteries. *Acs Appl. Mater. Interfaces* **2022**, *14*, 37793–37803. [CrossRef]
112. Dong, R.; Zheng, L.; Bai, Y.; Ni, Q.; Li, Y.; Wu, F.; Ren, H.; Wu, C. Elucidating the mechanism of fast na storage kinetics in ether electrolytes for hard carbon anodes. *Adv. Mater.* **2021**, *33*, e2008810. [CrossRef]
113. Zeng, F.; Xing, L.; Zhang, W.; Xie, Z.; Liu, M.; Lin, X.; Tang, G.; Mo, C.; Li, W. Innovative discontinuous-SEI constructed in ether-based electrolyte to maximize the capacity of hard carbon anode. *J. Energy Chem.* **2023**, *79*, 459–467. [CrossRef]
114. Hou, X.; Li, T.; Qiu, Y.; Jiang, M.; Zheng, Q.; Li, X. Weak coulomb interaction between anions and Na plus during solvation enabling desirable solid electrolyte interphase and superior kinetics for HC-based sodium ion batteries. *Chem. Eng. J.* **2023**, *453*, 139932. [CrossRef]
115. Zhou, D.; Tang, X.; Guo, X.; Li, P.; Shanmukaraj, D.; Liu, H.; Gao, X.; Wang, Y.; Rojo, T.; Armand, M.; et al. Polyolefin-based janus separator for rechargeable sodium batteries. *Angew. Chem. Int. Ed. Engl.* **2020**, *59*, 16725–16734. [CrossRef]
116. Kang, T.; Sun, C.; Li, Y.; Song, T.; Guan, Z.; Tong, Z.; Nan, J.; Lee, C.S. Dendrite-free sodium metal anodes via solid electrolyte interphase engineering with a covalent organic framework separator. *Adv. Energy Mater.* **2023**, *10*, 2204083. [CrossRef]
117. Niu, X.; Li, J.; Song, J.; Li, Y.; He, T. Interconnected porous poly(ether imide) separator for thermally stable sodium ion battery. *ACS Appl. Energy Mater.* **2021**, *4*, 11080–11089. [CrossRef]
118. Zhou, H.; Gu, J.; Wei, Y.; Zhang, W.; Kang, J.; Huang, J.-Q.; Zhang, B.; Hu, C.; Lin, X. Cellulose filter papers derived separator featuring effective ion transferring channels for sodium-ion batteries. *J. Power Sources* **2023**, *558*, 232649. [CrossRef]
119. Wang, J.; Zhao, J.; He, X.; Qiao, Y.; Li, L.; Chou, S.-L. Hard carbon derived from hazelnut shell with facile HCl treatment as high-initial-coulombic-efficiency anode for sodium ion batteries. *Sustain. Mater. Technol.* **2022**, *33*, e00446. [CrossRef]

120. Wang, P.; Guo, Y.-J.; Chen, W.-P.; Duan, H.; Ye, H.; Yao, H.-R.; Yin, Y.-X.; Cao, F.-F. Self-supported hard carbon anode from fungus-treated basswood towards sodium-ion batteries. *Nano Res.* **2022**, *11*, 5974. [CrossRef]
121. Zhen, Y.; Chen, Y.; Li, F.; Guo, Z.; Hong, Z.; Titirici, M.-M. Ultrafast synthesis of hard carbon anodes for sodium-ion batteries. *Proc. Natl. Acad. Sci. USA* **2021**, *118*, e111119118. [CrossRef] [PubMed]
122. Fan, C.; Zhang, R.; Luo, X.; Hu, Z.; Zhou, W.; Zhang, W.; Liu, J.; Liu, J. Epoxy phenol novolac resin: A novel precursor to construct high performance hard carbon anode toward enhanced sodium-ion batteries. *Carbon* **2023**, *205*, 353–364. [CrossRef]
123. Gan, Q.; Qin, N.; Gu, S.; Wang, Z.; Li, Z.; Liao, K.; Zhang, K.; Lu, L.; Xu, Z.; Lu, Z. Extra sodiation sites in hard carbon for high performance sodium ion batteries. *Small Methods* **2021**, *5*, 2100580. [CrossRef] [PubMed]
124. Pi, Y.; Gan, Z.; Yan, M.; Pei, C.; Yu, H.; Ge, Y.; An, Q.; Mai, L. Insight into pre-sodiation in $Na_3V_2(PO_4)_{(2)}F$-3/C @ hard carbon full cells for promoting the development of sodium-ion battery. *Chem. Eng. J.* **2021**, *413*, 127565. [CrossRef]
125. Kamiyama, A.; Kubota, K.; Igarashi, D.; Youn, Y.; Tateyama, Y.; Ando, H.; Gotoh, K.; Komaba, S. MgO-template synthesis of extremely high capacity hard carbon for na-ion battery. *Angew. Chem. Int. Ed.* **2021**, *60*, 5114–5120. [CrossRef]
126. Jin, Q.; Wang, K.; Feng, P.; Zhang, Z.; Cheng, S.; Jiang, K. Surface-dominated storage of heteroatoms-doping hard carbon for sodium-ion batteries. *Energy Storage Mater.* **2020**, *27*, 43–50. [CrossRef]
127. Yan, L.; Wang, J.; Ren, Q.; Fan, L.; Liu, B.; Zhang, L.; He, L.; Mei, X.; Shi, Z. In-Situ graphene-coated carbon microsphere as high initial coulombic efficiency anode for superior Na/K-ion full cell. *Chem. Eng. J.* **2022**, *432*, 133257. [CrossRef]

MDPI AG
Grosspeteranlage 5
4052 Basel
Switzerland
Tel.: +41 61 683 77 34

Molecules Editorial Office
E-mail: molecules@mdpi.com
www.mdpi.com/journal/molecules

www.ingramcontent.com/pod-product-compliance
Lightning Source LLC
Chambersburg PA
CBHW061619210926
51850CB00009B/2375
* 9 7 8 3 7 2 5 8 5 2 6 9 7 *